The Theory of Information and Coding
Second Edition

ENCYCLOPEDIA OF MATHEMATICS AND ITS APPLICATIONS

ENCYCLOPEDIA OF MATHEMATICS AND ITS APPLICATIONS

The Theory of Information and Coding
Second Edition

R. J. McELIECE

California Institute of Technology

CAMBRIDGE
UNIVERSITY PRESS

CAMBRIDGE
UNIVERSITY PRESS

University Printing House, Cambridge CB2 8BS, United Kingdom

One Liberty Plaza, 20th Floor, New York, NY 10006, USA

477 Williamstown Road, Port Melbourne, VIC 3207, Australia

314-321, 3rd Floor, Plot 3, Splendor Forum, Jasola District Centre, New Delhi - 110025, India

103 Penang Road, #05-06/07, Visioncrest Commercial, Singapore 238467

Cambridge University Press is part of the University of Cambridge.

It furthers the University's mission by disseminating knowledge in the pursuit of education, learning and research at the highest international levels of excellence.

www.cambridge.org
Information on this title: www.cambridge.org/9780521000956

Published by Addison-Wesley 1977, 1979, 1982

First published by Cambridge University Press 1985

Second edition 2002
Reprinted 2003

A catalogue record for this publication is available from the British Library

ISBN 978-0-521-00095-6 Hardback

Contents

v

Editor's statement

A large body of mathematics consists of facts that can be presented and described much like any other natural phenomenon. These facts, at times explicitly brought out as theorems, at other times concealed within a proof, make up most of the applications of mathematics, and are the most likely to survive changes of style and of interest.

This ENCYCLOPEDIA will attempt to present the factual body of all mathematics. Clarity of exposition, accessibility to the non-specialist, and a thorough bibliography are required of each author. Volumes will appear in no particular order, but will be organized into sections, each one comprising a recognizable branch of present-day mathematics. Numbers of volumes and sections will be reconsidered as times and needs change.

It is hoped that this enterprise will make mathematics more widely used where it is needed, and more accessible in fields in which it can be applied but where it has not yet penetrated because of insufficient information.

Information theory is a success story in contemporary mathematics. Born out of very real engineering problems, it has left its imprint on such far-flung endeavors as the approximation of functions and the central limit theorem of probability. It is an idea whose time has come.

Most mathematicians cannot afford to ignore the basic results in this field. Yet, because of the enormous outpouring of research, it is difficult for anyone who is not a specialist to single out the basic results and the relevant material. Robert McEliece has succeeded in giving a presentation that achieves this objective, perhaps the first of its kind.

GIAN-CARLO ROTA

Foreword

Transmission of information is at the heart of what we call communication. As an area of concern it is so vast as to touch upon the preoccupations of philosophers and to give rise to a thriving technology.

We owe to the genius of Claude Shannon[*] the recognition that a large class of problems related to encoding, transmitting, and decoding information can be approached in a systematic and disciplined way: his classic paper of 1948 marks the birth of a new chapter of Mathematics.

In the past thirty years there has grown a staggering literature in this fledgling field, and some of its terminology even has become part of our daily language.

The present monograph (actually two monographs in one) is an excellent introduction to the two aspects of communication: coding and transmission.

The first (which is the subject of Part two) is an elegant illustration of the power and beauty of Algebra; the second belongs to Probability Theory which the chapter begun by Shannon enriched in novel and unexpected ways.

MARK KAC
General Editor, Section on Probability

---

[*] C. E. Shannon, A Mathematical Theory of Communication, *Bell System Tech. J.* **27** (1948), Introduction: 379–382; Part one: Discrete Noiseless Systems, 382–405; Part two: The Discrete Channel with Noise (and Appendixes), 406–423; Part III: Mathematical Preliminaries, 623–636; Part IV: The Continuous Channel (and Appendixes), 637–656).

Preface to the first edition

This book is meant to be a self-contained introduction to the basic results in the theory of information and coding. It was written during 1972–1976, when I taught this subject at Caltech. About half my students were electrical engineering graduate students; the others were majoring in all sorts of other fields (mathematics, physics, biology, even one English major!). As a result the course was aimed at nonspecialists as well as specialists, and so is this book.

The book is in three parts: Introduction, Part one (Information Theory), and Part two (Coding Theory). It is essential to read the introduction first, because it gives an overview of the whole subject. In Part one, Chapter 1 is fundamental, but it is probably a mistake to read it first, since it is really just a collection of technical results about entropy, mutual information, and so forth. It is better regarded as a reference section, and should be consulted as necessary to understand Chapters 2–5. Chapter 6 is a survey of advanced results, and can be read independently. In Part two, Chapter 7 is basic and must be read before Chapters 8 and 9; but Chapter 10 is almost, and Chapter 11 is completely, independent from Chapter 7. Chapter 12 is another survey chapter independent of everything else.

The problems at the end of the chapters are very important. They contain verification of many omitted details, as well as many important results not mentioned in the text. It is a good idea to at least read the problems.

There are four appendices. Appendix A gives a brief survey of probability theory, essential for Part one. Appendix B discusses convex functions and Jensen's inequality. Appeals to Jensen's inequality are frequent in Part one, and the reader unfamiliar with it should read Appendix B at the first opportunity. Appendix C sketches the main results about finite fields needed in Chapter 9. Appendix D describes an algorithm for counting paths in directed graphs which is needed in Chapter 10.

A word about cross-references is in order: sections, figures, examples, theorems, equations, and problems are numbered consecutively by chapters, using double numeration. Thus "Section 2.3," "Theorem 3.4," and "Prob. 4.17" refer to section 3 of Chapter 2, Theorem 4 of Chapter 3, and Problem 17 of Chapter 4, respectively. The appendices are referred to by letter; thus "Equation (B.4)" refers to the fourth numbered equation in Appendix B.

The following special symbols perhaps need explanation: "□" signals the end of a proof or example; "iff" means *if and only if*; $\lfloor x \rfloor$ denotes the largest integer $\leq x$; and $\lceil x \rceil$ denotes the smallest integer $\geq x$.

Finally, I am happy to acknowledge my debts: To Gus Solomon, for introducing me to the subject in the first place; to John Pierce, for giving me the opportunity to teach at Caltech; to Gian-Carlo Rota, for encouraging me to write this book; to Len Baumert, Stan Butman, Gene Rodemich, and Howard Rumsey, for letting me pick their brains; to Jim Lesh and Jerry Heller, for supplying data for Figures 6.7 and 12.2; to Bob Hall, for drafting the figures; to my typists, Ruth Stratton, Lillian Johnson, and especially Dian Rapchak; and to Ruth Flohn for copy editing.

ROBERT J. McELIECE

Preface to the second edition

The main changes in this edition are in Part two. The old Chapter 8 ("BCH, Goppa, and Related Codes") has been revised and expanded into two new chapters, numbered 8 and 9. The old chapters 9, 10, and 11 have then been renumbered 10, 11, and 12. The new Chapter 8 ("Cyclic codes") presents a fairly complete treatment of the mathematical theory of cyclic codes, and their implementation with shift register circuits. It culminates with a discussion of the use of cyclic codes in burst error correction. The new Chapter 9 ("BCH, Reed–Solomon, and Related Codes") is much like the old Chapter 8, except that increased emphasis has been placed on Reed-Solomon codes, reflecting their importance in practice. Both of the new chapters feature dozens of new problems.

Introduction

In 1948, in the introduction to his classic paper, "A mathematical theory of communication," Claude Shannon[1,*] wrote:

> *"The fundamental problem of communication is that of reproducing at one point either exactly or approximately a message selected at another point."*

To solve that problem he created, in the pages that followed, a completely new branch of applied mathematics, which is today called *information theory* and/ or *coding theory*. This book's object is the presentation of the main results of this theory as they stand 30 years later.

In this introductory chapter we illustrate the central ideas of information theory by means of a specific pair of mathematical models, the *binary symmetric source* and the *binary symmetric channel*.

The binary symmetric source (the source, for short) is an object which emits one of two possible symbols, which we take to be "0" and "1," at a rate of R symbols per unit of time. We shall call these symbols *bits*, an abbreviation of *binary digits*. The bits emitted by the source are random, and a "0" is as likely to be emitted as a "1." We imagine that the source rate R is continuously variable, that is, R can assume any nonnegative value.

The binary symmetric channel (the BSC^2 for short) is an object through which it is possible to transmit one bit per unit of time. However, the channel is not completely reliable: there is a fixed probability p (called the *raw bit error probability*[3]), $0 \leqslant p \leqslant \frac{1}{2}$, that the output bit will not be the same as the input bit.

We now imagine two individuals, the sender and the receiver. The sender must try to convey to the receiver as accurately as possible the source output,

---

and the only communication link allowed between the two is the BSC described above. (However, we will allow the sender and receiver to get together before the source is turned on, so that each will know the nature of the data-processing strategies the other will be using.) We assume that both the sender and receiver have access to unlimited amounts of computing power, storage capacity, government funds, and other resources.

We now ask, For a given source rate R, how accurately can the sender communicate with the receiver over the BSC? We shall eventually give a very precise general answer to this question, but let's begin by considering some special cases.

Suppose $R = 1/3$. This means that the channel can transmit bits three times as fast as the source produces them, so the source output can be *encoded* before transmission by repeating each bit three times. For example, if the source's first five bits were 10100, the encoded stream would be 111000111000000. The receiver will get three versions of each source bit, but because of the channel "noise" these versions may not all be the same. If the channel garbled the second, fifth, sixth, twelfth, and thirteenth transmitted bits, the receiver would receive 101011111001100. A little thought should convince you that in this situation the receiver's best strategy for *decoding* a given source bit is to take the majority vote of the three versions of it. In our example he would decode the received message as 11100, and would make an error in the second bit. In general, a source bit will be received in error if either two or three of its three copies are garbled by the channel. Thus, if P_e denotes the *bit error probability*,

$$P_e = P \{2 \text{ channel errors}\} + P \{3 \text{ channel errors}\}$$

$$= 3p^2(1 - p) \qquad + p^3$$

$$= 3p^2 - 2p^3. \tag{0.1}$$

Since $p \leq \frac{1}{2}$, this is less than the raw bit error probability p; our simple coding scheme has improved the channel's reliability, and for very small p the relative improvement is dramatic.

It is now easy to see that even higher reliability can be achieved by repeating each bit more times. Thus, if $R = 1/(2n + 1)$ for some integer n, we could repeat each bit $2n + 1$ times before transmission (see Prob. 0.2) and use majority-vote decoding as before. It is simple to obtain a formula for the resulting bit error probability $P_e^{(2n+1)}$:

$$P_e^{(2n+1)} = \sum_{k=n+1}^{2n+1} P\,\{k \text{ channel errors out of } 2n+1 \text{ transmitted bits}\}$$

$$= \sum_{k=n+1}^{2n+1} \binom{2n+1}{k} p^k (1-p)^{2n+1-k}$$

$$= \binom{2n+1}{n+1} p^{n+1} + \text{terms of higher degree in } p. \qquad (0.2)$$

If $n > 1$, this approaches 0 much more rapidly as $p \to 0$ than the special case $n = 1$ considered above.[4] So in this rather weak sense the longer repetition schemes are more powerful than the shorter ones. However, we would like to make the stronger assertion that, for a fixed BSC with a fixed raw error probability $p < \frac{1}{2}$, $P_e^{(2n+1)} \to 0$ as $n \to \infty$, that is, by means of these repetition schemes the channel can be made as reliable as desired. It is possible but not easy to do this by studying formula (0.2) for $P_e^{(2n+1)}$. We shall use another approach and invoke the *weak law of large numbers*,[*] which implies that, if N bits are transmitted over the channel, then for any $\varepsilon > 0$

$$\lim_{N \to \infty} P\left\{\left|\frac{\text{number of channel errors}}{N} - p\right| > \varepsilon\right\} = 0. \qquad (0.3)$$

In other words, for large N, the fraction of bits received in error is unlikely to differ substantially from p. Thus we can make the following estimate of $P_e^{(2n+1)}$:

$$P_e^{(2n+1)} = P\left\{\text{fraction of transmitted bits received in error}\right.$$

$$\geq \frac{n+1}{2n+1} = \frac{1}{2} + \frac{1}{4n+2}\right\}$$

$$\leq P\{\text{fraction} > \tfrac{1}{2}\}$$

$$\leq P\{|\text{fraction} - p| > \tfrac{1}{2} - p\},$$

and so by (0.3) $P_e^{(2n+1)}$ does approach 0 as $n \to \infty$. We have thus reached the conclusion that if R is very small, it is possible to make the overall error probability very small as well, even though the channel itself is quite noisy. This is of course not particularly surprising.

---

[*] Discussed in Appendix A.

So much, temporarily, for rates less than 1. What about rates larger than 1? How accurately can we communicate under those circumstances?

If $R > 1$, we could, for example, merely transmit the fraction $1/R$ of the source bits and require the receiver to guess the rest of the bits, say by flipping an unbiased coin. For this not-very-bright scheme it is easy to calculate that the resulting bit error probability would be

$$P_e = \frac{1}{R} \times p + \frac{R-1}{R} \times \frac{1}{2}$$

$$= \tfrac{1}{2} - (\tfrac{1}{2} - p)/R. \tag{0.4}$$

Another, less uninspired method which works for some values of $R > 1$ will be illustrated for $R = 3$. If $R = 3$ there is time to transmit only one third of the bits emitted by the source over the channel. So the sender divides the source bits into blocks of three and transmits only the majority-vote of the three. For example if the source emits 101110101000101, the sender will transmit 11101 over the channel. The receiver merely triples each received bit. In the present case if the channel garbled the second transmitted bit he would receive 10101, which he would expand to 111000111000111, thereby making five bit errors. In general, the resulting bit error probability turns out to be

$$P_e = \tfrac{1}{4} \times (1 - p) + \tfrac{3}{4} \times p$$

$$= \tfrac{1}{4} + p/2. \tag{0.5}$$

Notice that this is less than $\tfrac{1}{3} + p/3$, which is what our primitive "coin-flipping" strategy gives for $R = 3$. The generalization of this strategy to other integral values of R is left as an exercise (see Prob. 0.4).

The schemes we have considered so far have been trivial, though perhaps not completely uninteresting. Let us now give an example which is much less trivial and in fact was unknown before 1948.

We assume now that $R = 4/7$, so that for every four bits emitted by the source there is just time to send three extra bits over the channel. We choose these extra bits very carefully: if the four source bits are denoted by x_0, x_1, x_2, x_3, then the extra or *redundant* or *parity-check* bits, labeled x_4, x_5, x_6, are determined by the equations

$$x_4 \equiv x_1 + x_2 + x_3 \quad (\bmod 2),$$

$$x_5 \equiv x_0 + x_2 + x_3 \quad (\bmod 2), \tag{0.6}$$

$$x_6 \equiv x_0 + x_1 + x_3 \quad (\bmod 2).$$

Thus, for example, if $(x_0, x_1, x_2, x_3) = (0110)$, then $(x_4, x_5, x_6) = (011)$, and the complete seven-bit *codeword* which would be sent over the channel is 0110011.

To describe how the receiver makes his estimate of the four source bits from a garbled seven-bit codeword, that is, to describe his *decoding algorithm*, let us rewrite the parity-check equations (0.6) in the following way:

$$
\begin{aligned}
x_1 + x_2 + x_3 + x_4 &= 0, \\
x_0 \quad + x_2 + x_3 \quad + x_5 &= 0, \\
x_0 + x_1 \quad + x_3 \quad\quad + x_6 &= 0.
\end{aligned}
\tag{0.7}
$$

(In (0.7) it is to be understood that the arithmetic is modulo 2.) Stated in a slightly different way, if the binary matrix H is defined by

$$
H = \begin{bmatrix} 0 & 1 & 1 & 1 & 1 & 0 & 0 \\ 1 & 0 & 1 & 1 & 0 & 1 & 0 \\ 1 & 1 & 0 & 1 & 0 & 0 & 1 \end{bmatrix},
$$

we see that each of the 16 possible codewords $\mathbf{x} = (x_0, x_1, x_2, x_3, x_4, x_5, x_6)$ satisfies the matrix-vector equation

$$
H\mathbf{x}^T = \begin{bmatrix} 0 \\ 0 \\ 0 \end{bmatrix}.
\tag{0.8}
$$

(In (0.8) the superscript T means "transpose.")

It turns out to be fruitful to imagine that the BSC adds (mod 2) either a 0 or a 1 to each transmitted bit, 0 if the bit is not received in error and 1 if it is. Thus if $\mathbf{x} = (x_0, x_1, \ldots, x_6)$ is transmitted, the received vector is $\mathbf{y} = (x_0 + z_0, x_1 + z_1, \ldots, x_6 + z_6)$, where $z_i = 1$ if the channel caused an error in the ith coordinate and $z_i = 0$ if not. Thus, if $\mathbf{z} = (z_0, \ldots, z_6)$ denotes the *error pattern*, then $\mathbf{y} = \mathbf{x} + \mathbf{z}$.

The receiver, who knows only $\mathbf{y}$ but wants to know $\mathbf{x}$, now does a very clever thing: he computes the following vector $\mathbf{s} = (s_0, s_1, s_2)$:

$$
\begin{aligned}
\mathbf{s}^T &= H\mathbf{y}^T \\
&= H(\mathbf{x} + \mathbf{z})^T \\
&= H\mathbf{x}^T + H\mathbf{z}^T \\
&= H\mathbf{z}^T \quad \text{(see (0.8))}.
\end{aligned}
\tag{0.9}
$$

Here $\mathbf{s}$ is called the *syndrome*[5] of $\mathbf{y}$; a 0 component in the syndrome indicates that the corresponding parity-check equation is satisfied by $\mathbf{y}$, a 1 indicates that it is not. According to (0.9), the syndrome does not depend on which codeword was sent, but only on the error pattern $\mathbf{z}$. However, since $\mathbf{x} = \mathbf{y} + \mathbf{z}$, if the receiver can find $\mathbf{z}$ he will know $\mathbf{x}$ as well, and so he focuses on the problem of finding $\mathbf{z}$. The equation $\mathbf{s}^T = H\mathbf{z}^T$ shows that $\mathbf{s}^T$ is the (binary) sum of those columns of H corresponding to 1's in $\mathbf{z}$, that is, corresponding to the bits of the codeword that were garbled by the channel:

$$\mathbf{s}^T = z_0 \begin{bmatrix} 0 \\ 1 \\ 1 \end{bmatrix} + z_1 \begin{bmatrix} 1 \\ 0 \\ 1 \end{bmatrix} + \cdots + z_6 \begin{bmatrix} 0 \\ 0 \\ 1 \end{bmatrix}. \tag{0.10}$$

The receiver's task, once he has computed $\mathbf{s}$, is to "solve" the equation $\mathbf{s}^T = H\mathbf{z}^T$ for $\mathbf{z}$. Unfortunately, this is only three equations in seven unknowns, and for any $\mathbf{s}$ there will always be 16 possibilities for $\mathbf{z}$. This is clearly progress, since there were a priori 128 possibilities for $\mathbf{z}$, but how can the receiver choose among the remaining 16? For example, suppose $\mathbf{y} = (0111001)$ was received. Then $\mathbf{s} = (101)$, and the 16 candidate $\mathbf{z}$'s turn out to be:

0	1	0	0	0	0	0		0	0	1	0	0	1	1
1	1	0	0	0	1	1		0	0	0	1	0	1	0
0	0	0	0	1	0	1		0	1	1	1	0	0	1
0	1	1	0	1	1	0		1	0	1	0	0	0	0
0	1	0	1	1	1	1		1	0	0	1	0	0	1
1	0	0	0	1	1	0		1	1	1	1	0	1	0
1	1	1	0	1	0	1		0	0	1	1	1	0	0
1	1	0	1	1	0	0		1	0	1	1	1	1	1

Faced with this set of possible error patterns, it is fairly obvious what to do: since the raw bit error probability p is $< \frac{1}{2}$, the fewer 1's (errors) in an error pattern, the more likely it is to have been the actual error pattern. In the current example, we're lucky: there is a unique error pattern (0100000) of least weight, the weight being the number of 1's. So in this case the receiver's best estimate of $\mathbf{z}$ (based both on the syndrome and on the channel statistics) is $\mathbf{z} = (0100000)$; the estimate of the transmitted codeword is $\mathbf{x} = \mathbf{y} + \mathbf{z} = (0011001)$; and finally, the estimate of the four source bits is (0011).

Of course we weren't really lucky in the above example, since we can show that for any syndrome $\mathbf{s}$ there will always be a unique solution to $H\mathbf{z}^T = \mathbf{s}^T$ of weight 0 or 1. To see this, notice that if $\mathbf{s} = (000)$, then $\mathbf{z} = (0000000)$ is the desired solution. But if $\mathbf{s} \neq (000)$, then $\mathbf{s}^T$ must occur as one of the columns of H; if $\mathbf{s}^T$ = the ith column of H, then the error pattern $\mathbf{z}$, which has one 1 in the ith position and 0's elsewhere, is the unique minimum-weight solution to $H\mathbf{z}^T = \mathbf{s}^T$.

We can now formally describe a *decoding algorithm* for this scheme, which is called the (7, 4) *Hamming code*. Given the received vector $\mathbf{y}$, the receiver executes the following steps:

1. Compute the syndrome $\mathbf{s}^T = H\mathbf{y}^T$.
2. If $\mathbf{s} = \mathbf{0}$, set $\hat{\mathbf{z}} = 0$; go to 4.
3. Locate the unique column of H which is equal to $\mathbf{s}$; call it column i; set $\hat{\mathbf{z}}$ = all 0's except for a single 1 in the ith coordinate.
4. Set $\hat{\mathbf{x}} = \mathbf{y} + \hat{\mathbf{z}}$. (This is the decoder's estimate of the transmitted code-word.)
5. Output $(\hat{x}_0, \hat{x}_1, \hat{x}_2, \hat{x}_3)$, the first four components of $\mathbf{x}$. (This is the decoder's estimate of the original source bits.)

It is of course possible that the vector $\hat{\mathbf{z}}$ produced by this algorithm will not be equal to the actual error pattern $\mathbf{z}$. However, if the channel causes at most one error, that is, if the weight of $\mathbf{z}$ is 0 or 1, then it follows from the above discussion that $\hat{\mathbf{z}} = \mathbf{z}$. Thus the Hamming code is a *single-error-correcting code*. In fact it is easy to see that the above decoding algorithm will fail to correctly identify the original codeword $\mathbf{x}$ iff the channel causes two or more errors. Thus, if P_E denotes the *block* error probability $P\{\hat{\mathbf{x}} \neq \mathbf{x}\}$,

$$P_E = \sum_{k=2}^{7} \binom{7}{k} p^k (1-p)^{7-k}$$

$$= 21p^2 - 70p^3 + \text{etc.}$$

Of course the block error probability P_E doesn't tell the whole story, for even if $\hat{\mathbf{x}} \neq \mathbf{x}$, some of the components of $\hat{\mathbf{x}}$ may nevertheless be right. If we denote the bit error probability $P\{\hat{x}_i \neq x_i\}$ by $P_e^{(i)}$, it is possible to show that, for all $0 \leq i \leq 6$,

$$P_e^{(i)} = 9p^2(1-p)^5 + 19p^3(1-p)^4 + 16p^4(1-p)^3$$

$$+ 12p^5(1-p)^2 + 7p^6(1-p) + p^7$$

$$= 9p^2 - 26p^3 + \text{etc.} \tag{0.11}$$

Comparing this to (0.1), we see that for BSC's with very small raw error probabilities the Hamming code performs at rate $4/7 = 0.571$ about as well as the crude repetition scheme at rate $1/3 = 0.333$.

We could also use the (7, 4) Hamming code to communicate at $R = 7/4$ by reversing the roles of sender and receiver. Here the sender would partition the sequence of source bits into blocks of seven, reduce each block of seven to only four via the above decoding algorithm (which in this context would become an "encoding algorithm"), and transmit these four bits over the channel. The receiver would decode the four received bits by adding three extra bits, computed by the parity-check rules (0.6). For this scheme the resulting bit error probability $P_e^{(i)} = P\{\hat{x}_i \neq x_i\}$ is not independent of i, but the average $P_e = \left(\sum_{i=0}^{6} P_e^{(i)} \right)/7$ is given by

$$P_e = \frac{1}{8}(1-p)^4 + \frac{53}{28}(1-p)^3 p + 3(1-p)^2 p^2 + \frac{59}{28}(1-p)p^3 + \frac{7}{8}p^4$$

$$= \frac{1}{8} + \frac{39}{28}p + \text{etc.} \tag{0.12}$$

For a noiseless ($p = 0$) BSC, this is much superior, for example, to the "coin-flipping" technique for $R = 7/4$, which from (0.4) gives $P_e = \frac{3}{14} = .214$.

Let us summarize what we know so far by specializing to a particular BSC, say $p = .1$, and for each of the communication schemes discussed so far placing a point on the (x, y) plane, with $x = R$, the rate, and $y = P_e$, the overall bit error probability, as shown in Fig. 0.1. Given sufficient patience and ingenuity, we could continue inventing ad hoc schemes and putting points on Fig. 0.1. Our eventual goal would be, of course, to learn which points are achievable and which are not. Incredibly, this goal has already been reached by Shannon. But before giving Shannon's result, let us formalize somewhat the concept of a rate R coding scheme with associated bit error probability P_e.

As suggested by Fig. 0.2, an (n, k) code is a scheme in which the source sequence is partitioned into blocks of k bits, and in which each k-bit source **u** block is mapped ("encoded") into an n-bit codeword **x**, which is transmitted over the channel and received, possibly garbled, as **y**. The decoder maps the n-bit noisy codeword **y** into a k-bit block **v**, which is an estimate of the original source sequence **u**. The *rate* of this communication system is $R = k/n$; the bit error probability is defined as

$$P_e = \frac{1}{k} \sum_{i=1}^{k} P_e^{(i)},$$

where

$$P_e^{(i)} = P\{v_i \neq u_i\}, \qquad i = 1, 2, \ldots, k.$$

(You should be able to see immediately how each of the schemes described so

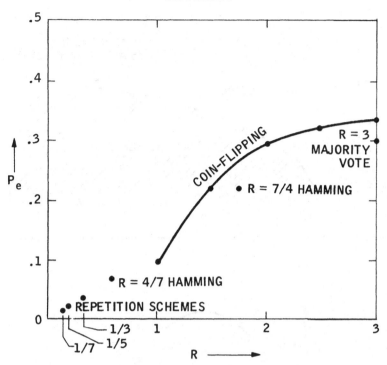

Figure 0.1 Some achievable (R, P_e) pairs for a BSC with $p = .1$.

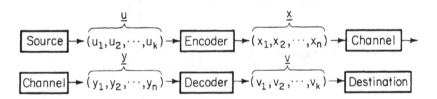

Figure 0.2 An (n, k) code for the binary symmetric source and BSC.

far, with the possible exception of the "coin-flipping" strategy for $R \geq 1$, fits this description; see Prob. 0.5.) We say that a point (x, y) in Fig. 0.1 is "achievable" if there exists such an (n, k) code with $k/n \geq x$, $P_e \leq y$. Not to prolong the suspense, Fig. 0.3 shows the set of achievable points for our special BSC $(p = .1)$. Of course the crucial thing to know about Fig. 0.3 is the description of the boundary between the achievable and nonachievable regions. In order to give the description, we need to introduce the important *binary entropy function*:

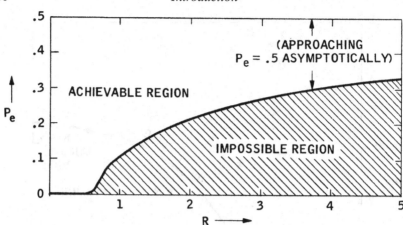

Figure 0.3 The achievable (R, P_e) pairs for a binary symmetric source and a BSC $(p = .1)$.

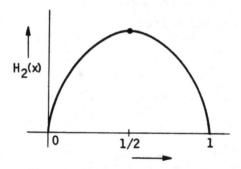

Figure 0.4 The binary entropy function.

$$H_2(x) = -x \log_2 x - (1 - x) \log_2(1 - x), \quad 0 < x < 1,$$

$$H_2(0) = H_2(1) = 0. \tag{0.13}$$

A graph of $y = H_2(x)$ is shown in Fig. 0.4. (Some important properties of $H_2(x)$ are described in Prob. 0.10.) We can now describe the boundary between the achievable and nonachieveable regions in Fig. 0.3. The curved part of the boundary is the set of points (R, P_e) satisfying

$$R = \frac{1 - H_2(.1)}{1 - H_2(P_e)}, \quad 0 \leqslant P_e < \tfrac{1}{2}. \tag{0.14}$$

The remainder of the boundary is a segment of the R axis, from $R = 0$ to

$R = 1 - H_2(.1) = 0.531$. For a general BSC, the story is just the same, except that (0.14) is replaced by

$$R = \frac{1 - H_2(p)}{1 - H_2(P_e)},\tag{0.15}$$

and the corresponding segment of the R axis runs from $R = 0$ to $R = 1 - H_2(p)$, as shown in Fig. 0.5.

There are many remarkable things about the results sketched in Figs. 0.3 and 0.5, but the most remarkable is this: If $R < 1 - H_2(p)$, then any positive P_e, however small, is achievable! For example, if $p = .1$, according to Fig. 0.3, there should exist a code with rate $\geqslant 0.5$ and overall error probability $< 10^{-500}$! The important number $1 - H_2(p)$ is called the *capacity* of the channel; this particular implication of Fig. 0.5 is called the *channel coding theorem*; it says that *arbitrarily reliable communication is possible at any rate below channel capacity*. (Compare this to our earlier result, proved via repetition codes, that as $R \to 0$, P_e could be made to approach 0 as well.)

It would not be much of an exaggeration to say that this entire book is devoted to the study of Fig. 0.5 and its generalizations. In Part one we shall describe (complete with proofs) the "achievable" region for a large class of source–channel pairs. (The description is given in Chapter 5, which the reader is encouraged to skim as soon as possible.) However, the proofs given are

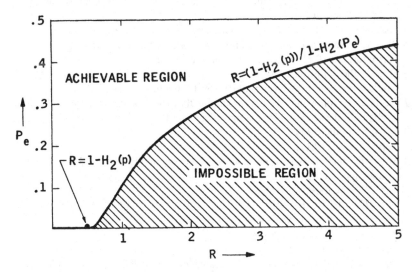

Figure 0.5 The achievable (R, P_e) pairs for a binary symmetric source and a general BSC.

somewhat unsatisfactory, since they only assert the existence of, but do not describe explicitly, the corresponding codes. In Part two this problem will be partially remedied, for there we shall describe in detail some of the important practical codes in use today.

Problems

0.1 Explain why the assumption $0 \leqslant p \leqslant \frac{1}{2}$ about the raw bit error probability of the BSC (see p. 1) involves no essential loss of generality.

0.2 Suppose $R = 1/2n$ for some integer n, and consider a "repeat each bit $2n$ times" strategy for communicating over the BSC.
  (a) Devise a suitable decoding strategy, and compute the resulting bit error probability P_e.
  (b) Show that the formula you obtained in part (a) is exactly the same as formula (0.2) for the performance of the $R = 1/(2n - 1)$ repetition scheme.

0.3 Verify formula (0.5).

0.4 Suppose $R = 2n + 1$ for some integer n, and consider a "send the majority vote of each $2n + 1$ successive source bits" strategy analogous to the $R = 3$ scheme described on p. 4.
  (a) Show that $P_e = (1 - p)Q + p(1 - Q)$, where

$$Q = \frac{1}{2} - \binom{2n}{n} 2^{-(2n+1)}.$$

  (b) Show that the performance of the analogous $R = 2n$ scheme yields exactly the same value of P_e.

0.5 Assuming $R \geqslant 1$ and R is rational, show how to convert the vague "coin-flipping" strategy on p. 4 into an (n, k) code in the sense of Fig. 0.2. Then verify formula (0.4) for the resulting bit error probability.

0.6 Decode the following vectors, assuming they are "noisy" versions of codewords from the $R = 4/7$ Hamming code: 1100000, 1010101, 0111100.

0.7 Decode the following vectors, assuming they are "noisy" versions of codewords from the $R = 7/4$ Hamming code: 1010, 0001, 0111.

0.8 Verify formula (0.11).

0.9 Verify formula (0.12).

0.10 Show that the binary entropy function $H_2(x)$ defined by Eq. (0.13), enjoys the following properties:
  (a) $H_2'(x) = \log_2(1 - x)/x$.
  (b) $H_2''(x) = -[x(1 - x) \log 2]^{-1}$.
  (c) $H_2(x) \leqslant 1$ with equality iff $x = \frac{1}{2}$.
  (d) $H_2(x) \geqslant 0$; $\lim_{x \to 0,1} H_2(x) = 0$.
  (e) $H_2(x) = H_2(1 - x)$.

0.11 Does the boundary between the achievable and impossible regions, call it $B(R)$, have a continuous derivative for all $R \geqslant 0$? Compute $\lim_{R \to \infty} B(R)$, and interpret your result in communications terms.

Notes

1 (p. 1). Claude Elwood Shannon was born in Petoskey, Michigan, on April 30, 1916. He did his undergraduate work at the University of Michigan, majoring in electrical engineering and mathematics, and got his B. S. in 1936. He earned his Ph.D. at the Massachusetts Institute of Technology in 1940, and after spending a year at the Institute for Advanced Study in Princeton, he joined the technical staff of the Bell Telephone Laboratories in Princeton, New Jersey.

In 1941 Shannon began a serious study of communications problems, motivated in part by the needs of the war effort, and this research culminated in the publication of "A mathematical theory of communication" in 1948. (This paper is reprinted in full in [25].)

With many profound scientific discoveries (for example Einstein's discovery in 1905 of the special theory of relativity) it is possible with the aid of hindsight to see that the times were ripe for a breakthrough. Not so with information theory. While of course Shannon was not working in a vacuum in the 1940's, his results were so breathtakingly original that even the communication specialists of the day were at a loss to understand their significance. Gradually, as Shannon's theorems were digested by the mathematical/ engineering community, it became clear that he had created a brand-new science, and others began to make first-rate contributions of their own. Slowly at first, and then more rapidly, the subject grew, until now hundreds of research papers in information theory are published each year.

Thus Shannon is universally acknowledged as the unique father of information theory, solely on the basis of his 1948 paper. But in addition, he is also universally acknowledged as the most important post-1948 contributor to the subject! Nearly every one of his papers since "A mathematical theory of communication" has proved to be a priceless source of research ideas for lesser mortals. (For example, in a 1973 collection of the key papers in the development of information theory [25], Shannon was the author or coauthor of 12 of the 49 papers cited. No other author appeared more than three times.)

2 (p. 1). In the Los Angeles, California, area channel 52 (KBSC) is the binary symmetric channel.

3 (p. 1). Sometimes p is also called the *transition* probability or the *crossover* probability for the BSC.

4 (p. 3). The derivation of this formula for P_e deserves further comment. The *binomial coefficient* $\binom{N}{K}$ ("N choose K") denotes the number of ways of choosing K objects without repetition out of a set of N objects. Numerically it is equal to $N(N-1)(N-2)\ldots(N-K+1)/K(K-1)\ldots 2\cdot 1$. (For further properties of binomial coefficients, see Knuth [7], Vol. 1.) So there are $\binom{2n+1}{k}$ different ways the channel could cause exactly k errors among the $2n+1$ transmitted bits. Each such pattern occurs with probability $p^k(1-p)^{2n+1-k}$, and so the term $\binom{2n+1}{k}p^k(1-p)^{2n+1-k}$ is equal to the probability that exactly k of the $2n+1$ copies of the source bit are received in error.

5 (p. 5). In medical parlance a syndrome is a pattern of symptoms that aids in the diagnosis of a disease. Here the "disease" is the error pattern, and a "symptom" is a parity-check failure. This felicitous coinage is due to Hagelbarger.

Part one
Information theory

1

Entropy and mutual information

1.1 Discrete random variables

Suppose X is a discrete random variable, that is, one whose range $R = \{x_1, x_2, \ldots\}$ is finite or countable. Let $p_i = P\{X = x_i\}$. (For probabilistic terminology consult Appendix A.) The *entropy* of X is defined by

$$H(X) = \sum_{i \geq 1} p_i \log \frac{1}{p_i}. \tag{1.1}$$

This definition needs elaboration. First, the base of the logarithm is purposely left unspecified. If necessary, however, we shall denote the base-b entropy by $H_b(X)$, and say that the entropy of X is being measured in base-b units. Base-2 units are called *bits* (binary igits), and base-e units are called *nats* (natural digits). Second, if $p_i = 0$, the term $p_i \log p_i^{-1}$ in (1.1) is indeterminate; we define it to be 0, however. (This convention is by no means arbitrary; see Prob. 1.1.) Finally, if R is infinite the sum (1.1) may not converge; in this case we set $H(X) = +\infty$.

Example 1.1 Let X represent the outcome of a single roll of a fair die. Then $R = \{1, 2, 3, 4, 5, 6\}$ and $p_i = \frac{1}{6}$ for each i. Here $H(X) = \log 6 = 2.58$ bits $= 1.79$ nats. $\square$

Example 1.2 Let $R = \{0, 1\}$, and define X by $P\{X = 0\} = p$, $P\{X = 1\} = 1 - p$. Then $H(X) = -p \log p - (1 - p) \log(1 - p)$, and so $H_2(X)$, as a function of $0 \leq p \leq 1$, is identical to the binary entropy function $H_2(p)$, which was defined in Eq. (0.13). In what follows, we will frequently represent the function $-p \log p - (1 - p) \log(1 - p)$, where the bases of the logarithms are unspecified, by $H(p)$, and call it the *entropy function*. Figure 1.1 gives its graph (cf. Fig. 0.4). More generally, if $\mathbf{p} = (p_1, \ldots, p_r)$ is any *probability*

17

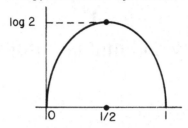

Figure 1.1 The entropy function $H(p)$.

vector, that is, $p_i \geq 0$ and $\sum p_i = 1$, we define $H(\mathbf{p}) = H(p_1, p_2, \ldots, p_r) = \sum p_i \log p_i^{-1}$. This notation is not quite consistent, since for $r = 2$ we have $H(p, 1 - p) = H(p)$. (Thus we use the symbol H in three slightly different ways: $H(X)$ is the entropy of the random variable X; $H(p) = -p \log p - (1 - p) \log(1 - p)$ for $0 \leq p \leq 1$; and $H(p_1, p_2, \ldots, p_r) = \sum p_i \log p_i^{-1}$ if $\mathbf{p}$ is a probability vector.)   □

Example 1.3 If the sum $\sum_{n=2}^{\infty}(n \log^2 n)^{-1}$ is denoted by A, and if the random variable X is defined by $P\{X = n\} = (An \log^2 n)^{-1}$ for $n = 2, 3, \ldots$, then $H(X) = +\infty$. (See Prob. 1.2.)   □

It turns out that $H(X)$ can be thought of as a measure of the following things about X:

(a) The amount of "information" provided by an observation of X.
(b) Our "uncertainty" about X.
(c) The "randomness" of X.

In the next few paragraphs we will discuss these properties informally, but the reader should be told immediately that $H(X)$ does in fact measure these things in a deep mathematical sense as well. Indeed there are many possible functions of a random variable X that share the properties to be discussed below, but only $H(X)$ will do for the study of communications problems.

For each $x \in R$ define $I(x) = -\log P\{X = x\}$. Then I is a new random variable, and $H(X)$ is its average. The function $I(x)$ (see Fig. 1.2) can be interpreted as the amount of information provided by the event $\{X = x\}$. According to this interpretation, the less probable an event is, the more information we receive when it occurs. A certain event (one that occurs with probability 1) provides no information, whereas an unlikely event provides a very large amount of information. For example, suppose you visited an oracle

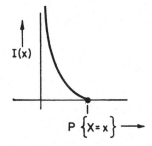

Figure 1.2 The function $I(x)$.

who could answer any "yes or no" question. If you asked, "Will I live to be 125?" and got a "no" answer, you would have gained very little information, since such extreme longevity is exceedingly improbable. Conversely, if you got a "yes," you would have learned much. If now millions of people visited the oracle and asked the same question, most would get a "no," a few would get a "yes," and the average amount of information provided would be $H(p)$, where $p = P\{\text{age at death} \geq 125\}$. Moreover, just before receiving the oracle's reply you would probably be slightly anxious; this reflects the fact that a small amount of uncertainty exists about the answer. $H(p)$ is equally a measure of this uncertainty.[1] Finally, if a dispassionate census worker were assigned to record the oracle's answers, he would become extremely bored and might begin to suspect the oracle of being a machine that always says "no." This reflects the fact that the random variable X representing the oracle's reply is not very random. Here $H(p)$ measures the randomness of X.

As a less transcendental example, define X by $P\{X = 0\} = P\{X = 1\} = \frac{1}{2}$. Then $I(0) = I(1) = H(X) = \log 2 = 1$ bit, that is, the observation of the "bit" X provides one "bit" of information.

Our first theorem concerns the maximum possible value for $H(X)$ in terms of the size of R.

Theorem 1.1 *Let X assume values in $R = \{x_1, x_2, \ldots, x_r\}$. Then $0 \leq H(X) \leq \log r$. Furthermore $H(X) = 0$ iff $p_i = 1$ for some i, and $H(X) = \log r$ iff $p_i = 1/r$ for all i.*

Proof Since each p_i is ≤ 1, each term $p_i \log p_i^{-1}$ in (1.1) is ≥ 0, so $H(X) \geq 0$. Furthermore $p \log p^{-1} = 0$ iff $p = 0$ or 1, and so $H(X) = 0$ iff each $p_i = 0$ or 1, i.e., one $p_i = 1$ and all the rest are 0.

Now by Jensen's inequality (see Appendix B), since $\log x$ is strictly convex $\cap$,

$$H(X) = \sum_{i=1}^{r} p_i \log \frac{1}{p_i} \leq \log \sum_{i=1}^{r} p_i \frac{1}{p_i} = \log r,$$

with equality iff p_i is a constant independent of i, i.e., $p_i = 1/r$ for all i.   □

Informally, Theorem 1.1 identifies a uniformly distributed random variable as the most "random" kind of random variable. Formally, it asserts that the maximum value of the function $H(p_1, p_2, \ldots, p_r)$, as $\mathbf{p} = (p_1, \ldots, p_r)$ ranges over the $r - 1$ dimensional simplex $\{p_i \geq 0, \sum p_i = 1\}$, is $\log r$ and is achieved uniquely at $\mathbf{p} = (1/r, 1/r, \ldots, 1/r)$.

Our next goal is to define, for a pair of random variables X and Y, a quantity $H(X|Y)$ called the *conditional entropy*[2] of X, given Y. In order to do this neatly, we introduce some streamlined notation. For x in the range of X, y in the range of Y, define:

$$p(x) = P\{X = x\},$$

$$p(y) = P\{Y = y\},$$

$$p(x, y) = P\{X = x, Y = y\}, \tag{1.2}$$

$$p(x|y) = P\{X = x | Y = y\} = p(x, y)/p(y),$$

$$p(y|x) = P\{Y = y | X = x\} = p(x, y)/p(x).$$

(This notation is occasionally ambiguous, and if absolutely necessary appropriate subscripts will be added, for example, $p_X(x)$, $p_{Y|X}(y, x)$. This need will arise, however, only when actual numbers are substituted for the letters x, y; see Example 1.6.) Our definition is:

$$H(X|Y) = E\left[\log \frac{1}{p(x|y)}\right]$$

$$= \sum_{x,y} p(x, y) \log \frac{1}{p(x|y)}. \tag{1.3}$$

(In (1.3) we observe the same conventions as we did for sum (1.1): $0 \log 0^{-1} = 0$; a divergent sum means $H(X|Y) = +\infty$.) Let us pause to motivate the definition via a simple model for a communications channel, called a *discrete memoryless channel* (DMC).

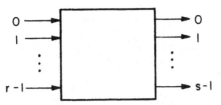

Figure 1.3 A discrete memoryless channel.

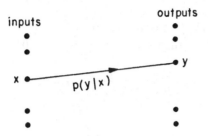

Figure 1.4 Another view of a DMC.

A DMC (Fig. 1.3) is an object that accepts, every unit of time, one of r input symbols, and in response expels one of s output symbols. (This channel is "discrete" because there are only finitely[3] many input and output symbols, "memoryless" because the current output depends only on the current input and not on any of the previous ones.) The precise labeling of the input and output symbols is of no real importance, but it is often convenient to let $\{0, 1, \ldots, r-1\}$ and $\{0, 1, \ldots, s-1\}$ represent the input and output alphabets.

The output is not a definite function of the input, however; rather the channel's behavior is governed by an $r \times s$ matrix of *transition probabilities* $(p(y|x))$. The number $p(y|x)$ represents the probability that y will be the output, given that x is the input. Clearly the number $p(y|x)$ must satisfy

$$p(y|x) \geq 0 \qquad \text{for all } x, y,$$

$$\sum_y p(y|x) = 1 \qquad \text{for all } x.$$

Sometimes. when r and s are not too big, the DMC is depicted graphically as shown in Fig. 1.4. In such a picture each pair (x, y) with $p(y|x) > 0$ is joined by a line labeled with the number $p(y|x)$.

Example 1.4 (the binary symmetric channel, already discussed in the introduction). Here $r = s = 2$, and the graph looks like this:

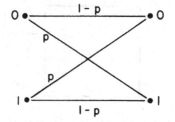

Example 1.5 (The binary erasure channel). Here $r = 2$, $s = 3$. The inputs are labeled "0" and "1," and the outputs are labeled "0," "1," and "?."

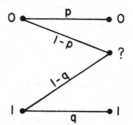

Such a channel might arise in practice for example if the inputs to a physical channel were the two squarewaves.

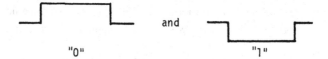

The detector at the output would receive a noisy version of these square waves, $r(t)$:

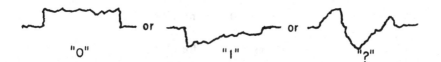

It might base its decision about whether "0" or "1" was sent on the value of the integral $\int r(t)\, dt = I$. If I is positive, the detector could decide "0" was sent; if negative, "1." However, if $|I|$ is very small, it might be best not to make a "hard decision" about the transmitted bit, but rather to output a special erasure symbol "?." If the channel is relatively quiet, the transitions $0 \to 1$ and $1 \to 0$ would be much less likely than $0 \to ?$ and $1 \to ?$, so the assumptions $P\{Y = 1|X = 0\} = P\{Y = 0|X = 1\} = 0$ might be reasonable. (For more on "hard decisions," see Prob. 4.15.) □

Suppose now that the inputs to a DMC are selected according to a probability distribution $p(x)$ on $\{0, 1, \ldots, r - 1\}$, that is, assume the input X to the channel is characterized by

$$P\{X = x\} = p(x), \qquad x \in \{0, 1, \ldots, r - 1\}.$$

Having specified X, we can now define a random variable Y which will represent the *output* of the channel. The joint distribution of X and Y is given by

$$p(x, y) = P\{X = x, Y = y\}$$
$$= P\{X = x\}P\{Y = y|X = x\}$$
$$= p(x)p(y|x),$$

and the marginal distribution of Y is

$$p(y) = P\{Y = y\}$$
$$= \sum_x P\{Y = y|X = x\}P\{X = x\}$$
$$= \sum_x p(y|x)p(x).$$

Similarly,

$$p(x|y) = p(x, y)/p(y)$$
$$= p(y|x)p(x)/\sum_{x'} p(y|x')p(x').$$

Hence corresponding to every DMC and input distribution there is a pair of random variables: X, the "input" to, and Y the "output" from, the channel. Conversely, given any pair (X, Y) of discrete random variables, there exist a DMC and input distribution such that X is the input and Y is the output: simply define the channel's transition probabilities by $p(y|x) = P\{Y = y|X = x\}$. In other words, given any ordered pair (X, Y) of random variables, it is possible to think of Y as a "noisy" version of X, that is, as the result of transmitting X through a certain DMC.

Example 1.6 Let X assume the values ± 1, ± 2, each with probability $\frac{1}{4}$, and let $Y = X^2$. The corresponding DMC looks like this:

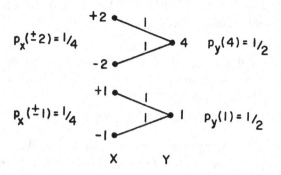

In this example X and Y are uncorrelated, and yet it is clear that Y provides a considerable amount of "information" about X (see Prob. 1.10). □

Given that we think of Y as a noisy version of X, and that $H(X)$ is a measure of our prior uncertainty about X, how can we measure our uncertainty about X after observing Y? Well, suppose we have observed that $Y = y$. Then, since the numbers $p(x|y) = P\{X = x | Y = y\}$ for fixed y represent the conditional distribution of X, given that $Y = y$, we define the *conditional entropy* of X, given $Y = y$:

$$H(X|Y = y) = \sum_x p(x|y) \log \frac{1}{p(x|y)}.$$

This quantity is itself a random variable defined on the range of Y; let us define the *conditional entropy* $H(X|Y)$ as its expectation:

$$H(X|Y) = \sum_y p(y) H(X|Y = y)$$

$$= \sum_y p(y) \sum_x p(x|y) \log \frac{1}{p(x|y)}$$

$$= \sum_{x,y} p(x, y) \log \frac{1}{p(x|y)},$$

in agreement with Eq. (1.3). Thus, for a given pair X, Y of random variables, $H(X|Y)$ *represents the amount of uncertainty remaining about X after Y has been observed*.

Example 1.7 Consider the following DMC, which is a particular case of the binary erasure channel of Example 1.5:

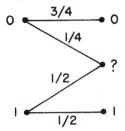

Here $p_X(0) = \frac{2}{3}$, $p_X(1) = \frac{1}{3}$. Then a simple calculation yields:

$$H_2(X) = 0.9183 \text{ bits},$$

$$H_2(X|Y = 0) = 0,$$

$$H_2(X|Y = 1) = 0,$$

$$H_2(X|Y = ?) = 1.$$

Thus, if $Y = 0$ or 1, there is no remaining uncertainty about X, but if $Y = ?$, we are more uncertain about X after receiving Y than before! However,

$$H_2(X|Y) = 0.3333 \text{ bits},$$

so that, on the average, at least, an observation of Y reduces our uncertainty about X. □

We now present a technical lemma on $H(X|Y)$ that will be useful later.

Theorem 1.2 *Let X, Y, Z be discrete random variables. Using obvious notation (see Eqs. (1.2)), define, for each z, $A(z) = \sum_{x,y} p(y) p(z|x, y)$. Then*

$$H(X|Y) \leqslant H(Z) + E(\log A).$$

Proof

$$H(X|Y) = E\left[\log \frac{1}{p(x|y)}\right]$$

$$= \sum_{x,y,z} p(x, y, z) \log \frac{1}{p(x|y)}.$$

$$= \sum_z p(z) \sum_{x,y} \frac{p(x, y, z)}{p(z)} \log \frac{1}{p(x|y)}.$$

For fixed z, $p(x, y, z)/p(z) = p(x, y|z)$ is a probability distribution, and so we can apply Jensen's inequality to the inner sum. The result is

$$H(X|Y) \leqslant \sum_z p(z) \log \left[\frac{1}{p(z)} \cdot \sum_{x,y} \frac{p(x, y, z)}{p(x|y)} \right]$$

$$= \sum_z p(z) \log \frac{1}{p(z)} + \sum_z p(z) \log \sum_{x,y} \frac{p(x, y, z)}{p(x|y)}.$$

But $p(x, y, z)/p(x|y) = p(x, y, z)p(y)/p(x, y) = p(y)p(z|x, y)$. $\square$

Corollary (*"Fano's inequality"*). *Let X and Y be random variables, each taking values in the set $\{x_1, x_2, \ldots, x_r\}$. Let $P_e = P\{X \neq Y\}$. Then*

$$H(X|Y) \leqslant H(P_e) + P_e \log(r - 1).$$

Proof In Theorem 1.2 define $Z = 0$ if $X = Y$ and $Z = 1$ if $X \neq Y$. Then $A(0) = 1$ and $A(1) = r - 1$. $\square$

[*Note:* The proof of Theorem 1.2 via our streamlined notation contains some subtle features; see Prob. 1.11.]

Fano's inequality has an interesting heuristic interpretation. Think of $H(X|Y)$ as the amount of information needed to determine X once Y is known. One way to determine X is to first determine whether or not $X = Y$; if $X = Y$, we are done. If, however, $X \neq Y$, there are $r - 1$ remaining possibilities for X. Determining whether or not $X = Y$ is equivalent to determining the random variable Z defined in the proof; since $H(Z) = H(P_e)$, it takes $H(P_e)$ bits to do this. If $X \neq Y$ (this happens with probability P_e), the amount of information needed to find out which of the remaining $r - 1$ values X has is, by Theorem 1.1. at most $\log(r - 1)$.

Example 1.8 We apply Fano's inequality to the channel of Example 1.7. Here $r = 3$, and $P\{X = Y\} = \frac{2}{3}$, $P_e = \frac{1}{3}$. Fano's bound is thus $H(X|Y) \leqslant H\left(\frac{1}{3}\right) + \frac{1}{3}\log 2 = \frac{1}{3}\log 3 + \frac{2}{3}\log \frac{3}{2} + \frac{1}{3}\log 2 = \log 3 - \frac{1}{3}\log 2 = 1.2520$ bits. (For examples where Fano's inequality does better, see Prob.1.11.) $\square$

Now since $H(X)$ represents our uncertainty about X before we know Y, and $H(X|Y)$ represents our uncertainty after, the difference $H(X) - H(X|Y)$ must represent the amount of information provided about X by Y. This important quantity is called the *mutual information* between X and Y, and is denoted by $I(X; Y)$:

$$I(X; Y) = H(X) - H(X|Y). \tag{1.4}$$

(In Example 1.7, $I_2(X; Y) = 0.9183 - 0.3333 = 0.5850$; thus, informally at least, the observation of a channel output provides 0.5850 bits of information about the input, on the average.) Using the notation of Eq. (1.2), we obtain several important alternative forms for $I(X; Y)$:

$$I(X; Y) = \sum_{x,y} p(x, y) \log \frac{p(x|y)}{p(x)}, \tag{1.5}$$

$$= \sum_{x,y} p(x, y) \log \frac{p(x, y)}{p(x)p(y)}, \tag{1.6}$$

$$= \sum_{x,y} p(x, y) \log \frac{p(y|x)}{p(y)}. \tag{1.7}$$

(The details are left as Prob. 1.14.)

We thus see that $I(X; Y)$ is the average, taken over the X, Y sample space, of the random variable[4]

$$I(x; y) = \log \frac{p(x|y)}{p(x)} = \log \frac{p(x, y)}{p(x)p(y)} = \log \frac{p(y|x)}{p(y)}.$$

Now $I(x; y)$ can be either positive or negative (e.g., in Example 1.7 $I(0; 0) = \log \frac{3}{2}$ and $I(0; ?) = \log \frac{3}{4}$); however, we shall now prove the important fact that $I(X; Y)$ cannot be negative. This is surely reasonable, given our heuristics: we do not expect to be misled (on the average) by observing the output of the channel.

Theorem 1.3 *For any discrete random variables X and Y, $I(X; Y) \geqslant 0$. Moreover $I(X; Y) = 0$ if and only if X and Y are independent.*

Proof We apply Jensen's inequality to Eq. (1.6):

$$-I(X; Y) = \sum_{x,y} p(x, y) \log \frac{p(x)p(y)}{p(x, y)}$$

$$\leqslant \log \sum_{x,y} p(x)p(y)$$

$$= \log 1 = 0.$$

Furthermore, in view of the strict convexity $\cap$ of $\log x$, equality holds iff $p(x)p(y) = p(x, y)$ for all x, y, that is, iff X and Y are independent. $\square$

(Although we shall not emphasize it, Theorem 1.3 shows that $I(X; Y)$ is a good measure of the dependence between X and Y, better for example than the covariance $\text{Cov}(X; Y)$. for example, recall Example 1.6. There, as is easily verified, $\text{Cov}(X; Y) = 0$ but $I_2(X; Y) = 1$ bit.)

Using Eqs. (1.4)–(1.7), it is possible to prove immediately several important facts about mutual information:

$$I(X; Y) = I(Y; X), \tag{1.8}$$

$$I(X; Y) = H(Y) - H(Y|X), \tag{1.9}$$

$$I(X; Y) = H(X) + H(Y) - H(X, Y), \tag{1.10}$$

where in (1.10) we have defined the *joint entropy* of X and Y by

$$H(X, Y) = \sum_{x,y} p(x, y) \log \frac{1}{p(x, y)}. \tag{1.11}$$

The proofs of these relationships are left as Prob. 1.14. They can be easily remembered by means of the Venn diagram shown in Fig. 1.5. It is a fruitful exercise to give informal interpretations of each of the relations implied by Fig. 1.5. For example, Eq. (1.8) expresses the "mutuality" of mutual information; $H(X, Y) = H(X) + H(Y|X)$ becomes "our uncertainty about X and Y is the sum of our uncertainty about X and our uncertainty about Y, once X is known," and so on.

Now if we are given three random variables X, Y, Z, we define the mutual information $I(X, Y; Z)$ ("the amount of information X and Y provide about Z"), analogously with Eq. (1.7), by

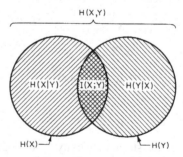

Figure 1.5 A mnemonic Venn diagram for Eqs. (1.4) and (1.8)–(1.10).

$$I(X, Y; Z) = E\left[\log \frac{p(z|x, y)}{p(z)}\right]$$

$$= \sum_{x,y,z} p(x, y, z) \log \frac{p(z|x, y)}{p(z)}.$$

We would not expect X and Y together to provide less information about Z than Y alone does, and indeed this is the case.

Theorem 1.4 $I(X, Y; Z) \geq I(Y; Z)$, *with equality iff* $p(z|x, y) = p(z|y)$ *for all* (x, y, z) *with* $p(x, y, z) > 0$.

Proof

$$I(Y; Z) - I(X, Y; Z) = E\left[\log \frac{p(z|y)}{p(z)} - \log \frac{p(z|x, y)}{p(z)}\right]$$

$$= E\left[\log \frac{p(z|y)}{p(z|x, y)}\right]$$

$$= \sum_{xy,z} p(x, y, z) \log \frac{p(z|y)}{p(z|x, y)}.$$

Applying Jensen's inequality, we have

$$I(Y; Z) - I(X, Y; Z) \leq \log \sum_{xy,z} p(x, y, z) \frac{p(z|y)}{p(z|x, y)}$$

$$= \log \sum_{xy,z} p(x, y) \cdot p(z|y)$$

$$= \log 1 = 0.$$

The conditions for equality follow from the discussion of Jensen's inequality in Appendix B. □

The condition for equality in Theorem 1.4 is very interesting; it says that the sequence (X, Y, Z) is a Markov chain, which for our purposes means simply that X, Y, *and* Z can be viewed as shown in Fig. 1.6. Here DMC 1 is characterized by the transition probabilities $p(y|x)$, and DMC 2 by the transition probabilities $p(z|y) = p(z|x, y)$. We have already observed that given any pair of random variables (X, Y), it is possible to devise a DMC with X as the input and Y as the output. However it is not true that if

(X, Y, Z) is any triple of random variables, there exists a pair of DMC's such that X, Y, Z have the relationship of Fig. 1.6. Indeed, it is clear that a necessary and sufficient condition for this is that (X, Y, Z) forms a Markov chain, that is, $p(z|y) = p(z|x, y)$ (i.e., Z depends on X only through Y).

Now let's assume that (X, Y, Z) is a Markov chain, as in Fig. 1.6. Then by Theorem 1.4, $I(X; Z) \leqslant I(X, Y; Z)$, and since (X, Y, Z) is a Markov chain, $I(X, Y; Z) = I(Y; Z)$. Hence $I(X; Z) \leqslant I(Y; Z)$. Now if (X, Y, Z) is a Markov chain, so is (Z, Y, X) (see Prob. 1.15), and hence $I(X; Z) \leqslant I(X; Y)$. Since this is an extremely important information-theoretic property of Markov chains, we display it as a theorem.

Theorem 1.5 *If (X, Y, Z) is a Markov chain, then*

$$I(X; Z) \leqslant \begin{cases} I(X; Y) \\ I(Y; Z). \end{cases} \qquad \square$$

Referring again to Fig. 1.6, we find that DMC's tend to "leak" information. If the DMC's are deterministic (i.e., if Y is a definite function of X and Z a definite function of Y), we can think of the casade in Fig. 1.6 as a kind of data-processing configuration. Paradoxically, Theorem 1.5 says that data processing can only destroy information! (For an important generalization of this, see Eq. (1.15).)

Example 1.9 Let X_1, X_2, X_3 be independent random variables; then $(X_1, X_1 + X_2, X_1 + X_2 + X_3)$ is a Markov chain, and so $I(X_1; X_1 + X_2 + X_3) \leqslant I(X_1; X_1 + X_2)$ (see Probs. 1.16 and 1.39). $\qquad \square$

Example 1.10 In Fig. 1.6 assume that X is described by $P\{X = 0\} = P\{X = 1\} = \frac{1}{2}$, and that both DMC's are binary symmetric channels with error probability p. Then

$$I(X; Y) = 1 - H_2(p) \qquad \text{bits,}$$

$$I(X; Z) = 1 - H_2[2p(1 - p)] \quad \text{bits.}$$

These two functions are plotted as follows: (For an extension of this

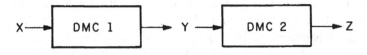

Figure 1.6 An information theorist's view of a Markov chain.

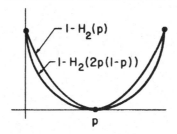

example, see Probs. 1.18 and 1.20.)   □

We conclude this section with two results about the convexity of $I(X; Y)$ when it is viewed as a function either of the input probabilities $p(x)$ or of the transition probabilities $p(y|x)$.

Theorem 1.6 $I(X; Y)$ *is a convex $\cap$ function of the input probabilities $p(x)$.*

Proof We think of the transition probabilities $p(y|x)$ as being fixed, and consider two input random variables X_1 and X_2 with probability distributions $p_1(x)$ and $p_2(x)$. If X's probability distribution is a convex combination $p(x) = \alpha p_1(x) + \beta p_2(x)$, we must show that

$$\alpha I(X_1; Y_1) + \beta I(X_2; Y_2) \leq I(X; Y),$$

where Y_1, Y_2 and Y are the channel outputs corresponding to X_1, X_2, and X, respectively. To do this consider the following manipulation, which uses obvious notational shorthand:

$$\alpha I(X_1; Y_1) + \beta I(X_2; Y_2) - I(X; Y)$$

$$= \sum_{x,y} \alpha p_1(x, y) \log \frac{p(y|x)}{p_1(y)} + \sum_{x,y} \beta p_2(x, y) \log \frac{p(y|x)}{p_2(y)} \qquad \text{(see Eq.(1.7))}$$

$$- \sum_{x,y} [\alpha p_1(x, y) + \beta p_2(x, y)] \log \frac{p(y|x)}{p(y)}$$

$$= \alpha \sum_{x,y} p_1(x, y) \log \frac{p(y)}{p_1(y)} + \beta \sum_{x,y} p_2(x, y) \log \frac{p(y)}{p_2(y)}. \qquad (1.12)$$

We now apply Jensen's inequality to each of the above sums. For example,

$$\sum_{x,y} p_1(x, y) \log \frac{p(y)}{p_1(y)} \leq \log \sum_{x,y} p_1(x, y) \frac{p(y)}{p_1(y)}.$$

But

$$\sum_{x,y} p_1(x,\ y)\frac{p(y)}{p_1(y)} = \sum_y \frac{p(y)}{p_1(y)} \sum_x p_1(x,\ y)$$

$$= \sum_y \frac{p(y)}{p_1(y)} \cdot p_1(y)$$

$$= 1.$$

Hence the first sum in (1.12) is ≤ 0; similarly, so is the second.   □

Corollary *The entropy function* $H(p_1,\ p_2,\ \ldots,\ p_r)$ *is convex* $\cap$.

Proof Let X be a random variable distributed according to $P\{X = i\} = p_i$. Then $I(X;\ X) = H(X) = H(p_1,\ p_2,\ \ldots,\ p_r)$. The result now follows from Theorem 1.6.   □

Theorem 1.7 $I(X;\ Y)$ *is convex* $\cup$ *in the transition probabilities* $p(y|x)$.

Proof Here the input probabilities $p(x)$ are fixed, but we are given two sets of transition probabilities $p_1(y|x)$ and $p_2(y|x)$ and a convex combination $p(y|x) = \alpha p_1(y|x) + \beta p_2(y|x)$. It is required to show that

$$I(X;\ Y) \leq \alpha I(X;\ Y_1) + \beta I(X;\ Y_2), \qquad (1.13)$$

where Y, Y_1, Y_2 are the channel outputs corresponding to the transition probabilities $p(y|x)$, $p_1(y|x)$, and $p_2(y|x)$. Again using obvious notation, the difference between the left and right sides of (1.13) is (see Eq. (1.5))

$$\sum_{x,y} [\alpha p_1(x,\ y) + \beta p_2(x,\ y)] \log \frac{p(x|y)}{p(x)}$$

$$- \sum_{x,y} \alpha p_1(x,\ y) \log \frac{p_1(x|y)}{p(x)} - \sum_{x,y} \beta p_2(x,\ y) \log \frac{p_2(x|y)}{p(x)}$$

$$= \alpha \sum_{x,y} p_1(x,\ y) \log \frac{p(x|y)}{p_1(x|y)} + \beta \sum_{x,y} p_2(x,\ y) \log \frac{p(x|y)}{p_2(x|y)}. \quad (1.14)$$

The first sum in (1.14) is, by Jensen's inequality,

$$\leq \alpha \log \left[\sum_{x,y} p_1(x, y) \frac{p(x|y)}{p_1(x|y)} \right]$$

$$= \alpha \log \left[\sum_{x,y} p(x|y) p_1(y) \right]$$

$$= \alpha \log \sum_y p_1(y) = 0.$$

Similarly the second sum is ≤ 0. $\qquad\qquad\qquad\qquad\qquad\qquad\qquad$ $\square$

1.2 Discrete random vectors

In Eq. (1.11) we defined the entropy $H(X, Y)$ of a pair of random variables, and on p. 28 we defined the mutual information $I(X, Y; Z)$ between a pair of random variables and a third random variable. In this section we will generalize those definitions and define $H(\mathbf{X})$, $H(\mathbf{X}|\mathbf{Y})$, and $I(\mathbf{X}; \mathbf{Y})$, where $\mathbf{X}$ and $\mathbf{Y}$ are arbitrary random vectors.

Our point of view is that a random vector $\mathbf{X} = (X_1, X_2, \ldots, X_n)$ is just a finite list of random variables X_i. The distribution of $\mathbf{X}$ (the joint distribution of $X_1, X_2, \ldots, X_n$) is the function $p(x_1, x_2, \ldots, x_n) = P\{X_1 = x_1, X_2 = x_2, \ldots, X_n = x_n\}$, where each x_i is in the range of X_i. A glance at the definitions in Section 1.1 should convince the reader that $H(X)$, $H(X|Y)$, $I(X; Y)$ depend only on the distribution functions $p(x)$, $p(y|x)$, etc., and not in any way on the fact that the values assumed by X and Y are real numbers. Hence we can immediately extend these definitions to arbitrary random vectors; for example the entropy of $\mathbf{X} = (X_1, \ldots, X_n)$ is defined as

$$H(\mathbf{X}) = \sum_{\mathbf{x}} p(\mathbf{x}) \log \frac{1}{p(\mathbf{x})},$$

where the summation is extended over all vectors $\mathbf{x}$ in the range of $\mathbf{X}$. And obviously Theorems 1.1–1.7 remain true.

The generalization of Theorem 1.5 to arbitrary random vectors has a particularly important application, which we now discuss. Consider the model for a communication system shown in Fig. 1.7 (cf. Figs. 0.2 and 5.1). In Fig. 1.7 the random vector $\mathbf{U}$ is a model for k consecutive source outputs; the encoder is a device that takes $\mathbf{U}$ and maps it into an n-tuple $\mathbf{X}$ for transmission over the channel; $\mathbf{Y}$ is the channel's noisy version of $\mathbf{X}$; and the decoder is a device that takes $\mathbf{Y}$ and maps it into a k-tuple $\mathbf{V}$, which is delivered to the destination and is supposed to reproduce $\mathbf{U}$, at least approximately.

The point of all this is that, for any realizable communication system, the sequence $(\mathbf{U}, \mathbf{X}, \mathbf{Y}, \mathbf{V})$ of random vectors forms a Markov chain (see Fig. 1.6). Informally this says that the output of each box in Fig. 1.7 depends only on its input and not on any of the earlier random vectors. Formally it gives many conditions on the various conditional probabilities, for example, $p(\mathbf{y}|\mathbf{x}, \mathbf{u}) = p(\mathbf{y}|\mathbf{x})$, $p(\mathbf{v}|\mathbf{y}, \mathbf{x}) = p(\mathbf{v}|\mathbf{y})$. (There is really no question of proving this part; it is one of the fundamental assumptions we make about a communication system.) Applying Theorem 1.5 to the sub-Markov chain $(\mathbf{U}, \mathbf{X}, \mathbf{V})$, we get $I(\mathbf{U}; \mathbf{V}) \leqslant I(\mathbf{X}; \mathbf{V})$. Similarly $I(\mathbf{X}; \mathbf{V}) \leqslant I(\mathbf{X}; \mathbf{Y})$. Hence for the random variables of Fig. 1.7,

$$I(\mathbf{U}; \mathbf{V}) \leqslant I(\mathbf{X}; \mathbf{Y}). \tag{1.15}$$

This result is called the *data-processing theorem*. Stated bluntly, it says that the information processing (the work done by the encoder and decoder of Fig. 1.7) can only destroy information! It says, for example, that the noisy channel output $\mathbf{Y}$ in Fig. 1.7 contains more information about the source sequence $\mathbf{U}$ than does the decoder's estimate $\mathbf{V}$. (While this is true theoretically, the data processing of the decoder is nevertheless required to render this information usable.)

We now come to a pair of inequalities involving $I(\mathbf{X}; \mathbf{Y})$ and $\sum_{i=1}^{n} I(X_i; Y_i)$, where $\mathbf{X} = (X_1, X_2, \ldots, X_n)$ and $\mathbf{Y} = (Y_1, Y_2, \ldots, Y_n)$ are a pair of n-dimensional random vectors.

Theorem 1.8 *If the components $(X_1, X_2, \ldots, X_n)$ of $\mathbf{X}$ are independent, then*

$$I(\mathbf{X}; \mathbf{Y}) \geqslant \sum_{i=1}^{n} I(X_i; Y_i).$$

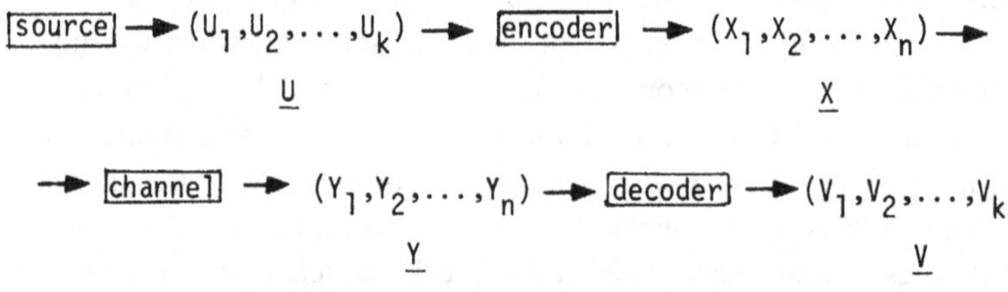

source $\longrightarrow$ $(U_1, U_2, \ldots, U_k)$ $\longrightarrow$ encoder $\longrightarrow$ $(X_1, X_2, \ldots, X_n)$ $\longrightarrow$

$\underline{U}$ $\qquad\qquad\qquad\qquad\qquad$ $\underline{X}$

$\longrightarrow$ channel $\longrightarrow$ $(Y_1, Y_2, \ldots, Y_n)$ $\longrightarrow$ decoder $\longrightarrow$ $(V_1, V_2, \ldots, V_k)$

$\underline{Y}$ $\qquad\qquad\qquad\qquad\qquad$ $\underline{V}$

Figure 1.7 A general communication system.

Proof Letting E denote expectation on the joint sample space of $\mathbf{X}$ and $\mathbf{Y}$, we have

$$I(\mathbf{X}; \mathbf{Y}) = E\left[\log\frac{p(\mathbf{x}|\mathbf{y})}{p(\mathbf{x})}\right] \quad (see\ Eq.\ (1.5))$$

$$= E\left[\log\frac{p(\mathbf{x}|\mathbf{y})}{p(x_1)p(x_2)\ldots p(x_n)}\right],$$

since $X_1, X_2, \ldots, X_n$ are assumed independent. On the other hand,

$$\sum_{i=1}^{n} I(X_i, Y_i) = \sum_{i=1}^{n} E\left[\log\frac{p(x_i|y_i)}{p(x_i)}\right]$$

$$= E\left[\log\frac{p(x_1|y_1)\ldots p(x_n|y_n)}{p(x_1)\ldots p(x_n)}\right].$$

Hence

$$\sum_{i=1}^{n} I(X_i, Y_i) - I(\mathbf{X}; \mathbf{Y})$$

$$= E\left[\log\frac{p(x_1|y_1)\ldots p(x_n|y_n)}{p(\mathbf{x}|\mathbf{y})}\right]$$

$$\leqslant \log E\left[\frac{p(x_1|y_1)\ldots p(x_n|y_n)}{p(\mathbf{x}|\mathbf{y})}\right] = 0$$

by Jensen's inequality, since this last expectation is

$$\sum_{\mathbf{x},\mathbf{y}} p(\mathbf{x}, \mathbf{y})\{\cdots\} = \sum_{\mathbf{x},\mathbf{y}} p(x_1|y_1)\ldots p(x_n|y_n)p(\mathbf{y})$$

$$= 1. \qquad \square$$

Example 1.11 Let $X_1, X_2, \ldots, X_n$ be independent identically distributed random variables with common entropy H. Also let π be a permutation of the set $\{1, 2, \ldots, n\}$, and let $Y_i = X_{\pi(i)}$. Then $I(\mathbf{X}; \mathbf{Y}) = nH$, but $\sum I(X_i; Y_i) = kH$, where k is the number of fixed points of π, that is, the number of integers i with $\pi(i) = i$. In particular if π has no fixed points, for example if $\pi(i) \equiv i + 1 \pmod{n}$, then $\sum I(X_i; Y_i) = 0$ (see Prob. 1.23). $\square$

If we think of $(Y_1, Y_2, \ldots, Y_n)$ as the n outputs of a noisy channel when the inputs are $X_1, X_2, \ldots, X_n$, Theorem 1.8 tells us that, if the inputs are independent, $\mathbf{Y}$ provides more information about $\mathbf{X}$ than the total amount of information provided about each X_i by the corresponding Y_i. The next theorem will tell us that if we drop the assumption of independence about the X_i and assume instead that the $(\mathbf{X}, \mathbf{Y})$ channel is memoryless, that is,

$$p(y_1, \ldots, y_n | x_1, \ldots, x_n) = \prod_{i=1}^{n} p(y_i | x_i), \qquad (1.16)$$

the situation is quite different!

Theorem 1.9 *If* $\mathbf{X} = (X_1, \ldots, X_n)$ *and* $\mathbf{Y} = (Y_1, \ldots, Y_n)$ *are random vectors and the channel is memoryless, that is, if* (1.16) *holds, then*

$$I(\mathbf{X}; \mathbf{Y}) \leq \sum_{i=1}^{n} I(X_i; Y_i).$$

Proof Again letting E denote expectation on the joint sample space of $\mathbf{X}$ and $\mathbf{Y}$, we have

$$I(\mathbf{X}; \mathbf{Y}) = E\left[\log \frac{p(\mathbf{y}|\mathbf{x})}{p(\mathbf{y})}\right] \qquad (see\ Eq.\ (1.7))$$

$$= E\left[\log \frac{p(y_1|x_1) \cdots p(y_n|x_n)}{p(\mathbf{y})}\right]$$

by (1.16). On the other hand,

$$\sum_{i=1}^{n} I(X_i; Y_i) = \sum_{i=1}^{n} E\left[\log \frac{p(y_i|x_i)}{p(y_i)}\right]$$

$$= E\left[\log \frac{p(y_1|x_1) \cdots p(y_n|x_n)}{p(y_1) \cdots p(y_n)}\right].$$

Hence

$$I(\mathbf{X}; \mathbf{Y}) - \sum_{i=1}^{n} I(X_i; Y_i)$$

$$= E\left[\log \frac{p(y_1) \cdots p(y_n)}{p(\mathbf{y})}\right]$$

$$\leq \log E\left[\frac{p(y_1) \cdots p(y_n)}{p(\mathbf{y})}\right]$$

$$= 0$$

by Jensen's inequality, since this last expectation is

$$\sum_y p(\mathbf{y})\{\cdots\} = \sum_y p(y_1) \cdots p(y_n)$$

$$= 1. \qquad \square$$

Example 1.12 Let X be a random variable with entropy H, and let $X_1 = X_2 = \cdots = X_n = Y_1 = \cdots = Y_n = X$. Then the hypotheses of Theorem 1.9 are satisfied, and $I(\mathbf{X}; \mathbf{Y}) = H$, $\sum I(X_i; Y_i) = nH$ (see Prob. 1.23). $\qquad \square$

Corollary *If* $X = (X_1, X_2, \ldots, X_n)$, *then*

$$H(\mathbf{X}) \leqslant \sum_{i=1}^{n} H(X_i).$$

Proof Define $Y_i = X_i$ and apply Theorem 1.9. $\qquad \square$

Note *Since the conclusions of Theorems 1.8 and 1.9 can both be satisfied simultaneously iff* $I(\mathbf{X}; \mathbf{Y}) = \sum I(X_i; Y_i)$, *it follows that a sufficient condition for equality to hold in either theorem is the hypothesis of the other. Interestingly enough, these conditions are also necessary (see Probs. 1.24 and 1.25).*

1.3 Nondiscrete random variables and vectors

In this section our goal is to extend the results of Sections 1.1 and 1.2 to random variables and vectors which can assume an uncountable number of values. A completely rigorous treatment of this subject proves to be very difficult, and so we shall content ourselves with a few elementary facts and refer the interested reader elsewhere[5] for details.

As we shall see, the entropy $H(X)$ can be defined for a nondiscrete random variable, but it always turns out to be infinite! However, the definition of the mutual information between a pair X, Y of random variables (or vectors) can be generalized to the nondiscrete case in a much more interesting and useful way. The key to this generalization is the notion of a *discrete quantization* of a random variable X.

If X is a random variable with distribution function $F(x) = P\{X \leqslant x\}$, and if $\{S_i, i = 1, 2, \ldots\} = P$ is a partition[6] of the real line R into a finite or countable number of Lebesgue-measurable subsets, the *quantization of X by*

P, denoted by $[X]_P$ or merely by $[X]$, is the discrete random variable defined by

$$P\{[X] = i\} = P\{X \in S_i\}$$

$$= \int_{S_i} dF(x).$$

If X and Y are a pair of random variables, we define their mutual information $I(X; Y)$ as

$$I(X; Y) = \sup_{P,Q} I([X]_P; [Y]_Q) \tag{1.17}$$

where the "sup" is taken over all pairs of partitions P and Q of the real line. Similarly if $\mathbf{X} = (X_1, \ldots, X_r)$ and $\mathbf{Y} = (Y_1, \ldots, Y_s)$ are a pair of random vectors,

$$I(\mathbf{X}; \mathbf{Y}) = \sup I([\mathbf{X}]; [\mathbf{Y}]), \tag{1.17'}$$

where the supremum is taken over all partitions $[\mathbf{X}] = ([X_1], \ldots, [X_r])$, $[\mathbf{Y}] = ([Y_1], \ldots, [Y_s])$ of the component random variables.

A partition P_1 is said to be a refinement of P_2 if every set in P_1 is a subset of some set in P_2. Notice that, if P_1 is a refinement of P_2, $[X]_{P_2}$ is a deterministic function of $[X]_{P_1}$, and so by Theorem 1.5 (also see Prob. 1.22) $I([X]_{P_2}; Y) \le I([X]_{P_1}; Y)$ for any random variable Y. This observation shows that the "sup" in definition (1.17) is in fact a kind of limit as the partitions P and Q become increasingly fine. Indeed, it is easy to show that, if P_1, P_2, Q_1, Q_2 are partitions and $[X]_1$, $[X]_2$, $[Y]_1$, $[Y]_2$ are the corresponding discrete random variables, there exist partitions P and Q corresponding to random variables $[X]$ and $[Y]$ such that

$$I([X]; [Y]) \ge I([X]_i; [Y]_i), \qquad i = 1, 2. \tag{1.18}$$

Now, if X and Y are already discrete with ranges $\{x_1, x_2, \ldots\}$ and $\{y_1, y_2, \ldots\}$, by selecting an X partition so that no set contains more than one x_i, and a Y partition so that no set contains more than one y_j, clearly $I([X]; [Y]) = I(X; Y)$. Equally clearly, no refinement could increase $I([X]; [Y])$, and so definition (1.17) reduces to our earlier definition.

Definition (1.17) of $I(X, Y)$ is fairly simple but it is not convenient for actually computing $I(X, Y)$ for a given pair of random variables. Ideally we would like to have a formula for $I(X, Y)$ analogous to Eq. (1.5), (1.6), or (1.7) involving integrals rather than sums, but unfortunately in general no such formula exists.[7] However if X and Y are sufficiently smooth such formulas do exist, as we shall now see.

Let us assume that X and Y have a continuous joint density, i.e. that there exists a continuous nonnegative function $p(x, y)$ defined for pairs (x, y) of real numbers such that if A and B are intervals on the real line,

$$P\{X \in A, Y \in B\} = \int_B \int_A p(x, y)\, dx\, dy.$$

In this situation the individual densitites of X and Y are given by

$$p(x) = \int_{-\infty}^{\infty} p(x, y)\, dy, \qquad q(y) = \int_{-\infty}^{\infty} p(x, y)\, dx,$$

and the conditional densities by

$$p(x|y) = \frac{p(x, y)}{q(y)}, \qquad p(y|x) = \frac{p(x, y)}{p(x)}.$$

Let us now assume that the integrals

$$h(X) = \int_{-\infty}^{\infty} p(x) \log \frac{1}{p(x)}\, dx \qquad (1.19)$$

$$h(X|Y) = \int_{-\infty}^{\infty} \int_{-\infty}^{\infty} p(x, y) \log \frac{1}{p(x|y)}\, dx\, dy \qquad (1.20)$$

both exist.

Theorem 1.10 *Under the preceding assumptions,*

$$I(X; Y) = h(X) - h(X|Y).$$

Proof Choose $0 < \varepsilon_1 < \varepsilon_2$ arbitrarily, and let $\cdots < x_{-1} < x_0 < x_1 < \cdots$ be a countable set of points such that $\Delta x_i = x_i - x_{i-1}$ satisfies $\varepsilon_1 < \Delta x_i < \varepsilon_2$ for all i. Similarly choose (y_j) such that $\varepsilon_1 < \Delta y_j < \varepsilon_2$. Let $[X]$ denote the quantization of X obtained from the partition of the real line consisting of the half-open intervals $[x_{i-1}, x_i)$, and let $[Y]$ denote the quantization of Y corresponding to the partition consisting of the intervals $[y_{j-1}, y_j)$. We now introduce the nonce notation:

$$p(i) = P\{[X] = i\} = \int_{x_{i-1}}^{x_i} p(x)\, dx$$

$$q(j) = P\{[Y] = j\} = \int_{y_{j-1}}^{y_j} q(y)\, dy$$

$$p(i, j) = P\{[X] = i, [Y] = j\} = \int_{y_{j-1}}^{y_j} \int_{x_{i-1}}^{x_i} p(x, y)\, dx\, dy$$

$$p(i|j) = P\{[X] = i|[Y] = j\} = p(i, j)/q(j).$$

Since all the densities involved are assumed to be continuous, by the mean value theorem for integrals[8]

$$p(i) = \Delta x_i p(s_i),$$

$$q(j) = \Delta y_j q(t_j), \tag{1.21}$$

for some $s_i \in [x_{i-1}, x_i]$, $t_j \in [y_{j-1}, y_j]$. Similary, by invoking a two-dimensional mean-value theorem,[8] we have

$$p(i, j) = \int_{y_{j-1}}^{y_j} \int_{x_{i-1}}^{x_i} p(x|y)q(y) \, dx \, dy$$

$$= p(s_{ij}|t_{ij}) \int_{y_{j-1}}^{y_j} \int_{x_{i-1}}^{x_i} q(y) \, dx \, dy$$

$$= p(s_{ij}|t_{ij})\Delta x_i \Delta y_j q(t_j) \tag{1.22}$$

for some point (s_{ij}, t_{ij}) in $[x_{i-1}, x_i] \times [y_{j-1}, y_j]$. Hence

$$p(i|j) = \Delta x_i p(s_{ij}|t_{ij}). \tag{1.23}$$

Now, by Eq. (1.4), $I([X]; [Y]) = H([X]) - H([X]|[Y]) = \sum_i p(i) \log p(i)^{-1} - \sum_{i,j} p(i, j) \log p(i|j)^{-1}$. By (1.21) we have

$$H([X]) = \sum_i \Delta x_i p(s_i) \log \frac{1}{p(s_i)} + \sum_i \Delta x_i p(s_i) \log \frac{1}{\Delta(x_i)}. \tag{1.24}$$

In (1.24) the first sum is an approximation to the integral for $h(X)$ (see (1.19)), and so for small enough ε_2 it will converge. The second sum converges (to a sum less than $\log \varepsilon_1^{-1}$) because $\Delta x_i > \varepsilon_1$. Similarly,

$$H([X]|[Y]) = \sum_{i,j} \Delta x_i \Delta y_j p(s_{ij}|t_{ij})q(t_j) \log \frac{1}{p(s_{ij}|t_{ij})}$$

$$+ \sum_i \Delta x_i \log \frac{1}{\Delta x_i} \sum_j \Delta y_j p(s_{ij}|t_{ij})q(t_j). \tag{1.25}$$

The first sum in (1.25) is the integral of a step-function approximation to $p(x, y) \log p(x|y)^{-1}$, and so for ε_2 small enough it will approximate $h(X|Y)$. The second sum is identical to the second sum in (1.24) since by (1.22)

$$\sum_j \Delta y_j p(s_{ij}|t_{ij})q(t_j) = \frac{1}{\Delta x_i} \sum_j \int_{y_{j-1}}^{y_j} \int_{x_{i-1}}^{x_i} p(x, y) \, dx \, dy$$

$$= \frac{1}{\Delta x_i} \cdot \Delta x_i p(s_i).$$

Thus we finally obtain

$$I([X]; [Y]) = \sum_i \Delta x_i p(s_i) \log \frac{1}{p(s_i)}$$

$$- \sum_{i,j} \Delta x_i \Delta y_j p(s_{ij}|t_{ij})q(t_j) \log \frac{1}{p(s_{ij}|t_{ij})}. \qquad (1.26)$$

But as we have already observed the first sum in (1.26) is an approximation to $h(x)$, and the second to $h(X|Y)$. So by making the X and Y partitions increasingly fine, i.e. as $\varepsilon_2 \to 0$, $I([X]; [Y])$ will approach $h(X) - h(X|Y)$, as asserted. (For an important application of Theorem 1.10, see Probs. 1.27 and 1.28.) $\qquad\qquad\square$

The definition of $h(X)$ bears a strong superficial resemblance to the definition (Eq. (1.1)) of the entropy of a random variable, and it is tempting to call $h(X)$ the entropy of X. This however would be a misnomer, because we should really define $H(X)$ in general as the supremum of $H([X])$ over all discrete quantizations of X. But it came out in the proof of Theorem 1.10 (Eq. (1.24)) that $H([X])$ is the sum of two terms; one is an approximating sum for $h(X)$, but the other, $\sum p(i) \log(\Delta x_i)^{-1}$, is a kind of measure of the mesh size and clearly approaches $+\infty$ as the Δx_i become smaller and smaller. Indeed, it is easy to show that $H([X])$ itself always approaches $+\infty$ (Prob. 1.31) even if the integral $h(X)$ converges to $-\infty$! Hence, although $h(X)$ is a useful quantity to know, it is not in any sense a measure of the randomness of X. Indeed, $h(X)$ is not even invariant under a coordinate transformation (see Prob. 1.33). For this reason $h(X)$ is usually called the *differential entropy* of X, to distinguish it from the ordinary, or *absolute entropy*.

As in the discrete case, we can easily extend the definitions of $I(X; Y)$ and $h(X)$ to random vectors. To define $I(\mathbf{X}; \mathbf{Y})$ in general, just take discrete quantizations of each component separately. For $h(\mathbf{X})$, assume the existence of an n-dimensional continuous density $p(\mathbf{x})$ and define $h(\mathbf{X}) = -\int p(\mathbf{x}) \log p(\mathbf{x}) \, d\mathbf{x}$.

Example 1.13 Let $\mathbf{X} = (X_1, X_2, \ldots, X_n)$, where the X_i are independent

Gaussian (normal) random variables with means μ_i and variances σ_i^2. Then the density for $\mathbf{X}$ is the function

$$g(\mathbf{x}) = \prod_{i=1}^{n} (2\pi\sigma_i^2)^{-1/2} \exp\left[-\frac{(x_i - \mu_i)^2}{2\sigma_i^2}\right]. \tag{1.27}$$

The differential entropy of $\mathbf{X}$ is easily computed:

$$h(\mathbf{X}) = \int g(\mathbf{x}) \log \frac{1}{g(\mathbf{x})} d\mathbf{x}$$

$$= \sum_{i=1}^{n} \int g(\mathbf{x}) \log\left\{(2\pi\sigma_i^2)^{1/2} \exp\left[\frac{(x_i - \mu_i)^2}{2\sigma_i^2}\right]\right\} d\mathbf{x}$$

$$= \sum_{i=1}^{n} \int g_i(x_i) \left[\tfrac{1}{2} \log(2\pi\sigma_i^2) + \frac{(x_i - \mu_i)^2}{2\sigma_i^2}\right] dx_i,$$

(where $g_i(x_i) = (1/\sqrt{2\pi\sigma_i^2}) \exp[-(x_i - \mu_i)^2/2\sigma_i^2]$)

$$= \sum_{i=1}^{n} \tfrac{1}{2}[\log(2\pi\sigma_i^2) + 1],$$

(since $\int g_i(x_i) \, dx_i = 1$ and $\int g_i(x_i)(x_i - \mu_i)^2 \, dx_i = \sigma_i^2$)

$$= \frac{n}{2} \log 2\pi e (\sigma_1^2 \ldots \sigma_n^2)^{1/n}.$$

In the special case $n = 1$, we have

$$h(X) = \tfrac{1}{2} \log 2\pi e \sigma^2. \qquad \square$$

It is an interesting and important fact that among all n-dimensional random variables with given variances, independent Gaussian random variables have the largest differential entropy.

Theorem 1.11 *If $\mathbf{X} = (X_1, X_2, \ldots, X_n)$ has density $p(\mathbf{x})$ and if $E[(X_i - \mu_i)^2] = \sigma_i^2$, $i = 1, 2, \ldots, n$, then $h(\mathbf{X}) \leq (n/2) \log 2\pi e (\sigma_1^2 \sigma_2^2 \ldots \sigma_n^2)^{1/n}$, with equality iff $p(\mathbf{x}) = g(\mathbf{x})$ (see Definition (1.27)) almost everywhere.*

Proof By hypothesis the marginal density $p_i(x)$ of X_i satisfies $\int p_i(x) \, dx = 1$, $\int p_i(x)(x - \mu_i)^2 \, dx = \sigma_i^2$. Hence it follows from the computation of Example 1.13 that

$$\int p(\mathbf{x}) \log \frac{1}{g(\mathbf{x})} d\mathbf{x} = \frac{n}{2} \log 2\pi e (\sigma_1^2 \ldots \sigma_n^2)^{1/n}.$$

Thus, if $\mathbf{Y}$ denotes an n-dimensional random vector distributed according to the Gaussian density $g(\mathbf{x})$,

$$h(\mathbf{X}) - h(\mathbf{Y}) = \int p(\mathbf{x}) \log \frac{g(\mathbf{x})}{p(\mathbf{x})} d\mathbf{x}.$$

By Jensen's inequality, however, the latter integral is

$$\leqslant \log \int g(\mathbf{x}) \, d\mathbf{x} = 0.$$

Furthermore, equality holds iff $g(\mathbf{x}) = p(\mathbf{x})$ for almost all $\mathbf{x}$. (For an extension of Theorem 1.11, see Probs. 1.34 and 1.35.) ☐

Having detoured briefly through the interesting topic of differential entropy, we return to our main concern, mutual information. Our aim is to show with some measure of credibility that Theorems 1.3, 1.4, 1.5, 1.8, and 1.9 remain true in the general case. Theorems 1.3, 1.8, and 1.9 present no serious difficulties; here are some remarks on their proofs.

Theorem 1.3. The fact that $I(\mathbf{X}; \mathbf{Y}) \geqslant 0$ follows immediately from the fact that $I([\mathbf{X}]; [\mathbf{Y}]) \geqslant 0$ for any discrete quantizations $[\mathbf{X}]$, $[\mathbf{Y}]$. Moreover, if $[\mathbf{X}]$ and $[\mathbf{Y}]$ are independent, so are $[\mathbf{X}]$ and $[\mathbf{Y}]$, and so $I([\mathbf{X}]; [\mathbf{Y}]) = 0$ for all quantizations; hence $I(\mathbf{X}; \mathbf{Y}) = 0$ by Eq. (1.17). Finally, if $\mathbf{X}$ and $\mathbf{Y}$ are dependent, it is possible to find quantizations $[\mathbf{X}]$ and $[\mathbf{Y}]$ which are also dependent (see Prob. 1.36), and so $I(\mathbf{X}; \mathbf{Y}) \geqslant I([\mathbf{X}]; [\mathbf{Y}]) > 0$.

Theorem 1.8. Generalizing the observation made above, we have that, if $X_1, \ldots, X_n$ are independent, so are $[X_1], [X_2], \ldots, [X_n]$, and so by Theorem 1.8

$$I([\mathbf{X}]; [\mathbf{Y}]) \geqslant \sum_{i=1}^{n} I([X_i]; [Y_i]). \tag{1.28}$$

But as the quantizations of the $[X_i]$ and $[Y_i]$ become increasingly fine, the left side of (1.28) approaches $I(\mathbf{X}; \mathbf{Y})$, and the right side approaches $\sum_{i=1}^{n} I(X_i; Y_i)$.

Theorem 1.9. Here the main problem is to appropriately generalize the hypothesis

$$p(\mathbf{y}|\mathbf{x}) = \prod_{i=1}^{n} p(y_i|x_i) \tag{1.29}$$

(see Eq. (1.16)). To do this properly would lead us far afield into a study of conditional probability distributions (see, e.g., Feller [4], Vol. II, §V.10). However, we avoid these problems by *defining* achannel to be memoryless if (1.29) holds for every quantization of the inputs to and outputs from the channel. This condition is usually easy to verify (see Prob. 4.27) and has the advantage that Theorem 1.9 immediately implies

$$I([\mathbf{X}]; [\mathbf{Y}]) \leq \sum_{i=1}^{n} I([X_i]; [Y_i]). \tag{1.30}$$

for every partitioning of the inputs and outputs, and so Theorem 1.9 follows for arbitrary random variables by taking increasingly fine quantizations in (1.30).

Finally we come to Theorems 1.4 and 1.5. First observe that Theorem 1.5 follows as a direct corollary of Theorem 1.4, so we focus on Theorem 1.4. There is no problem in proving that $I(X, Y; Z) \geq I(Y; Z)$ in general, since for every quantization of X, Y and Z we get $I([X], [Y]; [Z]) \geq I([Y]; [Z])$ directly from Theorem 1.4. However, it is not so easy to prove the second part of Theorem 1.4, i.e. that $I(X, Y; Z) = I(Y; Z)$ iff (X, Y, Z) forms a Markov chain. The problem is that if (X, Y, Z) is a Markov chain, and if $[X]$ and $[Z]$ are partitions of X and Z, it may not be possible to find a partition $[Y]$ of Y such that $([X], [Y], [Z])$ is a Markov chain (see Probs. 1.37–1.39). We prefer not to burden the reader with a detailed proof, but refer instead to Pinsker [24], Chapter 3, where the cited result is proved with full rigor.[9]

Problems

1.1 Show that the convention $0 \log 1/0 = 0$ adopted in definition (1.1) of $H(X)$ is forced if we make either of the following two assumptions:
   (a) $H(X)$ is a continuous function of the probability vector $\mathbf{p} = (p_1, p_2, \ldots)$.
   (b) If X and Y are random variables defined on the same sample space, and if $X = Y$ a.e., then $H(X) = H(Y)$.

1.2 If X is defined as in Example 1.3, show that $H(X) = +\infty$.

1.3 Let X be a discrete random variable, and let f be a real-valued function defined on the range of X. Show that $H(X) \geq H[f(X)]$, with equality iff f is one to one on the set $\{x: P\{X = x\} > 0\}$.

1.4 Find two distinct distributions $p_1 \geq p_2 \geq \cdots \geq p_n > 0$ and $q_1 \geq q_2 \geq \cdots \geq q_m > 0$ such that $H(p_1, p_2, \ldots, p_n) = H(q_1, q_2, \ldots, q_m)$.

1.5 What is the *minimum* value of $H(p_1, \ldots, p_n) = H(\mathbf{p})$ as $\mathbf{p}$ ranges over the set of n-dimensional probability vectors? Find all $\mathbf{p}$'s which achieve this minimum.

1.6 Given a probability distribution $(p_1, p_2, \ldots, p_n)$ and an integer m, $0 \leq m \leq n$, define $q_m = 1 - \sum_{j=1}^{m} p_j$. Show that $H(p_1, \ldots, p_n) \leqslant H(p_1, \ldots, p_m, q_m)$ $+ q_m \log(n - m)$. When does equality hold?

1.7 Let $f(x)$ be any function defined for all $x \geqslant 1$. If X is a discrete random variable with range $R = \{x_1, \ldots, x_n\}$, define the *f-entropy* of X by $H_f(X) = \sum_{i=1}^{n} p_i f(1/p_i)$, where $p_i = P\{X = x_i\}$. (Ordinary entropy: $f(x) = \log x$.)

(a) If $f(x)$ is convex $\cap$, find the best possible upper bound on $H(X)$ that depends only on n. Show by example that the probability distribution that achieves this maximum may not be unique.

(b) If $f(x) = \log x/x$, show that $H_f(X) < \log(e)/e$. [N.B. $f(x)$ is not convex $\cap$.]

(c) Work harder on part (b) and show that in fact $H_f(X) \leqslant \log(3)/3$, with equality iff exactly three of the p_i's equal $\frac{1}{3}$, and the rest are 0.

1.8 Let $\mathbf{a} = (a_1, a_2, \ldots)$ be a sequence of nonnegative real numbers such that the sum $Z(s) = \sum_n e^{-a_n s}$ converges for all sufficiently large s. (For example, if there are only finitely many a's, this condition is satisfied.) For each $\alpha > 0$ define $\Phi(\alpha) = \sup \{H(\mathbf{p}): \sum p_n a_n = \alpha\}$. Show that $\Phi(\alpha)$ is given parametrically by $\Phi(\alpha) = \log Z(s) - sZ'(s)/Z(s)$, where $Z'(s)/Z(s) = -\alpha$. [*Hint*: If $\sum p_n a_n = \alpha$, define a new probability distribution by $q_n = e^{-a_n s}/Z(s)$, and apply Jensen's inequality to the sum $\sum p_n \log(q_n/p_n)$.] If $\mathbf{a} = (a_1, a_2, \ldots, a_N)$ has only finitely many components, show that a graph of $\Phi(\alpha)$ versus α looks like this:

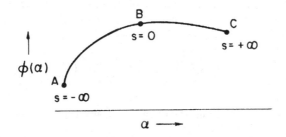

Give the coordinates of the points A, B, and C.

1.9 Let $\mathbf{p} = (p_1, p_2, \ldots)$ be a countable probability distribution, that is, $p_n \geqslant 0$ for $n = 1, 2, \ldots$, and $\sum_{n=1}^{\infty} p_n = 1$. Show that if $\sum_{n=1}^{\infty} p_n \log n$ converges, then $H(\mathbf{p})$ is finite. Conversely if $H(\mathbf{p})$ is finite and $\mathbf{p}$ is monotone (i.e., $p_1 \geqslant p_2 \geqslant \cdots$). show that $\sum p_n \log n < \infty$. Show by example that the latter assertion is not true if the assumption of monotonicity is dropped.

1.10 Let X and Y be the random variables defined in Example 1.6. Show that $\text{Cov}(X, Y) = 0$, $I(X; Y) = 1$ bit. Find the DMC for which X is a noisy version of Y.

1.11 Find necessary and sufficient conditions on the joint probability distribution $p(x, y, z)$ for equality to hold in Theorem 1.2. Using this result, for any $r \geqslant 2$ and any value $0 \leqslant P_e \leqslant 1$, construct random variables X and Y with ranges $\{1, 2, \ldots, r\}$ for which Fano's inequality (the corollary to Theorem 1.2) is sharp.

1.12 Show that Fano's inequality implies both an upper bound and a lower bound on P_e in terms of $H(X|Y)$. Interpret the upper bound heuristically.

1.13 Suppose X and Y are random variables which both assume values in a finite group. Define $Z = Y - X$. using Theorem 1.2, show that $H(X|Y) \leqslant H(Z)$. If X and Z are independent, show that $H(X|Y) = H(Z)$.

1.14 Verify Eqs. (1.5)–(1.10).

1.15 Show that (X, Y, Z) is a discrete Markov chain iff (Z, Y, X) is.

1.16 Verify Example 1.9.

1.17 (a) Suppose that X is a random variable uniformaly distributed on $\{0, 1, 2\}$. Among all random variables Y assuming only the two values $\{0, 1\}$, how large can $I(X; Y)$ be?

   (b) Now let X be a random varable uniformly distributed on $\{0, 1, \ldots, n-1\}$. Among all random variables Y assuming the values $\{0, 1, \ldots, m-1\}$, how large can $I(X; Y)$ be? [*Note*: Only the case $m < n$ is interesting.]

1.18 Show that a cascade of n identical BSC's.

$$X_0 \longrightarrow \boxed{\text{BSC \#1}} \longrightarrow X_1 \longrightarrow \cdots \longrightarrow \boxed{\text{BSC \#n}} \longrightarrow X_n \ ,$$

each with raw error probability p, is equivalent to a single BSC with error probability $\frac{1}{2}[1 - (1 - 2p)^n]$, and hence that $\lim_{n \to \infty} I(X_0; X_n) = 0$, if $p \neq 0, 1$.

1.19 If X, Y, and Z are three discrete random variables defined on the same sample space, the conditional mutual information $I(X; Y|Z)$ ("the mutual information between X and Y, given Z") is defined by

$$I(X; Y|Z) = \sum_{x,y,z} p(x, y, z) \log \frac{p(x, y|z)}{p(x|z)p(y|z)},$$

where we adopt the usual conventions about the sum. Prove:

   (a) $I(X; Y|Z) = I(Y; X|Z)$.

   (b) $I(X; Y, Z) = I(X; Z) + I(X; Y|Z)$.

   (c) $I(X; Y|Z) \geqslant 0$ with equality iff (X, Z, Y) is a Markov chain.

1.20 (Finite Markov chains). Let $A = (a_{ij})$ be an $r \times r$ stochastic matrix, that is $a_{ij} \geqslant 0$ and $\sum_{j=0}^{r-1} a_{ij} = 1$ for $i = 0, 1, \ldots, r-1$. Regard A as the matrix of transition probabilities for a DMC, i.e. $P\{Y = j | X = i\} = a_{ij}$. Consider a semi-infinite cascade of copies of this channel:

$$X_0 \longrightarrow \boxed{\begin{array}{c}\text{DMC}\\1\end{array}} \longrightarrow X_1 \longrightarrow \boxed{\begin{array}{c}\text{DMC}\\2\end{array}} \longrightarrow \cdots \longrightarrow X_{n-1} \longrightarrow \boxed{\begin{array}{c}\text{DMC}\\n\end{array}} \longrightarrow \cdots$$

Now let $(p_0, p_1, \ldots, p_{r-1}) = \mathbf{p}$ be a probability vector which is stable with respect to A, i.e. $p_j = \sum_{i=0}^{r-1} p_i a_{ij}$, $j = 0, 1, \ldots, r-1$, and suppose X_0 in the above sketch is distributed according to $\mathbf{p}$.

   (a) Show that $X_0, X_1, X_2, \ldots$ are identically distributed random variables. Under what conditions are they independent? (The sequence

$X_0, X_1, X_2, \ldots$ is called a finite Markov chain; see Feller [4], Vol. 1, Chapters XV and XVI.)

(b) The (per symbol) *entropy* of the Markov chain $X_0, X_1, \ldots$ is defined as

$$H = \lim_{n \to \infty} \frac{1}{n} H(X_0, X_1, \ldots X_{n-1}).$$

Show that

$$H = \sum_{i,j} p_i a_{ij} \log \frac{1}{a_{ij}}.$$

(c) Compute the stable probability distributions and the entropy of the Markov chains whose transition probability matrices are given by

$$\begin{bmatrix} \alpha & 1-\alpha \\ \beta & 1-\beta \end{bmatrix}, \quad \begin{bmatrix} 0 & 0 & 0 & 1 \\ 0 & 0 & 0 & 1 \\ \frac{1}{2} & \frac{1}{2} & 0 & 0 \\ 0 & 0 & 1 & 0 \end{bmatrix}, \quad \begin{bmatrix} q_0, & q_1, & \cdots, & q_{r-1} \\ q_1, & q_2, & \cdots, & q_0 \\ \vdots & & & \\ q_{r-1}, & q_0, & \cdots, & q_{r-2} \end{bmatrix}.$$

(In the last example each row is a left cyclic shift of the preceding one.)

1.21 Is $I(X; Y)$ strictly convex $\cap$ in $p(x)$? Is it strictly convex $\cup$ in $p(y|x)$?

1.22 Let $\mathbf{X} = (X_1, \ldots, X_n)$ and $\mathbf{Y} = (Y_1, \ldots, Y_m)$ be discrete random vectors, and let f and g be real-valued functions of n and m real variables, respectively. Show that $I[f(\mathbf{X}); g(\mathbf{Y})] \leq I(\mathbf{X}; \mathbf{Y})$.

1.23 Supply the missing details in Examples 1.11 and 1.12.

1.24 Show that equality holds in Theorem 1.8 iff $p(\mathbf{x}|\mathbf{y}) = \prod_{i=1}^{n} p(x_i|y_i)$ for all $(\mathbf{x}, \mathbf{y})$.

1.25 Show that equality holds in Theorem 1.9 iff $Y_1, Y_2, \ldots, Y_n$ are independent, and hence that equality holds in the corollary to Theorem 1.9 iff $X_1, \ldots, X_n$ are independent.

1.26 Prove or disprove: If $\mathbf{X}$ and $\mathbf{Y} = (Y_1, Y_2, \ldots, Y_n)$ are discrete random vectors, $I(\mathbf{X}; \mathbf{Y}) \leq \sum_{i=1}^{n} I(\mathbf{X}; Y_i)$.

1.27 Let X and Z be independent random variables with continuous density functions, and let $Y = X + Z$. If $h(Y)$ and $h(Z)$ exist, show that $I(X; Y) = h(Y) - h(Z)$. Show that the same formula holds if X and Z are random vectors.

1.28 (Continuation). Let X and Z be independent random variables, X discrete and Z with continuous density, and let $Y = X + Z$. Show that Y has a continuous density, and that if $h(Y)$ and $h(Z)$ exist, then $I(X; Y) = h(Y) - h(Z)$. Show that the same result holds if X and Z are random vectors.

1.29 If $\mathbf{X} = (X_1, \ldots, X_n)$ are independent Gaussian random variables with means μ_i and variances σ_i^2 and $\mathbf{X}' = (X_1', \ldots, X_n')$ are ditto with means μ_i' and variances $\sigma_i'^2$, show that $\mathbf{X} + \mathbf{X}' = (X_1 + X_1', \ldots, X_n + X_n')$ are ditto with means $\mu_i + \mu_i'$ and variances $\sigma_i^2 + \sigma_i'^2$.

1.30 If $X_1, X_2, \ldots, X_n$ are independent random variable with differential entropies $h(X_i) = h_i$, show that

$$h(X_1 + \cdots + X_n) \geq \tfrac{1}{2} \log \left(\sum_{i=1}^{n} e^{2h_i} \right),$$

with equality iff the X_i are Gaussian with variances $\sigma_i^2 = e^{2h_i}/2\pi e$.

1.31 If X has a continuous density, show that sup $H([X]) = +\infty$, where the supremum is taken over all discrete quantizations of X.

1.32 Construct a random variable whose density $p(x)$ is continuous for all real x such that $h(X) = -\infty$.

1.33 Let $\mathbf{X}$ be a random n-dimensional vector with continuous n-dimensional density $p(\mathbf{x})$ and differential entropy $h(\mathbf{X})$. Let f be a continuously differentiable one-to-one mapping of Euclidean n-space E^n onto itself. Show that $h[f(\mathbf{X})] = h(\mathbf{X}) + \int p(\mathbf{x}) \log|J| \, d\mathbf{x}$, where $J = J(x_1, \dots, x_n)$ is the Jacobian of the transformation f.

1.34 Let $\mathbf{X} = (X_1, \dots, X_n)$ be a random variable with density, and suppose $E[(X_i - \mu_i)(X_j - \mu_j)] = \rho_{ij}$ for all i, j. Prove that

$$h(\mathbf{X}) \leq \frac{n}{2} \log 2\pi e (\sigma^2)^{1/n},$$

where σ^2 is the determinant of the matrix (ρ_{ij}). Furthermore, show that equality holds iff $\mathbf{X}$ is an n-dimensional normal random variable whose covariance matrix is (ρ_{ij}). (See Feller [4], Vol. 2, Section III.6.)

1.35 Let $f(x)$ be a continuous real-valued function defined on an interval I. The object of this problem is to find out how large the differential entropy $h(X)$ can be if X has a density function that satisfies $p(x) = 0$ if $x \notin I$, and $\int_I p(x)f(x)\,dx = A$, where $\inf(f) < A < \sup(f)$. (For example, if $I = (-\infty, \infty)$ and $f(x) = (x - \mu)^2$, we will arrive by a new route at the one-dimensional case of Theorem 1.11.) Define $G(s) = \int_I e^{-sf(x)}\,dx$. Show that there exists s_0 with $G'(s_0)/G(s_0) = -A$, and define $q(x) = e^{-s_0 f(x)}/G(s_0)$ for $x \in I$, $q(x) = 0$ for $x \notin I$. Show that $h(X) \leq \log G(s_0) + s_0 A$ with equality iff X's density $= q(x)$ almost everywhere. [*Hint*: See Prob. 1.8.] Apply this general technique to three cases:
(a) $f(x) = \log x$, $I = (1, \infty)$.
(b) $f(x) = x$, $I = (0, \infty)$.
(c) $f(x) = |x|$, $I = (-\infty, \infty)$.

1.36 If X and Y are dependent random variables, show that there exist discrete quantizations $[X]$ and $[Y]$ which are also dependent.

The next three problems are for readers with some knowledge of general Markov chains; see, for example, Feller [4], Vol. 2, Section VI.II.

1.37 Let (X, Y, Z) be a Markov chain such that for any quantizations $[X]$ and $[Z]$ there exists a quantization $[Y]$ such that $([X], [Y], [Z])$ is a discrete Markov chain. Prove that $I(X; Y, Z) = I(X; Y)$.

1.38 Consider Markov chains of the form $(X, f(X), Z)$ or $(X, Y, f(Y))$. Show that such chains have the property described in Prob. 1.37, and hence that Theorem 1.4 holds for them. [*Hint*: For the $(X, f(X), Z)$ case, if $\{S_i\}$ is the X partition, let $f(S_i)$ be the Y partition.]

1.39 Let X_1, X_2, X_3 be independent, identically distributed Gaussian random variables with mean 0 and variance 1. Show that the sequence (X, Y, Z), where $X = X_1$, $Y = X_1 + X_2$, and $Z = X_1 + X_2 + X_3$, is a Markov chain. (See Example 1.9.) But show also that, if $[X]$ and $[Z]$ are quantizations of X and Z which both assume more than two values, $([X], [Y], [Z])$ is never a Markov chain, for any quantization $[Y]$ of Y.

Notes

1 (p. 19). Suppose you could ask question Q_n: "Will I live to be n?" Think about how your uncertainty (anxiety!) would increase as n decreased.

2 (p. 20). $H(X|Y)$ is also sometimes called the *equivocation* (of Y about X).

3 (p. 21). In orthodox mathematical jargon "discrete" means finite or *countable*. There would be no problem in letting our DMC's have countable input and output alphabets, but in order not to conflict with accepted information-theoretic usage we shall usually not do so.

4 (p. 27). These definitions make sense only when $p(x)p(y) \neq 0$. If $p(x)p(y) = 0$, then $p(x, y) = 0$ also, and we may define $I(x; y)$ arbitrarily without affecting the values of $I(X; Y)$.

5 (p. 37). The most complete treatment of this subject is given in the first three chapters of Pinsker's book [24].

6 (p. 37). A partition of the real line R is a collection of subsets (S_i) with $\cup S_i = R$ and $S_i \cap S_j = \phi$ if $i \neq j$.

7 (p. 38). This statement is not strictly true. In Pinsker [24] it is shown that $I(X; Y)$ is either infinite or equal to the expected value in the joint sample space of X, Y of the logarithm of the Radon-Nikodym derivative $d\mu_{xy}/d\mu_x \times \mu_y$, where μ_{xy} is the joint probability measure of (X, Y) and $\mu_x \times \mu_y$ is the product of the marginal X and Y measures. Thus symbolically

$$I(X; Y) = \int \left(\log \frac{d\mu_{xy}}{d\mu_x \times \mu_y} \right) d\mu_{xy}.$$

8 (p. 40). The general theorem being applied here is that, if $f(x)$ is continuous, $\int_S f(x)g(x)\, dx = f(x_0) \int_S g(x)\, dx$ for some $x_0 \in S$. (See Apostol [2], Theorem 14.16, for example.)

9 (p. 44). Perhaps a brief guide to Pinsker's proof is in order. The desired result is that $I(X, Y; Z) = I(Y; Z)$ iff (X, Y, Z) is a Markov chain. Thus, if the *conditional mutual information* $I(X; Z|Y)$ is defined by $I(X; Z|Y) = I(X, Y; Z) - I(Y; Z)$, the required result is $I(X; Z|Y) = 0$ iff (X, Y, Z) is a Markov chain. Chapter 3 in Pinsker is wholly devoted to the properties of conditional mutual information. He first gives an abstract measure-theoretic definition of $I(X; Z|Y)$, not the one given above, then proves various elementary properties like $I(X; Z|Y) \geqslant 0$ (equivalent to Theorem 1.4) and $I(X; Z|Y) = I(Z; X|Y)$ (see Prob. 1.19), and finally proves that $I(X; Z|Y) = 0$ iff (X, Y, Z) is Markov. Finally, in Section 3.6 he proves that $I(X; Z|Y) = I(X, Y; Z) - I(Y; Z)$, and attributes the result to Kolmogorov.

2
Discrete memoryless channels and their capacity–cost functions

2.1 The capacity–cost function

A *discrete memoryless channel* (DMC; see p. 20 ff.) is characterized by two finite sets: A_X, the *input alphabet*, and A_Y, the *output alphabet*, and a set of *transition probabilities* $p(y|x)$, defined for each $x \in A_X$ and $y \in A_Y$, which satisfy $p(y|x) \geq 0$, $\sum_y p(y|x) = 1$, all $x \in A_X$. If A_X has r elements and A_Y has s elements, the transition probabilities are conveniently displayed in an $r \times s$ stochastic matrix[1] $Q = (q_{xy})$, whose rows are indexed by A_X and columns by A_Y. Furthermore, associated with each input x there is a nonnegative number $b(x)$, the "cost"[2] of x. Usually A_X will be taken as $\{0, 1, \ldots, r-1\}$ and A_Y as $\{0, 1, \ldots, s-1\}$.

Example 2.1 $A_X = A_Y = \{0, 1\}$,

$$Q = \begin{array}{c} 0 \\ 1 \end{array} \begin{array}{c} \\[-1.2em] \overset{\begin{array}{cc} 0 & 1 \end{array}}{\left[\begin{array}{cc} q & p \\ p & q \end{array}\right]} \end{array},$$

where $0 \leq p \leq \frac{1}{2}$, $q = 1 - p$ (this is the binary symmetric channel; see p. 21), and $b(0) = 0$, $b(1) = 1$. □

Example 2.2 $A_X = \{0, \frac{1}{2}, 1\}$, $A_Y = \{0, 1\}$,

$$Q = \begin{array}{c} 0 \\ \frac{1}{2} \\ 1 \end{array} \begin{array}{c} \\[-1.2em] \overset{\begin{array}{cc} 0 & 1 \end{array}}{\left[\begin{array}{cc} 1 & 0 \\ \frac{1}{2} & \frac{1}{2} \\ 0 & 1 \end{array}\right]} \end{array},$$

and $b(0) = b(1) = 1$, $b\left(\frac{1}{2}\right) = 0$. □

50

Example 2.3 $A_X = A_Y = \{0, 1, 2\}$,

$$Q = \begin{bmatrix} 1 & 0 & 0 \\ 0 & 1 & 0 \\ 0 & 0 & 1 \end{bmatrix},$$

and $b(0) = b(1) = 1$, $b(2) = 4$. (We shall return to these examples later in the chapter.) $\square$

The motivation for this definition is that we are supposed to imagine a device—the "channel"—which accepts, every unit of time, one symbol $x \in A_X$, and in response expels a symbol $y \in A_Y$. The channel is unreliable in the sense that the output is not a definite function of the input; $p(y|x)$ is the probability that y is the output, given that x is the input. Furthermore, use of the channel is not free; the "cost" of using input x is $b(x)$.

More generally, imagine that this channel is used n consecutive times, with inputs $x_1, x_2, \ldots, x_n$ and corresponding outputs $y_1, y_2, \ldots, y_n$. The *memoryless assumption* is that the output y_i at time i depends only on the input x_i at time i, i.e., the probability that $y_1, \ldots, y_n$ is the output, given that $x_1, \ldots, x_n$ is the input, is the product $\prod_{i=1}^{n} p(x_i|x_i)$. The cost of sending $x_1, \ldots, x_n$ is defined as

$$b(\mathbf{x}) = \sum_{i=1}^{n} b(x_i). \tag{2.1}$$

If the n inputs are described probabilistically by the random variables $\mathbf{X} = (X_1, X_2, \ldots, X_n)$ with joint distribution function $p(\mathbf{x}) = p(x_1, \ldots, x_n)$, the *average cost* is defined by

$$E[b(\mathbf{X})] = \sum_{i=1}^{n} E[b(X_i)]$$

$$= \sum_{\mathbf{x}} p(\mathbf{x})b(\mathbf{x}). \tag{2.2}$$

For each $n = 1, 2, \ldots$, we define the nth *capacity–cost function* $C_n(\beta)$ of the channel by

$$C_n(\beta) = \max\{I(\mathbf{X}; \mathbf{Y}) : E[b(\mathbf{X})] \leq n\beta\}, \tag{2.3}$$

where the maximum in (2.3) is extended over all pairs $(\mathbf{X}, \mathbf{Y}) = ((X_1, \ldots, X_n), (Y_1, \ldots, Y_n))$ of n-dimensional random vectors for which (i) the conditional probabilities $P\{\mathbf{Y}|\mathbf{X}\}$ are consistent with the given channel transition probabilities, i.e. $P\{Y_1 = y_1, \ldots, Y_n = y_n | X_1 = x_1, \ldots,$ $X_n = x_n\} = \prod_{i=1}^{n} p(y_i | x_i)$, and (ii) the input vector $\mathbf{X}$ satisfies $E[b(\mathbf{X})] \leq n\beta$. We shall call an input vector $\mathbf{X}$ a *test source*; if it satisfies $E[b(\mathbf{X})] \leq n\beta$, we shall call it *$\beta$-admissible*. Hence the maximization in (2.3) is taken over all *n-dimensional β-admissible test sources*.

Several remarks about the function $C_n(\beta)$ are in order. First, note that for a given matrix of transition probabilities $(p(y|x))$ the function $I(\mathbf{X}; \mathbf{Y})$ is a continuous function of the input distribution $p(\mathbf{x})$. The set of such distributions satisfying $\sum p(\mathbf{x})b(\mathbf{x}) = E[b(\mathbf{X})] \leq n\beta$ is a compact subset of r^n-dimensional Euclidean space, and so the function $I(\mathbf{X}; \mathbf{Y})$ actually achieves its maximum value.[3] This is the reason "max" rather than "sup" appears in (2.3). Second, notice that if we define $\beta_{\min}$ by

$$\beta_{\min} = \min_{x \in A_X} b(x), \qquad (2.4)$$

then $E[b(\mathbf{X})] \geq n \cdot \beta_{\min}$, and so $C_n(\beta)$ is defined only for $\beta \geq \beta_{\min}$. Finally, observe that if $\beta_1 > \beta_2$, the set of test sources satisfying $E[b(\mathbf{X})] \leq n\beta_2$ is a subset of the set of test sources satisfying $E[b(\mathbf{X})] \leq n\beta_1$, and so $C_n(\beta_1) \geq C_n(\beta_2)$, i.e. $C_n(\beta)$ is an increasing function of $\beta \geq \beta_{\min}$.

The *capacity–cost* function[4] of the channel is now defined by

$$C(\beta) = \sup_n \frac{1}{n} C_n(\beta). \qquad (2.5)$$

The number $C(\beta)$ will turn out to represent the maximum amount of information that can be transmitted reliably over the channel per unit of time, if the channel must be used in such a way that the average cost is $\leq \beta$ per unit of time. The precise statement of this result, Shannon's *channel coding theorem*, will be proved in Section 2.2. Our object in this section is to develop techniques for computing $C(\beta)$ for a given DMC and cost function.

We begin by showing that the functions $C_n(\beta)$ are all *convex*.

Theorem 2.1 $C_n(\beta)$ *is a convex $\cap$ function of $\beta \geq \beta_{\min}$.*

Proof Let $\alpha_1, \alpha_2 \geq 0$, $\alpha_1 + \alpha_2 = 1$. We must show that, for $\beta_1, \beta_2 \geq \beta_{\min}$,

$$C_n(\alpha_1 \beta_1 + \alpha_2 \beta_2) \geq \alpha_1 C_n(\beta_1) + \alpha_2 C_n(\beta_2).$$

To do this, let $\mathbf{X}_1, \mathbf{X}_2$ be *n*-dimensional test sources with distributions $p_1(\mathbf{x})$, $p_2(\mathbf{x})$ which achieve $C_n(\beta_1)$, $C_n(\beta_2)$, respectively, that is, if $\mathbf{Y}_1, \mathbf{Y}_2$ denote the outputs corresponding to $\mathbf{X}_1, \mathbf{X}_2$, then

$$E[b(\mathbf{X}_i)] \leqslant n\beta_i \tag{2.6}$$

$$\left. \begin{array}{l} \end{array} \right\}, \quad i = 1, 2.$$

$$I(\mathbf{X}_i; \mathbf{Y}_i) = C_n(\beta_i) \tag{2.7}$$

Define a further test source $\mathbf{X}$ with distribution $p(\mathbf{x}) = \alpha_1 p_1(\mathbf{x}) + \alpha_2 p_2(\mathbf{x})$, and let $\mathbf{Y}$ be the corresponding output. Then $E[b(\mathbf{X})] = \sum_x p(\mathbf{x})b(\mathbf{x}) = \alpha_1 \sum_x p_1(\mathbf{x})b(\mathbf{x}) + \alpha_2 \sum_x p_2(\mathbf{x})b(\mathbf{x}) = \alpha_1 E[b(\mathbf{X}_1)] + \alpha_2 E[b(\mathbf{X}_2)] \leqslant n(\alpha_1\beta_1 + \alpha_2\beta_2)$ (see (2.6)), and so $\mathbf{X}$ is $(\alpha_1\beta_1 + \alpha_2\beta_2)$-admissible. Thus $I(\mathbf{X}; \mathbf{Y}) \leqslant C_n(\alpha_1\beta_1 + \alpha_2\beta_2)$. But, since $I(\mathbf{X}; \mathbf{Y})$ is a convex $\cap$ function of the input probability distribution $p(\mathbf{x})$ (Theorem 1.6), $I(\mathbf{X}; \mathbf{Y}) \geqslant \alpha_1 I(\mathbf{X}_1; \mathbf{Y}_1) + \alpha_2 I(\mathbf{X}_2; \mathbf{Y}_2) = \alpha_1 C_n(\beta_1) + \alpha_2 C_n(\beta_2)$ (see Eq. (2.7)). $\qquad\square$

Our next result shows that the apparently formidable definition (2.5) is not so bad after all.[5]

Theorem 2.2 *For any* DMC, $C_n(\beta) = nC_1(\beta)$ *for all* $n = 1, 2, \ldots$ *and all* $\beta \geqslant \beta_{\min}$.

Proof Let $\mathbf{X} = (X_1, \ldots, X_n)$ be a β-admissible test source that achieves $C_n(\beta)$, that is,

$$E[b(\mathbf{X})] \leqslant n\beta, \tag{2.8}$$

$$I(\mathbf{X}; \mathbf{Y}) = C_n(\beta), \tag{2.9}$$

where $\mathbf{Y} = (Y_1, \ldots, Y_n)$ is the corresponding channel output. By Theorem 1.9,

$$I(\mathbf{X}; \mathbf{Y}) \leqslant \sum_{i=1}^{n} I(X_i; Y_i). \tag{2.10}$$

If we define $\beta_i = E[b(X_i)]$, it follows that

$$\sum_{i=1}^{n} \beta_i = \sum_{i=1}^{n} E[b(X_i)]$$

$$= E\{b(\mathbf{X})\}$$

$$\leqslant n\beta. \tag{2.11}$$

Furthermore, from definition (2.3) of $C_1(\beta_i)$,

$$I(X_i; Y_i) \leqslant C_1(\beta_i). \tag{2.12}$$

Since by Theorem 2.1 $C_1(\beta)$ is a convex $\cap$ function of β, Jensen's inequality impies that

$$\frac{1}{n}\sum_{i=1}^{n} C_1(\beta_i) \leq C_1\left(\frac{1}{n}\sum_{i=1}^{n}\beta_i\right) = C_1\left\{\frac{1}{n}E[b(\mathbf{X})]\right\}.$$

But, since $(1/n)E[b(\mathbf{X})] \leq \beta$ and $C_1(\beta)$ is an increasing function of β,

$$\sum_{i=1}^{n} C_1(\beta_i) \leq nC_1(\beta). \tag{2.13}$$

Combinging (2.9), (2.10), (2.12), and (2.13), we obtain $C_n(\beta) \leq nC_1(\beta)$. For the opposite inequality, let (X, Y) be a pair of random variables that achieve $C_1(\beta)$:

$$E[b(X)] \leq \beta, \tag{2.14}$$

$$I(X; Y) = C_1(\beta). \tag{2.15}$$

Let $X_1, X_2, \ldots, X_n$ be independent, identically distributed random variables with distribution functions the same as X, and let $Y_1, Y_2, \ldots, Y_n$ be the corresponding channel outputs. Then $E[b(\mathbf{X})] = \sum E[b(X_i)] \leq n\beta$ by (2.14), and $I(\mathbf{X}; \mathbf{Y}) = \sum_{i=1}^{n} I(X_i; Y_i)$ (Theorems 1.8 and 1.9) $= nC_1(\beta)$. Thus $C_n(\beta) \geq nC_1(\beta)$ and well.   □

Corollary *For a memoryless channel $C(\beta) = C_1(\beta)$ (see definition (2.5)). [N.B. This result is not true for channels with memory; see Prob. 2.10.]*

Let us now discuss the general properties of the function $C(\beta)$ for a given DMC and cost function. We know that it is an increasing, convex $\cap$ function if $\beta \geq \beta_{\min}$. Its convexity implies that it is also continuous for $\beta > \beta_{\min}$; see Appendix B. (It is also continuous at $\beta = \beta_{\min}$; see Prob. 2.5.) We shall now argue that $C(\beta)$ is actually constant for sufficiently large β. Let us define $C_{\max} = \max\{C(\beta) : \beta \geq \beta_{\min}\}$, that is,

$$C_{\max} = \max\{I(X; Y)\}, \tag{2.16}$$

where the maximum is taken over all (one-dimensional) test sources X, with no bound on $E[b(X)]$. $C_{\max}$ is called the *capacity* of the channel. If we define

$$\beta_{\max} = \min\{E[b(X)] : I(X; Y) = C_{\max}\}, \tag{2.17}$$

then clearly $C(\beta) = C_{\max}$ for all $\beta \geq \beta_{\max}$, and $C(\beta) < C_{\max}$ for $\beta < \beta_{\max}$. It follows from the facts that $C(\beta)$ is increasing and convex $\cap$ for $\beta \geq \beta_{\min}$ and is constant for $\beta \geqslant \beta_{max}$ that $C(\beta)$ is actually strictly increasing for $\beta_{min} \leqslant \beta \leqslant \beta_{max}$ (see Prob. 2.6). Hence in this interval $C(\beta)$ could be defined by

$$C(\beta) = \max\{I(X; Y) : E[b(X)] = \beta\}, \quad \beta_{min} \leqslant \beta \leqslant \beta_{max}. \quad (2.18)$$

Finally, we turn to the computation of $C(\beta_{min}) = C_{min}$. A test source X is β_{min}-admissible iff it has $p(x) = 0$ when $b(x) > \beta_{min}$, that is, if it uses only the cheapest inputs. Hence C_{min} is the capacity of the reduced channel, which is derived from the original one by deleting those inputs x for which $b(x) > \beta_{min}$.

Combining all these facts, we see that a typical $C(\beta)$ curve might look as shown in Fig. 2.1.

Example 2.1 (continued)

$$Q = \begin{bmatrix} q & p \\ p & q \end{bmatrix},$$

$b(0) = 0$, $b(1) = 1$. Here $\beta_{min} = 0$, the reduced channel has only one input, 0, and so $C_{min} = C(0) = 0$. Let X be a test source that achieves $C(\beta)$ for some $0 \leqslant \beta \leqslant \beta_{max}$. Then (see (2.18)) we must have $P\{X = 1\} = \beta$, $P\{X = 0\} = \alpha = 1 - \beta$, and $C(\beta) = I(X; Y) = H(Y) - H(Y|X) = H(\alpha q + \beta p) - H(p)$. Since $H(x)$ attains its maximum value, $\log 2$, at $x = \frac{1}{2}$ (cf. Fig. 1.1), it follows that $H(\alpha q + \beta p)$ attains its maximum, $\log 2$, at $\beta = \frac{1}{2}$. Hence $\beta_{max} = \frac{1}{2}$, and the entire $C(\beta)$ curve (see Fig. 2.2) is given analytically by

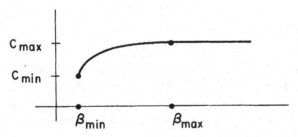

Figure 2.1 A typical $C(\beta)$ curve.

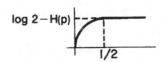

Figure 2.2 $C(\beta)$ for Example 2.1.

$$C(\beta) = \begin{cases} H[(1-\beta)q + \beta p] - H(p), & 0 \leqslant \beta \leqslant \frac{1}{2}, \\ \log 2 - H(p), & \beta \geqslant \frac{1}{2}. \end{cases}$$ $\square$

Example 2.2 (continued).

$$Q = \begin{bmatrix} 1 & 0 \\ \frac{1}{2} & \frac{1}{2} \\ 0 & 1 \end{bmatrix},$$

$b(0) = b(1) = 1$, $b(\frac{1}{2}) = 0$. Here $\beta_{min} = 0$ and $C(0) = 0$ as before. Let X denote a test source that achieves $C(\beta)$ for some $0 \leqslant \beta \leqslant \beta_{max}$, and let $p(x) = P\{X = x\}$ for $x = 0, \frac{1}{2}, 1$. Now $I(X; Y)$ is convex $\cap$ in $p(0)$, $p(\frac{1}{2})$, $p(1)$, and $E[b(X)] = (p(0) + p(1))$ is symmetric in $p(0)$ and $p(1)$, and so we must have $p(0) = p(1) = \beta/2$. Then $I(X; Y) = H(Y) - H(Y|X) = \log 2 - (1 - \beta)\log 2 = \beta \log 2$. Thus $\beta_{max} = 1$, and the $C(\beta)$ curve (see Fig. 2.3) is given by

$$C(\beta) = \begin{cases} \beta \log 2, & 0 \leqslant \beta \leqslant 1, \\ \log 2, & \beta \geqslant 1. \end{cases}$$

This illustrates the fact that $C(\beta)$ need not be strictly convex for $\beta_{min} \leqslant \beta \leqslant \beta_{max}$. $\square$

Example 2.3 (continued).

$$Q = \begin{bmatrix} 1 & 0 & 0 \\ 0 & 1 & 0 \\ 0 & 0 & 1 \end{bmatrix},$$

$b(0) = b(1) = 1$, $b(2) = 4$. Let X be a test source achieving $C(\beta)$, and let $\alpha_i = P\{X = i\}$, $i = 0, 1, 2$. Then $C(\beta) = H(Y) - H(Y|X) = H(Y) = H(X) = H(\alpha_0, \alpha_1, \alpha_2)$, and for $\beta_{min} \leqslant \beta \leqslant \beta_{max}$, $E[b(X)] = \alpha_0 + \alpha_1 + 4\alpha_2 = \beta$. Clearly, $\beta_{min} = 1$, and the reduced channel has only the inputs 0, 1, with transition probabilities

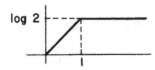

Figure 2.3 $C(\beta)$ for Example 2.2.

$$Q' = \begin{bmatrix} 1 & 0 \\ 0 & 1 \end{bmatrix}.$$

The capacity of this channel (see Example 2.1) is $\log 2$, so $C(1) = \log 2$. Also $C_{\max} = \max\{H(\alpha_0, \alpha_1, \alpha_2)\} = \log 3$, where $\alpha_0 = \alpha_1 = \alpha_2 = \frac{1}{3}$ (Theorem 1.1), and so $\beta_{\max} = \frac{1}{3} + \frac{1}{3} + \frac{4}{3} = 2$, $C_{\max} = \log 3$. For $1 \leq \beta \leq 2$, we must maximize $H(\alpha_0, \alpha_1, \alpha_2)$ subject to $\alpha_0 + \alpha_1 + 4\alpha_2 = \beta$. This is symmetric in α_0 and α_1, and so we take $\alpha_0 = \alpha_1 = \alpha$, $\alpha_2 = 1 - 2\alpha$. This forces $\alpha = 2/3 - \beta/6$, and so $C(\beta) = H(2/3 - \beta/6, 2/3 - \beta/6, \beta/3 - 1/3)$ for $1 \leq \beta \leq 2$. the graph of $C(\beta)$ is shown in Fig. 2.4. A generalization of Examples 2.1 and 2.3 is given in Prob. 2.3. $\qquad\square$

We conclude this section with a theorem which tells how to compute the capacity (i.e., $C_{\max}$) of a channel whose transition matrix has a high degree of symmetry. We shall call a stochastic matrix Q *symmetric* if every row of Q is a permutation of every other row, and every column is a permutation of every other column. A DMC is *symmetric* if its transition probability matrix is symmetric.

For example,

$$Q = \begin{bmatrix} \frac{1}{3} & \frac{1}{3} & \frac{1}{6} & \frac{1}{6} \\ \frac{1}{6} & \frac{1}{6} & \frac{1}{3} & \frac{1}{3} \end{bmatrix}$$

is symmetric, but

$$Q' = \begin{bmatrix} \frac{1}{3} & \frac{1}{3} & \frac{1}{6} & \frac{1}{6} \\ \frac{1}{6} & \frac{1}{3} & \frac{1}{6} & \frac{1}{3} \end{bmatrix}$$

is not. (But see Prob. 2.2.)

Theorem 2.3 *If a symmetric DMC has r inputs and s outputs, its capacity is*

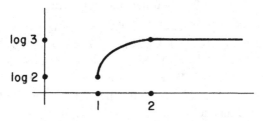

Figure 2.4 $C(\beta)$ for Example 2.3.

achieved with equiprobable inputs, i.e., $p(x) = 1/r$, $x \in \{0, 1, \ldots, r-1\}$, *and the capacity is*

$$C_{\max} = \log s - H(q_0, q_1, \ldots, q_{s-1}),$$

where $(q_0, q_1, \ldots, q_{s-1})$ is any row of the transition matrix.

Proof[6] $I(X; Y) = H(Y) - H(Y|X)$, $H(Y|X) = \sum_x p(x)H(Y|X = x)$. But since every row of the matrix is a permutation of every other row,

$$H(Y|X = x) = \sum_y p(y|x) \log \frac{1}{p(y|x)} = H(q_0, q_1, \ldots, q_{s-1})$$

is independent of x. Also, by Theorem 1.1, $H(Y) \leq \log s$ with equality iff $p(y) = 1/s$ for $y \in \{0, 1, \ldots, s-1\}$. But the condition on the columns of the transition matrix guarantees that, if $p(x) = 1/r$ for all x, then $p(y) = 1/s$ for all y.   □

For example, the channel with symmetric transition matrix Q given above has $C_{\max} = \log 4 - H(\frac{1}{3}, \frac{1}{3}, \frac{1}{6}, \frac{1}{6}) = \log(2^{5/3}3^{-1}) = 0.0817$ bit. As a more general example, the *r-ary symmetric* channel has as transition probability matrix the $r \times r$ matrix $q_{xy} = \varepsilon$ if $x \neq y$, and $q_{xy} = 1 - (r-1)\varepsilon$ if $x = y$, where $0 \leq \varepsilon \leq 1/(r-1)$. For $r = 2$ this is the familiar BSC; $r = 4$, is as follows:

$$Q = \begin{bmatrix} 1 - 3\varepsilon & \varepsilon & \varepsilon & \varepsilon \\ \varepsilon & 1 - 3\varepsilon & \varepsilon & \varepsilon \\ \varepsilon & \varepsilon & 1 - 3\varepsilon & \varepsilon \\ \varepsilon & \varepsilon & \varepsilon & 1 - 3\varepsilon \end{bmatrix}.$$

By Theorem 2.3, the capacity of the *r*-ary symmetric channel is $\log r - H[1 - (r-1)\varepsilon, \varepsilon, \ldots, \varepsilon] = \log r + (r-1)\varepsilon \log \varepsilon + (1 - r\varepsilon + \varepsilon) \log (1 - r\varepsilon + \varepsilon)$. (A method for computing the entire capacity–cost function $C(\beta)$ for an *r*-ary symmetric channel is given in Prob. 2.3.)

2.2 The channel coding theorem

According to the results of Section 2.1, if $\mathbf{X} = (X_1, X_2, \ldots, X_n)$ represents n consecutive inputs to a given DMC, and if $\mathbf{Y} = (Y_1, \ldots, Y_n)$ is the corresponding output, then $n^{-1}I(\mathbf{X}; \mathbf{Y}) \leq C_{\max}$. This suggests, according to the informal interpretation of $I(\mathbf{X}; \mathbf{Y})$ given in Chapter 1, that the channel can transmit *at most* $C_{\max}$ bits[7] of information per unit of time. On the other hand, if X is a test source that achieves $C(\beta_{\max}) = C_{\max}$, then $I(X; Y) = C_{\max}$, and this suggests that, if used properly, the channel can transmit *at least* $C_{\max}$ bits of information per unit of time. Thus C_{max} ought to represent the maximum rate at which information can be transmitted over the channel. More generally, for an arbitrary $\beta \geqslant \beta_{min}$, $C(\beta)$ ought to represent the maximum rate at which information can be transmitted over the channel, if the average input cost must be $\leqslant \beta$. It is our object in this section to make these vague allegations precise.

Consider the following thought experiment. We model an information source as a sequence $\mathbf{U} = (U_1, U_2, \ldots, U_k)$ of independent, identically distributed random variables with common distribution function $P\{U = 0\} = P\{U = 1\} = \frac{1}{2}$. Our object is to transmit these k "bits" of information over the channel, using the channel n times, with an average cost $\leqslant \beta$. Let $\mathbf{X} = (X_1, \ldots, X_n)$ be the corresponding channel input, $\mathbf{Y} = (Y_1, Y_2, \ldots, Y_n)$ the channel output, and $\hat{\mathbf{U}} = (\hat{U}_1, \ldots, \hat{U}_k)$ the receiver's estimate of $\mathbf{U}$, which we assume depends only on $\mathbf{Y}$ (Fig. 2.5).

We assume this is a pretty good system, say $P\{\hat{U}_i \neq U_i\} < \varepsilon$ for all i, where ε is some small number. Then, by Theorem 1.8, $I(\mathbf{U}; \hat{\mathbf{U}}) \geqslant \sum_{u=1}^{k} I(U; \hat{U}_i)$, and $I(U_i; \hat{U}_i) = H(U_i) - H(U_i|\hat{U}_i) = \log 2 - H(U_i|\hat{U}_i) \geqslant \log 2 - H(\varepsilon)$ by Fano's inequality (corollary to Theorem 1.2). Thus we have $I(\mathbf{U}; \hat{\mathbf{U}}) \geqslant k[1 - H_2(\varepsilon)]$. Also, by the data-processing theorem (Eq. (1.15)), $I(\mathbf{U}; \mathbf{V}) \leqslant I(\mathbf{X}; \mathbf{Y})$. Finally, by Eq. (2.3), $I(\mathbf{X}; \mathbf{Y}) \leqslant C_n(\beta) = nC(\beta)$. Combining these three inequalities, we have (implied logs are base -2):

$$\frac{k}{n} \leqslant \frac{C(\beta)}{1 - H_2(\varepsilon)}. \tag{2.19}$$

The ratio k/n, which we shall call the *rate* of the system, represents the number of *bits per channel use* being transmitted by our imaginary communication system. The bound (2.19) is an increasing function of the bit error probability ε; this is not surprising: it says merely that the more reliably we want to communicate, the slower we must communicate. Quantitatively, Eq. (2.19) says that, if the channel is being used so that the average input cost is $\leqslant \beta$, and if we wish to design a system with rate $r > C(\beta)$, then the resulting error probability ε is bounded below by $\varepsilon \geqslant H^{-1}[1 - C(\beta)/r] > 0$. Even if

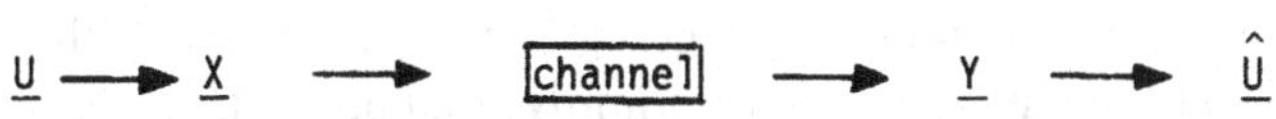

Figure 2.5 An imaginary communication system.

there is no cost constraint, if $r > C_{max}$, then $\varepsilon \geq H_2^{-1}(1 - C_{max}/r) > 0$. In words, *we cannot communicate reliably at rates above channel capacity.*

What about rates below $C(\beta)$? Here (2.19) is no help, since if $k/n < C(\beta)$, (2.19) is satisfied for all $\varepsilon \geq 0$. This is not a weakness in our bound, for we shall now prove that, if $R < C(\beta)$ and $\varepsilon > 0$, it is possible to design a system of the above sort with average cost $\leq \beta$, $k/n \geq R$, and $P\{U_i \neq \hat{U}_i\} < \varepsilon$ for all *i*! This stunning result, due to Claude Shannon, is called the *channel coding theorem.* The key to this result is the concept of a code, which we now describe.

Given an integer *n*, a (channel) *code of length n* over A_X is a subset $C = \{\mathbf{x}_1, \mathbf{x}_2, \ldots, \mathbf{x}_M\}$ of A_X^n.[8] The rate of the code is defined as $r = (1/n) \log M$—if logs are base 2, the rate is in units of *bits per* (channel input) *symbol.* The code is *β-admissible* if $b(\mathbf{x}_i) = \sum_{j=1}^{n} b(x_{ij}) \leq n\beta$ for all *i*, where $\mathbf{x}_i = (x_{i1}, \ldots, x_{in})$ is the expansion of $\mathbf{x}_i$ into components.

A *decoding rule* for the code *C* is a mapping $f: A_Y^n \to C \cup \{?\}$. The special symbol "?" denotes decoder failure; its significance will emerge below.

A code can be used to design a communication system, in the sense of Fig. 2.5, as follows. Let *k* be an integer such that $k \leq \log_2 M$. Then it will be possible to assign a *distinct* codeword $\mathbf{x}_i$ to each of the 2^k possible source sequences.[9] A one-to-one mapping from the set of all possible source sequences into the code *C* is called an *encoding rule.* If the source sequence to be transmitted is $\mathbf{u} = (u_1, \ldots, u_k)$, the idea is to *encode* **u** into one of the codewords $\mathbf{x}_i$ via the encoding rule, and transmit $\mathbf{x}_i$ over the channel. At the other end of the channel appears a noisy version of $\mathbf{x}_i$; call it **y**. The receiver decodes **y** into a codeword $\mathbf{x}_j$ (or "?") via the decoding rule *f*; the estimate $\hat{\mathbf{u}}$ of **u** is then the unique source sequence $\hat{\mathbf{u}}$ (if there is one) that corresponds to the codeword $\mathbf{x}_j$. The error probability for the system, given that $\mathbf{x}_i$ is transmitted, denoted by $P_E^{(i)}$, is then given by

$$P_E^{(i)} = P\{f(\mathbf{y}) \neq \mathbf{x}_i\}$$

$$= \sum \{p(\mathbf{y}|\mathbf{x}_i) : f(\mathbf{y}) \neq \mathbf{x}_i\} \tag{2.20}$$

where $p(\mathbf{y}|\mathbf{x}_i) = \prod_{j=1}^{n} p(y_i|x_{ij})$ is the appropriate product of entries from the channel's matrix *Q* of transition probabilities.

Before we state and prove the coding theorem, we shall illustrate the concept of coding with three rather primitive examples. (It is the object of Part two to give nonprimitive examples!)

Example 2.4 $A_X = A_Y = \{0, 1\}$,

$$Q = \begin{bmatrix} q & p \\ p & q \end{bmatrix},$$

$b(0) = b(1) = 0$. (This is the BSC with no input cost constraint.) Code: $n = 3$, $M = 2$, $C = \{(000), (111)\}$, rate $= \frac{1}{3}$ bit per symbol. Decoding rule: $f(y_1 y_2 y_3) = (xxx)$, where $x =$ "majority vote" of y_1, y_2, and y_3. It is an easy exercise to prove that $P_E^{(1)} = P_E^{(2)} = 3p^2 - 2p^3$, which is less than the raw error probability p if $p < \frac{1}{2}$ and much less if p is very small (see Eq. (0.1)). □

Example 2.5 $A_X = \{0, \frac{1}{2}, 1\}$, $A_Y = \{0, 1\}$,

$$Q = \begin{bmatrix} 1 & 0 \\ \frac{1}{2} & \frac{1}{2} \\ 0 & 1 \end{bmatrix},$$

$b(0) = b(1) = 1$, $b(\frac{1}{2}) = 0$. (This is the same as Example 2.2.) Code: $C = \{(x_1, x_2, \ldots, x_k, \frac{1}{2}, \frac{1}{2}, \ldots, \frac{1}{2}) : x_j = 0 \text{ or } 1, i = 1, 2, \ldots, k\}$. Here k is some fixed integer $\leqslant n$ and $M = 2^k$. The code's rate is k/n bits per symbol. This code is β-admissible for all $\beta \geqslant k/n$. Decoding rule: $f(y_1, \ldots, y_n) = (y_1, \ldots, y_k, \frac{1}{2}, \frac{1}{2}, \ldots, \frac{1}{2})$. Here $P_E^{(i)} = 0$ for all i (see Prob. 2.11). □

Example 2.6 $A_X = \{0, 1\}$, $A_Y = \{0, 1, 2, 3\}$,

$$Q = \begin{bmatrix} \frac{1}{3} & \frac{1}{3} & \frac{1}{6} & \frac{1}{6} \\ \frac{1}{6} & \frac{1}{6} & \frac{1}{3} & \frac{1}{3} \end{bmatrix},$$

$b(0) = b(1) = 0$. (This is the "symmetric" channel of p. 57.) Code: $n = 2$, $M = 2$; $C = \{(00), (11)\}$, rate $r = \frac{1}{2}$ bit per symbol. Decoding rule given by the table below:

		y_2			
		0	1	2	3
	0	00	00	00	?
y_1	1	00	00	?	11
	2	00	?	11	11
	3	?	11	11	11

$f(y_1 y_2) = (y_1, y_2)$th entry in the decoding matrix.

Here it turns out (see Prob. 2.12) that $P_E^{(i)} = \frac{4}{9}$, $i = 1, 2$. □

Returning now to the bound (2.19), recall that we set out to design, for given values of β, $R < C(\beta)$, and $\varepsilon > 0$, a system with $k/n \geqslant R$ and $P\{U_i \neq \hat{U}_i\} < \varepsilon$ for all i. From the preceding discussion, we see that it will be possible to do this if we can find a code C of length n, together with a decoding rule, for which $M \geqslant 2^{\lceil Rn \rceil}$ and $P_E^{(i)} < \varepsilon$ for all ε. The following theorem asserts that it is possible to do this.

Theorem 2.4 *Let a* DMC *with capacity–cost function* $C(\beta)$ *be given. Then, for any* $\beta_0 \geqslant \beta_{\min}$ *and real numbers* $\beta > \beta_0$, $R < C(\beta_0)$, $\varepsilon > 0$, *for all sufficiently large n there exists a code* $C = \{\mathbf{x}_1, \mathbf{x}_2, \ldots, \mathbf{x}_M\}$ *of length n and a decoding rule such that:*

(a) *each codeword* $\mathbf{x}_i$ *is* β-*admissible,*
(b) $M \geqslant 2^{\lceil Rn \rceil}$,
(c) $P_E^{(i)} < \varepsilon$ *for all* $i = 1, 2, \ldots, M$.

Corollary *(the channel coding theorem for DMC's[10]). For any* $R < C_{\max}$ *and* $\varepsilon > 0$, *there exists a code* $C = \{\mathbf{x}_1, \ldots, \mathbf{x}_M\}$ *of length n and a decoding rule such that:*

(a) $M \geqslant 2^{\lceil Rn \rceil}$,
(b) $P_E^{(i)} < \varepsilon$ *for all* $i = 1, 2, \ldots, M$.

Proof of corollary Let $\beta_0 = \beta_{\max}$ in Theorem 2.4.   ☐

Proof of Theorem 2.4 Throughout the proof, think of n as a large integer. It will be specified more precisely later.

Consider the set Ω of all pairs $(\mathbf{x}, \mathbf{y})$ consisting of one channel input sequence $\mathbf{x} = (x_1, \ldots, x_n)$ and one channel output sequence $\mathbf{y} = (y_1, \ldots, y_n)$, both of length n. In symbols, $\Omega = A_X^n \times A_Y^n$. We make Ω into a sample space by defining

$$p(\mathbf{x}, \mathbf{y}) = p(\mathbf{x})p(\mathbf{y}|\mathbf{x}); \tag{2.21}$$

$p(\mathbf{x}) = p(x_1) \ldots p(x_n)$, where $p(x)$ is a probability distribution on A_X that achieves $C(\beta_0)$; and $p(\mathbf{y}|\mathbf{x}) = p(y_1|x_1) \ldots p(y_n|x_n)$, where $p(y|x)$ are the channel transition probabilities.

Let us now choose R' to satisfy $R < R' < C(\beta_0)$ and define a subset $T \subseteq \Omega$ as follows:

$$T = \{(\mathbf{x}, \mathbf{y}) : I(\mathbf{x}; \mathbf{y}) \geqslant nR'\}, \tag{2.22}$$

where $I(\mathbf{x}; \mathbf{y}) = \log_2[p(\mathbf{y}|\mathbf{x})/p(\mathbf{y})]$. The set T can be thought of as the set of pairs which are close together (see Prob. 2.17) in a certain sense. Also, we define a subset $B \subseteq A_X^n$:

$$B = \{\mathbf{x} : b(\mathbf{x}) \leqslant \beta n\}. \tag{2.23}$$

Here B is the set of β-admissible codewords. Finally, we define the set $T^* \subseteq T$ by

$$T^* = \{(\mathbf{x}, \mathbf{y}) : (\mathbf{x}, \mathbf{y}) \in T \text{ and } \mathbf{x} \in B\}. \tag{2.24}$$

Now let $C = \{\mathbf{x}_1, \mathbf{x}_2, \ldots, \mathbf{x}_M\}$ be any code, good or bad, of length n. We define a decoding rule as follows. If $\mathbf{y}$ is received, we examine the set

$$S(\mathbf{y}) = \{\mathbf{x} : (\mathbf{x}, \mathbf{y}) \in T^*\} \subseteq B$$

(which could be thought of as a "sphere" around $\mathbf{y}$). If $S(\mathbf{y})$ contains *exactly one* codeword $\mathbf{x}_i$, we set $f(\mathbf{y}) = \mathbf{x}_i$. Otherwise, either because $S(\mathbf{y})$ contains no codewords or more than one codeword, we set $f(\mathbf{y}) = ?$, that is, we deliberately make an error.[11] This decoding rule is depicted in Fig. 2.6.

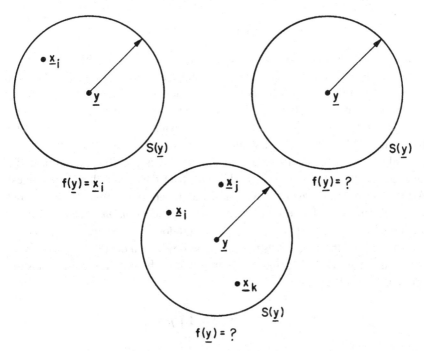

Figure 2.6 The decoding rule used in the proof of Theorem 2.4.

If we use the code C with the decoding rule just described, if $\mathbf{x}_i$ is transmitted and if $\mathbf{y}$ is received, an error can occur iff either $\mathbf{x}_i \notin S(\mathbf{y})$ or $\mathbf{x}_j \in S(\mathbf{y})$ for some $j \neq i$. Hence (see Eq. (2.20))

$$P_E^{(i)} \leqslant P\{\mathbf{x}_i \notin S(\mathbf{y})\} + \sum_{\substack{j=1 \\ j \neq i}}^{M} P\{\mathbf{x}_j \in S(\mathbf{y})\}. \tag{2.25}$$

To put (2.25) into a more convenient form, let us define the indicator functions of the set T^* as follows:

$$\Delta(\mathbf{x}, \mathbf{y}) = \begin{cases} 1 & \text{if} \quad (\mathbf{x}, \mathbf{y}) \in T^*, \\ 0 & \text{if} \quad (\mathbf{x}, \mathbf{y}) \notin T^*, \end{cases}$$

$$\overline{\Delta}(\mathbf{x}, \mathbf{y}) = \begin{cases} 0 & \text{if} \quad (\mathbf{x}, \mathbf{y}) \in T^*, \\ 1 & \text{if} \quad (\mathbf{x}, \mathbf{y}) \notin T^*. \end{cases} \tag{2.26}$$

Now (2.25) can be rewritten as

$$P_E^{(i)} \leqslant \sum_{\mathbf{y}} \overline{\Delta}(\mathbf{x}_i, \mathbf{y}) p(\mathbf{y}|\mathbf{x}_i) + \sum_{j \neq i} \sum_{\mathbf{y}} \Delta(\mathbf{x}_j, \mathbf{y}) p(\mathbf{y}|\mathbf{x}_i)$$

$$= Q_i(\mathbf{x}_1, \ldots, \mathbf{x}_M). \tag{2.27}$$

Our goal is to somehow find a code $\{\mathbf{x}_1, \ldots, \mathbf{x}_M\}$ for which Q_i is very small for all i simultaneously. Unfortunately, however, Q_i is an extermely complicated function and cannot be evaluated explicitly (or even closely estimated) except for the simplest possible codes—certainly not for big, complex codes with large n and huge M! So what is the point in introducing the bound $P_E^{(i)} \leqslant Q_i$ in the first place? The point is that, although it is not possible to estimate Q_i for a particular code, it is possible to estimate the average value of Q_i as $\{\mathbf{x}_1, \ldots, \mathbf{x}_M\}$ ranges over the set of all possible codes! Astonishingly,[12] this average will turn out to approach 0 if $M = 2^{Rn}$ and $n \to \infty$! This remarkable proof technique is called *random coding*, because we choose the code $\{\mathbf{x}_1, \ldots, \mathbf{x}_M\}$ "at random" according to a certain probability distribution. Let us now see how it works in detail.

The first step is to describe the appropriate probability distribution on the set of all possible codes. It is

$$p(\mathbf{x}_1, \ldots, \mathbf{x}_M) = \prod_{i=1}^{M} p(\mathbf{x}_i),$$

where if $\mathbf{x}_i = (x_{i1}, x_{i2}, \ldots, x_{in})$, $p(\mathbf{x}_i) = \prod_{k=1}^{n} p(x_{ik})$. This probability distribution corresponds to the experiment of choosing the code "randomly" by picking the individual codeword coordinates independently, according to the probability distribution $p(x)$ that achieves $C(\beta_0)$.

We now view $Q_i(\mathbf{x}_1, \ldots, \mathbf{x}_M)$ as a random variable on the sample space of all possible codes; its expected value is (see Eq. (2.27))

$$E(Q_i) = E\left[\sum_{\mathbf{y}} \overline{\Delta}(\mathbf{x}_i, \mathbf{y}) p(\mathbf{y}|\mathbf{x}_i)\right] + \sum_{j \neq i} E\left[\sum_{\mathbf{y}} \Delta(\mathbf{x}_j, \mathbf{y}) p(\mathbf{y}|\mathbf{x}_i)\right]$$

$$= E_1 + \sum_{j \neq i} E_2^{(j)}. \tag{2.28}$$

First, we bound E_1:

$$E_1 = \sum_{\mathbf{x}_1,\ldots,\mathbf{x}_M} p(\mathbf{x}_1) \ldots p(\mathbf{x}_M) \sum_{\mathbf{y}} \overline{\Delta}(\mathbf{x}_i, \mathbf{y}) p(\mathbf{y}|\mathbf{x}_i)$$

$$= \sum_{\mathbf{x}_i,\mathbf{y}} p(\mathbf{x}_i) p(\mathbf{y}|\mathbf{x}_i) \overline{\Delta}(\mathbf{x}_i, \mathbf{y})$$

$$= \sum_{\mathbf{x},\mathbf{y}} p(\mathbf{x}, \mathbf{y}) \overline{\Delta}(\mathbf{x}, \mathbf{y}) \quad \text{(see (2.21))}$$

$$= P\{(\mathbf{x}, \mathbf{y}) \notin T^*\} \quad \text{(see (2.26))}$$

$$= P\{(\mathbf{x}, \mathbf{y}) \notin T \text{ or } \mathbf{x} \notin B\} \quad \text{(see (2.24))}$$

$$\leqslant P\{(\mathbf{x}, \mathbf{y}) \notin T\} + P\{\mathbf{x} \notin B\}. \tag{2.29}$$

Hence

$$E_1 \leqslant P\{I(\mathbf{x}; \mathbf{y}) < nR'\} + P\{b(\mathbf{x}) > \beta n\} \quad \text{(see (2.22), (2.23), (2.29))}$$

But

$$I(\mathbf{x}; \mathbf{y}) = \log \frac{p(\mathbf{y}|\mathbf{x})}{p(\mathbf{y})}$$

$$= \log \prod_{k=1}^{n} \frac{p(y_k|x_k)}{p(y_k)}$$

$$= \sum_{k=1}^{n} \log \frac{p(y_k|x_k)}{p(y_k)}$$

$$= \sum_{k=1}^{n} I(x_k; y_k).$$

Hence $I(\mathbf{x}; \mathbf{y})$ is the sum of n independent identically distributed random variables $I(x_k; y_k)$. By definition, $E[I(x_k; y_k)] = I(X; Y) = C(\beta_0)$, so that each $I(x_k; y_k)$ has mean $C(\beta_0)$. Since $R' < C(\beta_0)$, it follows from the weak law of large numbers (see Appendix A) that

$$\lim_{n \to \infty} P\{I(\mathbf{x}; \mathbf{y}) < nR'\} = 0. \tag{2.30}$$

Similarly, $b(\mathbf{x}) = \sum_{k=1}^{n} b(x_k)$ is the sum of n independent, identically distributed random variables with mean $\leq \beta_0$. Since $\beta > \beta_0$, it follows that

$$\lim_{n \to \infty} P\{b(\mathbf{x}) > n\beta\} = 0. \tag{2.31}$$

Combining (2.29), (2.30), (2.31), we see that E_1 can be made as small as we wish by choosing n large enough.

Next, consider the term $E_2^{(j)}$ in (2.28):

$$E_2^{(j)} = \sum_{\mathbf{x}_1,\ldots,\mathbf{x}_M} p(\mathbf{x}_1) \cdots p(\mathbf{x}_M) \sum_{\mathbf{y}} \Delta(\mathbf{x}_j, \mathbf{y}) p(\mathbf{y}|\mathbf{x}_i)$$

$$= \sum_{\mathbf{x}_j, \mathbf{y}} p(\mathbf{x}_j) \Delta(\mathbf{x}_j, \mathbf{y}) \sum_{\mathbf{x}_i} p(\mathbf{x}_i) p(\mathbf{y}|\mathbf{x}_i)$$

$$= \sum_{\mathbf{x}_j, \mathbf{y}} p(\mathbf{x}_j) \Delta(\mathbf{x}_j, \mathbf{y}) p(\mathbf{y}).$$

Hence, by (2.26), (2.24), and (2.22),

$$E_2^{(j)} \leq \sum_{(\mathbf{x}, \mathbf{y}) \in T} p(\mathbf{x}) p(\mathbf{y}). \tag{2.32}$$

Now for $(\mathbf{x}, \mathbf{y}) \in T$, $p(\mathbf{x}) p(\mathbf{y}) \leq p(\mathbf{x}, \mathbf{y}) \cdot 2^{-R'n}$ (see (2.22)). Hence the bound (2.32) can be continued:

$$E_2^{(j)} \leq \sum_{(\mathbf{x},\mathbf{y})\in T} p(\mathbf{x})p(\mathbf{y})$$

$$\leq 2^{-R'n} \sum_{(\mathbf{x},\mathbf{y})\in T} p(\mathbf{x}, \mathbf{y})$$

$$\leq 2^{-R'n}. \tag{2.33}$$

Finally, by combining (2.28), (2.29), and (2.33), we get

$$E(Q_i) \leq P\{I(\mathbf{x}; \mathbf{y}) < nR'\} + P\{b(\mathbf{X}) > n\beta\} + M \cdot 2^{-R'n}. \tag{2.34}$$

If $M = 2 \cdot 2^{\lceil Rn \rceil}$, the last term in (2.34) is $\leq 4 \cdot 2^{-n(R'-R)}$. Since $R' > R$, it follows that for sufficiently large n, this term can be made as small as desired. We have already observed (see (2.30) and (2.31)) that the other two terms can be made arbitrarily small. Hence it is possible to choose n large enough so that, with $M = 2 \cdot 2^{\lceil Rn \rceil}$,

$$E(Q_i) < \varepsilon/2, \tag{2.35}$$

which is almost what we need.

The last step in the proof is to define a function $P_E(\mathbf{x}_1, \ldots, \mathbf{x}_M)$ by

$$P_E(\mathbf{x}_1, \ldots, \mathbf{x}_M) = \frac{1}{M} \sum_{i=1}^{M} P_E^{(i)}(\mathbf{x}_1, \ldots, \mathbf{x}_M). \tag{2.36}$$

Here P_E is a kind of overall error probability, assuming that each of the M codewords is transmitted with probability $1/M$. If we regard P_E as a random variable defined on the sample space of all possible codes, then from (2.27) and (2.35) we have, for $M = 2 \cdot 2^{\lceil Rn \rceil}$ and n large enough,

$$E(P_E) < \varepsilon/2.$$

Thus the average value of P_E is $< \varepsilon/2$, and so there must be a particular code $(\mathbf{x}_1, \ldots, \mathbf{x}_M)$ with $P_E(\mathbf{x}_1, \ldots, \mathbf{x}_M) < \varepsilon/2$. This code may not satisfy the conclusions of Theorem 2.4, because it may contain codewords $\mathbf{x}_i$ for which $b(\mathbf{x}_i) > n\beta$ and/or $P_E^{(i)} > \varepsilon$. But, if more than half of the codewords $\mathbf{x}_i$ had $P_E^{(i)} \geq \varepsilon$, from (2.36) we would have $P_E \geq \varepsilon/2$, a contradiction. So if we delete the codewords with $P_E^{(i)} \geq \varepsilon$ from the code, we obtain a code with $\geq 2^{\lceil Rn \rceil}$ codewords for which $P_E^{(i)} < \varepsilon$ for all i.[13] This code therefore satisfies conclusions (b) and (c) of Theorem 2.4. Finally, notice that, if $b(\mathbf{x}_i) > n\beta$, the decoding sphere $S(\mathbf{y}) = \{\mathbf{x} : (\mathbf{x}, \mathbf{y}) \in T \text{ and } b(\mathbf{x}) \leq n\beta\}$ cannot contain $\mathbf{x}_i$, that is, $P_E^{(i)} = 1$. Hence this new code cannot contain any codewords which are not β-admissible, and so (a) is satisfied also. $\square$

Problems

2.1 Compute the complete capacity–cost functions of the following channels:

(a) $Q = \begin{bmatrix} 1 & 0 & 0 \\ 0 & q & p \\ 0 & p & q \end{bmatrix}$, $b(0) = 1$, $b(1) = b(2) = 0$.

(b) $Q = \begin{bmatrix} q & p & 0 \\ 0 & p & q \end{bmatrix}$, $b(0) = 0$, $b(1) = 1$.

(The channel in part (b) is called a *binary erasure channel*; it can be depicted like this:

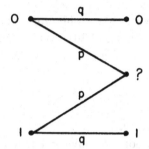

The input symbols can be received correctly or *erased*, that is, received as "?.")

2.2 Let us call a channel *weakly symmetric* if the columns of its transition matrix Q can be partitioned into subsets C_i such that, for each i, in the matrix Q_i formed by the columns in C_i each row is a permutation of every other row, and the same is true of columns. (For example, Q' on page 57 is weakly symmetric but not symmetric.) Show that the capacity of a weakly symmetric channel is achieved with equiprobable inputs.

2.3 Let us define a *strongly symmetric* channel of order r as one whose transition matrix Q has the form illustrated below for $r = 4$:

$$\begin{bmatrix} q & p & p & p \\ p & q & p & p \\ p & p & q & p \\ p & p & p & q \end{bmatrix}$$

where $p \leqslant 1/r$ and $(r-1)p + q = 1$. The object of this exercise is for you to compute $C(\beta)$ for this channel, for arbitrary input costs $b(x)$, by supplying the details in the following argument. (Assume $\beta_{min} = 0$; see Prob. 2.4.)

If X is a test source achieving $C(\beta)$, let $f(x) = P\{X = x\}$, $g(x) = P\{Y = x\}$, for $x \in \{0, 1, \ldots, r-1\}$. Then $f(x) = [g(x) - p]/(q - p)$, and the constraint $\sum b(x)f(x) = \beta$ is equivalent to $\sum b(x)g(x) = \beta(q - p) + Bp$, where $B = \sum b(x)$. $I(X; Y) = H(Y) - H(q) - (1 - q)\log(r - 1)$, so the object is to maximize $H(Y)$ subject to $\sum b(x)g(x) = \beta(q - p) + Bp$. To do this we can use the results of Prob. 1.8 and obtain the parametric solution:

$$C(\beta) = \log A(\lambda) - \lambda A'(\lambda)/A(\lambda) - H(q) - (1 - q)\log(r - 1),$$

$$\beta = -\frac{1}{q - p}(A'(\lambda)/A(\lambda) + B_P)$$

as λ ranges over the interval $\lambda \in [\lambda_0, 0]$ and β ranges over $[0, (1/n)B]$, λ_0 being the unique solution to the equation $A'(\lambda_0)/A(\lambda_0) = -Bp$.

2.4 The object of this exercise is to show that there is no essential loss of generality in assuming $\beta_{\min} = 0$. If we are given a channel whose capacity–cost function $C(\beta)$ has $\beta_{\min} > 0$, let $C'(\beta)$ denote the capacity–cost function of the same channel in which all the costs have been decreased by $\beta_{\min}$. Show that $C(\beta) = C'(\beta - \beta_{\min})$.

2.5 The object of this exercise is to prove that $C(\beta)$ is continuous at $\beta = \beta_{\min}$. To do this, supply the details in the following argument.

Let $(\mathbf{p}_1, \mathbf{p}_2, \ldots)$ be a sequence of probability vectors representing the input distributions of test sources achieving $C(\beta_1), C(\beta_2), \ldots$, where $\lim_{n\to\infty}\beta_n = \beta_{\min}$. Then there is a subsequence $(\mathbf{p}_{n_1}, \mathbf{p}_{n_2}, \ldots)$ which converges to a probability vector $\mathbf{p}$. If X is a test source with input distribution $\mathbf{p}$, then $E[b(X)] = \beta_{\min}$ and $I(X; Y) = \lim_{\beta\to\beta_{\min}^+} C(\beta)$. Hence $C(\beta_{\min}) \geq \lim_{\beta\to\beta_{\min}^+} C(\beta)$. Since $C(\beta)$ is an increasing function of $\beta \geq \beta_{\min}$, this proves the required continuity.

2.6 Let $f(x)$ be an increasing, convex $\cap$ function of $x \geq x_1$. Assume $f(x)$ is constant for $x \geq x_2$, where $x_1 < x_2$. Show that $f(x)$ is strictly increasing for $x_1 \leq x \leq x_2$.

2.7 Characterize as fully as you can the DMC's with $C_{\max} = 0$.

2.8 The *sum* of DMC's with transition matrices $Q_1, Q_2, \ldots, Q_{\hat{m}}$ is defined as the channel whose matrix is

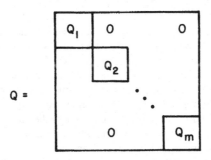

If $C_{\max}^{(i)}$ denotes the capacity of the ith channel, show that the capacity of the sum channel is given by

$$C_{\max} = \log_2 \sum 2^{C_{\max}^{(i)}} \text{ bits.}$$

2.9 Consider two DMC's with input alphabets $A_X^{(i)}$, output alphabets $A_Y^{(i)}$, transition probabilities $p^{(i)}(y|x)$, and cost functions $b^{(i)}(x)$, $i = 1, 2$. Their *product* is defined as a channel with input alphabet $A_X^{(1)} \times A_X^{(2)}$, output alphabet $A_Y^{(1)} \times A_Y^{(2)}$, transition probabilities $p((y_1, y_2)| (x_1, x_2)) = p^{(1)}(y_1|x_1) \cdot p^{(2)}(y_2|x_2)$, and cost function $b(x_1, x_2) = b^{(1)}(x_1) + b^{(2)}(x_2)$. (Physically, the product channel is the pair of channels being used in parallel.) Show that the capacity–cost function of the product channel is given by

$$C(\beta) = \max_t \{C_1(t) + C_2(\beta - t)\}.$$

Thus show that the graph of $C(\beta)$ is obtained by adding both coordinates of the curves $C_1(\beta)$ and $C_2(\beta)$ at points on the two curves having the same slope.

2.10 (The capacity of a simple channel with memory). Let $Z_1, Z_2, \ldots$ be identically distributed random variables, a Markov chain, taking values in $\{0, 1, \ldots, r-1\}$ (cf. Prob. 1.20). The Z_i's can be used to define an additive channel with $A_X = A_Y = \{0, 1, \ldots, r-1\}$, by the rule $Y_i \equiv X_i + Z_i \pmod{r}$ where X_i and Y_i are the input and output from the channel at time i. The n-*dimensional capacity* is defined as $C_{\max}^{(n)} = \max\{I(X; Y)\}$ as X ranges over all n-dimensional test sources, and the *capacity* is $C_{\max} = \sup(1/n)C_{\max}^{(n)}$. Show that $C_{\max}^{(n)} = n\log r - H(Z_1, \ldots, Z_n) = n\log r - H(\mathbf{p}) - (n-1)H$, where $\mathbf{p}$ is a probability vector describing the common distribution of the Z_i's, and H is the entropy of the chain. Conclude that $C_{\max} = \log r - H$. Apply this result to the case where $r = 2$ and the chain's transition probabilities are given by the stochastic matrix

$$Q = \begin{bmatrix} q & p \\ p & q \end{bmatrix}.$$

2.11 For the channel and cost function in Example 2.5, show explicitly how to achieve what is promised by the coding theorem, that is, give a precise description of a code C that has properties (a), (b), and (c) of Theorem 2.4.

2.12 Complete the details of Example 2.6, and show how the decoding function could be improved to achieve $P_E^{(i)} = \frac{1}{3}$ for $i = 1, 2$.

2.13 In the proof of Theorem 2.4 we gave an explicit decoding rule for a given code $C = \{\mathbf{x}_1, \ldots, \mathbf{x}_M\}$ (see pp. 63). This rule, while adequate for our purposes, is not optimal and is rarely used in practice. In this exercise we shall introduce and study two other decoding rules, which are better than the one in Theorem 2.4 but are more difficult to analyze. Thus suppose the probability that $\mathbf{x}_i$ will be transmitted is p_i, $i = 1, 2, \ldots, M$. Given the received vector $\mathbf{y}$, *minimum error probability decoding* (MED) chooses the codeword for which the conditional probability $p(\mathbf{x}_i|\mathbf{y})$ is largest. *Maximum likelihood decoding* (MLD) chooses the one for which $p(\mathbf{y}|\mathbf{x}_i)$ is largest.

(a) Show that MED lives up to its name, that is, the probability of decoder error, given that $\mathbf{y}$ is received, is minimized by MED, and hence also that the average error probability $P_E = \sum_{i=1}^{M} p_i P_E^{(i)}$ is minimized.

(b) Show that if $p_i = 1/M$ for all i, then MLD performs identically to MED.

(c) On a BSC, define the *Hamming distance* $d_H(\mathbf{x}, \mathbf{y})$ between $\mathbf{x}$ and $\mathbf{y}$ as the number of components in which they disagree. Show that a MLD decoder always picks a codeword $\mathbf{x}_i$ which is closest to $\mathbf{y}$ with respect to this distance (see Prob. 2.17 and also Section 7.3).

2.14 (Continuation). Consider the following code of length 4 over $\{0, 1\}$: $\mathbf{x}_1 = 0000$, $\mathbf{x}_2 = 0011$, $\mathbf{x}_3 = 1100$, $\mathbf{x}_4 = 1111$. Suppose the codewords are sent over a BSC (error probability p) with unequal probabilities, $P\{\mathbf{x}_1\} = \frac{1}{2}$, $P\{\mathbf{x}_2\} = P\{\mathbf{x}_3\} = \frac{1}{8}$, $P\{\mathbf{x}_4\} = \frac{1}{4}$. Find a decoding rule that minimizes $P_E = \frac{1}{2}P_E^{(1)} + \frac{1}{8}P_E^{(2)} + \frac{1}{8}P_E^{(3)} + \frac{1}{4}P_E^{(4)}$.

2.15 Show that, if $\beta_0 > \beta_{\min}$, it is possible to replace conclusion (a) in Theorem 2.4 with "each codeword is β_0-admissible"; and if $\beta_0 < \beta_{\max}$, to replace conclusion (b) with "$M \geqslant 2^{\lceil C(\beta_0)n \rceil}$."

2.16 Instead of requiring the average cost per channel use to be $\leqslant \beta$, we could require that the maximum allowable cost be $\leqslant \beta$, that is, define $\overline{C}(\beta) = \sup\{I(X; Y) : b(X) \leqslant \beta \text{ with probability } 1\}$. Compute $\overline{C}(\beta)$ for the channels of Examples 2.1, 2.2, and 2.3. (pp. 55–56). Does the coding theorem remain valid for $\overline{C}(\beta)$?

2.17 For a BSC, show that the set T defined in Eq. (2.22) is of the form $T = \{(\mathbf{x}; \mathbf{y}) : d_H(\mathbf{x}; \mathbf{y}) \leqslant r\}$, where $d_H(\mathbf{x}, \mathbf{y})$ is the *Hamming distance* between $\mathbf{x}$ and $\mathbf{y}$ (i.e., the number of components in which $\mathbf{x}$ and $\mathbf{y}$ differ), and find r in terms of n, R', and ε, the crossover probability of the channel. (Assume $p(x = 0) = p(x = 1) = \frac{1}{2}$.)

2.18 Let Ω be a finite set, and let f be a function mapping Ω into the real numbers. Prove that there exists an element $\omega \in \Omega$ such that $f(\omega) < y$ iff it is possible to define a probability distribution on Ω such that $E[f(\omega)] < y$.

2.19 Consider the DMC whose matrix of transition probabilities is

$$Q = \begin{bmatrix} \frac{1}{2} & \frac{1}{2} & 0 & 0 & 0 \\ 0 & \frac{1}{2} & \frac{1}{2} & 0 & 0 \\ 0 & 0 & \frac{1}{2} & \frac{1}{2} & 0 \\ 0 & 0 & 0 & \frac{1}{2} & \frac{1}{2} \\ \frac{1}{2} & 0 & 0 & 0 & \frac{1}{2} \end{bmatrix}.$$

(There is no cost constraint, i.e., $b(x) = 0$ for all x.)

(a) Compute $C_{\max}$.

(b) Find a code of length 1 with rate $\log 2$, and $P_E^{(i)} = 0$ for all i.

(c) Find a code of length 2 with rate $\frac{1}{2}\log 5$, and $P_E^{(i)} = 0$ for all i.

(d) If $\{\mathbf{x}_1, \ldots, \mathbf{x}_M\}$ is any code of length n for this channel with $P_E^{(i)} = 0$ for all i, show that the code's rate is $< \log \frac{5}{2}$.[14]

2.20 The binary erasure channel has matrix of transition probabilities

$$Q = \begin{matrix} 0 \\ 1 \end{matrix} \begin{bmatrix} q & p & 0 \\ 0 & p & q \end{bmatrix} \begin{matrix} 0 & ? & 1 \end{matrix},$$

where $0 \leqslant p \leqslant \frac{1}{2}$, $p + q = 1$ (Assume no input cost, i.e., $b(x) = 0$ for all x.) Its capacity is $C_{\max} = 1 - p$ (see Prob. 2.1b). Assume that this channel is equipped with *noiseless, delayless feedback*, that is, the receiver is able to communicate back to the sender the symbol he has received. The sender adopts the following "coding" strategy for communicating the output of a binary symmetric source over this channel: he merely repeats each symbol until it is finally received correctly.

(a) Calculate the average number of channel symbols required per source symbol if this strategy is used.

(b) Using the results of part (a), design, for any $R < C_{\max}$, $\varepsilon > 0$, a code of length n with M codewords such that $M \geqslant 2^{\lceil Rn \rceil}$, $P_E^{(i)} < \varepsilon$ for each i (see corollary to Theorem 2.4).[15]

In the next six problems we will sketch a proof of a coding theorem for a DMC without cost constraint that is in some ways stronger, and in other ways weaker, than the channel coding theorem (corollary to Theorem 2.4). First we must make some definitions. For each pair $x_1, x_2 \in A_X$, we define

$$J(x_1, x_2) = \sum_{y \in A_Y} \sqrt{p(y|x_1)p(y|x_2)}$$

and

$$J_0 = \min\{E(J(X_1, X_2))\},$$

where the minimization is taken over all independent, identically distributed random variables assuming values in A_X. Finally, the important quantity R_0 is defined by

$$R_0 = -\log_2 J_0.$$

The theorem to be proved is this:

R_0 *theorem for a* DMC. For any $R < R_0$ there exists a code $\{\mathbf{x}_1, \mathbf{x}_2, \ldots, \mathbf{x}_M\}$ with at least $M = \lceil 2^{Rn} \rceil$ codewords of length n, and an appropriate decoding rule, such that if $P_E = (\sum_{i=1}^M P_E^{(i)})/M$ denotes the average decoding error probability, then $P_E < 2^{-n(R_0 - R)}$. (This theorem is stronger than the channel coding theorem in that it gives an explicit estimate of how small P_E can be made as a function of n. It is weaker in that $R_0 < C_{\max}$, so that for rates $R_0 < R < C_{\max}$ it does not imply that $P_E \to 0$ is possible at all.)

2.21 Consider a code of length n containing only two codewords, $\mathbf{x}_1 = (x_{11}, x_{12}, \ldots, x_{1n})$ and $\mathbf{x}_2 = (x_{21}, x_{22}, \ldots, x_{2n})$. Assume that, given a received n-tuple $\mathbf{y}$, the decoder outputs $\mathbf{x}_1$ if $p(\mathbf{y}|\mathbf{x}_1) > p(\mathbf{y}|\mathbf{x}_2)$ and $\mathbf{x}_2$ if $p(\mathbf{y}|\mathbf{x}_2) > p(\mathbf{y}|\mathbf{x}_1)$. Let $Y_1 = \{\mathbf{y} : p(\mathbf{y}|\mathbf{x}_1) > p(\mathbf{y}|\mathbf{x}_2)\}$, $Y_2 = \{\mathbf{y}: p(\mathbf{y}|\mathbf{x}_2) > p(\mathbf{y}|\mathbf{x}_1)\}$. If $P_E^{(i)}$ denotes the probability of decoder error, given that $\mathbf{x}_i$ is sent, show that

$$P_E^{(i)} \leqslant \prod_{k=1}^{n} J(x_{1k}, x_{2k}), \quad i = 1, 2,$$

by verifying the following steps:

$$P_E^{(i)} \leqslant \sum_{y \in Y_i} p(\mathbf{y}|\mathbf{x}_i)$$

$$\leqslant \sum_{y \in A_Y^n} \sqrt{p(\mathbf{y}|\mathbf{x}_1)p(\mathbf{y}|\mathbf{x}_2)}$$

$$= \prod_{k=1}^{n} J(x_{1k}, x_{2k}).$$

2.22 If now $\{x_1, \ldots, x_M\}$ is a code containing M codewords of length n, and the decoder picks the codeword x_i for which $p(y|x_i)$ is largest, denote by $P_E^{(i)}$ the probability of decoder error, given that x_i is transmitted, and show that

$$P_E^{(i)} \leq \sum_{\substack{j=1 \\ j \neq i}}^{M} \prod_{k=1}^{n} J(x_{ik}, x_{jk}).$$

2.23 By averaging the expression obtained in Prob. 2.22 over all codes, where the individual codewords are chosen independently according to the probability distribution that achieves J_0, obtain the estimate

$$E[P_E^{(i)}] < M \cdot 2^{-R_0 n}.$$

2.24 Now complete the proof of the R_0 coding theorem.

2.25 Show that for a BSC,

$$R_0 = 1 - \log_2\left[1 + 2\sqrt{p(1-p)}\right],$$

where p is the raw bit error probability. Furthermore show that $\frac{1}{2}C_{\max} < R_0 < C_{\max}$, where $C_{\max} = 1 - H_2(p)$ is the capacity of the BSC.

2.26 Compute R_0 for the binary erasure channel (see Prob. 2.1b).

Notes

1 (p. 50). A stochastic matrix is a matrix of nonnegative real numbers with the property that the sum of the entries in each row is 1 (cf. Prob. 1.20).

2 (p. 50). We emphasize at once that the most important example of a cost function is $b(x) = 0$ for all $x \in A_X$. The main reasons for introducing $b(x)$ at all are to underscore the duality between the results of Chapters 2 and 3, and to prepare for the important input-constrained Gaussian channel of Chapter 4.

3 (p. 52). We are invoking the theorem that a continuous real-valued function defined on a compact subset of a metric space achieves its supremum and infimum; see Apostol [2], Theroem 4.28.

4 (p. 52). This terminology is not standard; in the special case $b(x) \equiv 0$ (cf. Note 2) $C(\beta)$ is obviously constant for all $\beta \geq 0$, and is called merely the channel *capacity*; see p. 54.

5 (p. 53). Given the conclusion of Theorem 2.2, the reader may wonder why we gave the complicated definition (2.5) in the first place. The answer is that this definition allows a simple proof of the *converse* to the coding theorem (see Chapter 5, especially Eq. (5.9)); also, for channels with memory, Theorem 2.2 is false (see Prob. 2.10).

6 (p. 58). A close inspection of the proof of Theorem 2.3 reveals that all we need is for the transition probability matrix to have constant row entropy and column sum.

7 (p. 58). Assuming the implied logarithms are base 2.

8 (p. 60). If A is a set, the symbol A^n denotes the set $\{(a_1, a_2, \ldots, a_n): a_i \in A\}$ of ordered n-tuples from A.

9 (p. 60). More generally, the code C could be used to communicate the output of any source, binary, memoryless, or otherwise, provided the number of possible source

outputs is $\leq M$. This robustness makes the coding theorem (Theorem 2.4) much more valuable than our discussion indicates, since few real sources are accurately modeled by a binary symmetric source.

10 (p. 62). This corollary, which states that *we can communicate arbitrarily reliably at rates below channel capacity*, is called the *channel coding theorem*, since, as mentioned in Note 2, the cost constraint is usually absent, and so the more general result of Theorem 2.4 is relatively unimportant.

11 (p. 63). This decoding rule is not the best possible one; see Probs. 2.13 and 2.14. However it is relatively easy to analyze and is sufficiently close to optimal to allow us to prove the coding theorem.

12 (p. 64). But see Prob. 2.18.

13 (p. 67). Note that deleting codewords from C cannot change $P_E^{(i)}$ for the remaining codewords, since $P_E^{(i)} = \sum \{p(\mathbf{y}|\mathbf{x}_i): \mathbf{y} \notin f^{-1}(\mathbf{x}_i)\}$.

14 (p. 71). This is a famous example of a channel with a nonzero *zero-error capacity*. It is known that no codes exist for this channel with $P_E = 0$ and rate $> \frac{1}{2} \log 5$; see the paper by Shannon [25], pp. 112–113, and Lovász [51].

15 (p. 72). This result becomes more interesting in light of the theorem, due to Shannon, that feedback cannot increase the capacity of a DMC! (See Shannon [25], p. 120) What is true however is that feedback in general makes the design of encoders and decoders much simpler. There is a substantial literature on channels with feedback; see [25], pp. 373–436.

3

Discrete memoryless sources and their rate-distortion functions

3.1 The rate-distortion function

Consider an information source that produces, every unit of time, a symbol u from a finite set A_U, called the *source alphabet*. Suppose that the sequence of symbols produced by the source can be modeled by a sequence $U_1, U_2, \ldots$ of independent, identically distributed random variables with common distribution function $P\{U = u\} = p(u)$. Such an information source is called a *discrete memoryless source* (DMS), and the numbers $p(u)$ are called the *source statistics*.

Now suppose that we are required to transmit the source's output over a channel to a certain destination. We assume that a transmitted source symbol $u \in A_U$ will be reproduced at the destination as a symbol v which is an element of another finite set A_V, called the *destination alphabet*. (A_V will usually, but not always, contain A_U as a subset.) Also, suppose that for each pair (u, v) there is a nonnegative number $d(u, v)$ which measures the error, or *distortion*, caused when the source symbol u is reproduced as the destination symbol v. The function d is called a *distortion measure*.[1]

Finally, suppose that the function d is extended to pairs $(\mathbf{u}, \mathbf{v}) = (u_1, u_2, \ldots, u_k; v_1, v_2, \ldots, v_k)$ from $A_U^k \times A_V^k$ by

$$d(\mathbf{u}, \mathbf{v}) = \sum_{i=1}^{k} d(u_i, v_i). \tag{3.1}$$

It is usually, but not always, convenient to take $A_U = \{0, 1, \ldots, r-1\}$, $A_V = \{0, 1, \ldots, s-1\}$, and to arrange the distortions $d(u, v)$ into an $r \times s$ matrix D.

Example 3.1 $A_U = A_V = \{0, 1\}$; source statistics $p(0) = p$, $p(1) = q = 1 - p$, where $p \leqslant \frac{1}{2}$; and distortion matrix

$$D = \begin{bmatrix} 0 & 1 \\ 1 & 0 \end{bmatrix}. \qquad \qquad \square$$

Example 3.2[2] $A_U = \{-1, 0, +1\}$, $A_V = \{-\frac{1}{2}, +\frac{1}{2}\}$; source statistics $(\frac{1}{3}, \frac{1}{3}, \frac{1}{3})$ and distortion matrix

$$D = \begin{bmatrix} 1 & 2 \\ 1 & 1 \\ 2 & 1 \end{bmatrix}.$$

(These two examples will be explored further on p. 81 ff.) $\square$

Let k be a fixed, positive integer. Consider the independent random variables $U_1, U_2, \ldots, U_k$ which model the first k symbols emitted by the source, and let $V_1, V_2, \ldots, V_k$ be *any* collection of k random variables taking values in the destination alphabet A_V, and defined on the same sample space as the U_i's. We can calculate the mutual information $I(\mathbf{U}; \mathbf{V})$ between the random vectors $\mathbf{U} = (U_1, \ldots, U_k)$ and $\mathbf{V} = (V_1, \ldots, V_k)$, and also the *average distortion* $E(d) = E[d(\mathbf{U}, \mathbf{V})]$, defined by

$$E(d) = \sum_{\mathbf{u},\mathbf{v}} p(\mathbf{u}, \mathbf{v})d(\mathbf{u}, \mathbf{v})$$

$$= \sum_{\mathbf{u},\mathbf{v}} p(\mathbf{u})p(\mathbf{v}|\mathbf{u})d(\mathbf{u}, \mathbf{v}). \qquad (3.2)$$

(In (3.2) the summation is extended over the $r^k s^k$ pairs $(\mathbf{u}, \mathbf{v}) = (u_1, \ldots, u_k, v_1, \ldots, v_k)$, where $u_i \in A_U$, $v_i \in A_V$, and $p(\mathbf{u}, \mathbf{v}) = P\{\mathbf{U} = \mathbf{u}, \mathbf{V} = \mathbf{v}\}$, $p(\mathbf{v}|\mathbf{u}) = P\{\mathbf{V} = \mathbf{v}|\mathbf{U} = \mathbf{u}\}$.)

We now define[3] the function $R_k(\delta)$, which is a function of the source statistics $(p(u))$, the distortion matrix D, and the real number δ, by

$$R_k(\delta) = \min\{I(\mathbf{U}; \mathbf{V}): E(d) \le k\delta\}. \qquad (3.3)$$

In (3.3) the minimization is extended over all pairs $(\mathbf{U}, \mathbf{V}) = ((U_1, \ldots, U_k), (V_1, \ldots, V_k))$ of k-dimensional random vectors assuming values in $A_U^k \times A_V^k$ for which $U_1, \ldots, U_k$ are independent with common distribution function $P\{U = u\} = p(u)$, where $(p(u))$ are the given source statistics, and the average distortion $E(d)$ defined in (3.2) is $\le k\delta$. Since the source statistics $(p(u))$ are fixed, in computing $R_k(\delta)$ for a fixed δ we must vary the conditional probabilities $p(\mathbf{v}|\mathbf{u})$ that define $\mathbf{V}$. These probabilities can be thought of as transition probabilities defining a channel with $\mathbf{U}$ as input, $\mathbf{V}$ as output. In the present context this channel is usually called a k-dimensional *test channel*, and the minimization (3.3) is said to take place over all test channels whose average distortion is $\leq k\delta$.

Several preliminary remarks about the function $R_k(\delta)$ are in order. First note that for fixed $(p(u))$ the function $I(U; V)$ is a continuous function of the $r^k s^k$ transition probabilities $p(v|u)$. The subset of the set of transition probabilities where $E(d) \leq k\delta$ is a compact region of $r^k s^k$-dimensional Euclidean space, and so the function $I(U; V)$ actually achieves its minimum value on this region.[4] This is the reason we write "min" rather than "inf" in Eq. (3.3). Second, notice that the minimum possible value of $E(d)$ is given by $k \cdot \delta_{\min}$, where

$$\delta_{\min} = \sum_{u \in A_U} p(u) \cdot \min_v d(u, v), \qquad (3.4)$$

as may be seen from (3.2), since $E(d) \geq \sum_{(u,v)} p(u, v) \min_v d(u, v) = k\delta_{\min}$. Thus $R_k(\delta)$ is defined only for $\delta \geq \delta_{\min}$. Finally, observe that if $\delta_1 > \delta_2$, the set of k-dimensional test channels satisfying $E(d) \leq \delta_2$ is a subset of those for which $E(d) \leq \delta_1$, and so $R_k(\delta_1) \leq R_k(\delta_2)$, that is, $R_k(\delta)$ is a decreasing function of $\delta \geq \delta_{\min}$.

The *rate-distortion* function of the source is now defined as

$$R(\delta) = \inf_k \frac{1}{k} R_k(\delta). \qquad (3.5)$$

The number $R(\delta)$ turns out to represent the minimum number of bits (assuming all logarithms are base 2) needed to represent a source symbol if we are willing to tolerate an average distortion of δ. The precise statement of this result, Shannon's *source coding theorem*, will be stated and proved in Section 3.2. Our object in the rest of this section is to develop techniques for computing $R(\delta)$ for a given DMS and distortion measure.

Our first result is that the functions $R_k(\delta)$ are convex.

Theorem 3.1 $R_k(\delta)$ *is a convex* $\cup$ *function of* $\delta \geq \delta_{\min}$.

Proof Suppose $\alpha_1, \alpha_2 \geq 0$, $\alpha_1 + \alpha_2 = 1$. We must show that, for δ_1, $\delta_2 \geq \delta_{\max}$,

$$R_k(\alpha_1 \delta_1 + \alpha_2 \delta_2) \leq \alpha_1 R_k(\delta_1) + \alpha_2 R_k(\delta_2). \qquad (3.6)$$

To do this, let $p_i(u|v)$ be the transition probabilities for a test channel achieving $R_k(\delta_i)$, $i = 1, 2$. Then if V_1, V_2 denote the test channel outputs,

$$I(U; V_i) = R_k(\delta_i), \qquad (3.7)$$

$$E(d_i) \leqslant k\delta_i, \qquad i = 1, 2, \tag{3.8}$$

where $d_i = d(\mathbf{U}, \mathbf{V}_i)$ denotes the average distortion in the ith test channel. Now define a new test channel with transition probabilities $p(\mathbf{v}|\mathbf{u}) = \alpha_1 p_1(\mathbf{v}|\mathbf{u}) + \alpha_2 p_2(\mathbf{v}|\mathbf{u})$. If $\mathbf{V}$ denotes the output from this test channel, then from (3.2) and (3.8), $E[d(\mathbf{U}; \mathbf{V})] = \alpha_1 E[d(\mathbf{U}; \mathbf{V}_1)] + \alpha_2 E[d(\mathbf{U}; \mathbf{V}_2)] \leqslant (\alpha_1 \delta_1 + \alpha_2 \delta_2)$. Hence this test channel is admissible for the calculation of $R_k(\alpha_1 \delta_1 + \alpha_2 \delta_2)$, and so $I(\mathbf{U}; \mathbf{V}) \geqslant R_k(\alpha_1 \delta_1 + \alpha_2 \delta_2)$.

On the other hand since $I(\mathbf{U}; \mathbf{V})$ is a convex $\cup$ function of the transition probabilities $p(\mathbf{v}|\mathbf{u})$ (Theorem 1.7), $I(\mathbf{U}; \mathbf{V}) \leqslant \alpha_1 I(\mathbf{U}; \mathbf{V}_1) + \alpha_2 I(\mathbf{U}; \mathbf{V}_2) = \alpha_1 R_k(\delta_1) + \alpha_2 R_k(\delta_2)$. The last two inequalities combine to give (3.6) and so prove Theorem 3.1. $\qquad\square$

The next result shows that for a DMS the computation of $R(\delta)$ is considerably easier than would appear from definition (3.5).

Theorem 3.2[5] *For a DMS, $R_k(\delta) = kR_1(\delta)$ for all k and $\delta \geqslant \delta_{\min}$.*

Proof Let $p(\mathbf{v}|\mathbf{u})$ be the transition probabilities for a k-dimensional test channel achieving $R_k(\delta)$. Then

$$I(\mathbf{U}; \mathbf{V}) = R_k(\delta), \tag{3.9}$$

$$E[d(\mathbf{U}, \mathbf{V})] \leqslant k\delta. \tag{3.10}$$

Since $U_1, U_2, \ldots, U_k$ are independent, by Theorem 1.8 we have

$$I(\mathbf{U}; \mathbf{V}) \geqslant \sum_{i=1}^{k} I(U_i; V_i). \tag{3.11}$$

If we define $\delta_i = E[d(U_i, V_i)$, we have

$$I(U_i; V_i) \geqslant R_1(\delta_i), \qquad i = 1, 2, \ldots, k, \tag{3.12}$$

$$E[d(\mathbf{U}, \mathbf{V})] = \sum_{i=1}^{k} \delta_i \leqslant k\delta. \tag{3.13}$$

Combining (3.11) and (3.12), we have $I(\mathbf{U}; \mathbf{V}) \geqslant \sum_{i=1}^{k} R_1(\delta_i)$. But since R_1 is convex $\cup$,

$$\sum_{i=1}^{k} R_1(\delta_i) \geqslant k \cdot R_1\left(\frac{\delta_1 + \cdots + \delta_k}{k}\right)$$

$$\geqslant kR_1(\delta),$$

by (3.13) and the fact that R_1 is a decreasing function of δ. Hence $R_k(\delta) = I(\mathbf{U}; \mathbf{V}) \geq kR_1(\delta)$. To prove the opposite inequality, let $p(v|u)$ be a one-dimensional test channel that achieves $R_1(\delta)$, and define $p(\mathbf{v}|\mathbf{u}) = \prod_{i=1}^{k} p(v_i|u_i)$. It is easy to verify that the memoryless test channel so defined has $E(d) \leq k\delta$ and $I(\mathbf{U}; \mathbf{V}) = kR_1(\delta)$ (see Prob. 3.3). Thus $R_k(\delta) \leq kR_1(\delta)$ as well, and Theorem 3.2 is proved. $\qquad\square$

Corollary

$$R(\delta) = R_1(\delta)$$

$$= \min\{I(U; V): E(d) \leq \delta\}.$$

Proof This follows immediately from Theorem 3.2 and definitions (3.3) and (3.5). $\qquad\square$

Let us now pause for a general description of the function $R(\delta)$. We already know that $R(\delta)$ is a decreasing, convex $\cup$ function of $\delta \geq \delta_{\min}$. The convexity implies immediately (see Appendix B) that $R(\delta)$ is continuous for $\delta > \delta_{\min}$. $R(\delta)$ is also continuous at $\delta = \delta_{\min}$, but we leave the proof of this fact as Prob. 3.4. Furthermore, we shall now show that $R(\delta) = 0$ for all sufficiently large δ. Indeed, if we define $\delta_{\max}$ by

$$\delta_{\max} = \min_v \sum_u p(u)d(u, v), \qquad (3.14)$$

then $R(\delta) = 0$ iff $\delta \geq \delta_{\max}$. To see this, observe that a test channel which maps every input u deterministically onto a v for which $\sum p(u)d(u, v) = \delta_{\max}$ will have $I(U; V) = 0$ and $E(d) = \delta_{\max}$. This shows that $R(\delta) = 0$ for $\delta \geq \delta_{\max}$. Conversely, if $R(\delta) = 0$, then a test channel achieving this must have U and V independent (Theorem 1.3), and so

$$E(d) = \sum p(u)p(v)d(u, v) = \sum_v p(v) \sum_u p(u)d(u, v)$$

$$\geq \sum_v p(v) \cdot \delta_{\max} = \delta_{\max}$$

by (3.14). Thus if $R(\delta) = 0$, then $\delta \geq \delta_{\max}$.

Since $R(\delta)$ is decreasing and convex $\cup$ for $\delta \geq \delta_{\min}$, and is constant for $\delta \geq \delta_{\max}$, it follows that $R(\delta)$ is strictly decreasing for $\delta_{\min} \leq \delta \leq \delta_{\max}$ (see Prob. 2.6), and so in this range $R(\delta)$ is given by

$$R(\delta) = \min\{I(U; V): E(d) = \delta, \qquad \delta_{\min} \leq \delta \leq \delta_{\max}. \qquad (3.15)$$

Thus a typical $R(\delta)$ curve will look something like Fig. 3.1. (However $R(\delta)$ need not be strictly convex $\cup$; see Prob. 3.2.) In Prob. 3.8, however, it is shown that there is no essential loss of generality in assuming that $\delta_{\min} = 0$, i.e. that for every source symbol u there exists at least one destination symbol for which $d(u, v) = 0$, in which case the $R(\delta)$ curve looks like Fig. 3.2.

The value of $R(0)$ (or $R(\delta_{\min})$ if $\delta_{\min} > 0$) is in general somewhat difficult to compute (see Prob. 3.7). However, if in addition to assuming that each row of the matrix D has at least one 0, we assume that each column has at most one 0, $R(0)$ is easy to compute. For in this case a test channel with $E(d) = 0$ must map each u into the set of "perfect" representatives of u, $G_u = \{v | d(u, v) = 0\}$. But by our assumption on the columns of D the sets G_u are disjoint, and so in the test channel V completely determines U. Thus $R(0) = I(U; V) = H(U) - H(U|V) = H(U)$, the source entropy. In this rather typical situation, the $R(\delta)$ curve looks like Fig. 3.3 (cf. Prob. 3.5).

We shall now actually compute $R(\delta)$ for Examples 3.1 and 3.2.

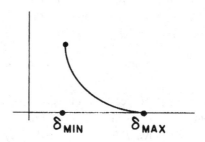

Figure 3.1 A typical $R(\delta)$ curve.

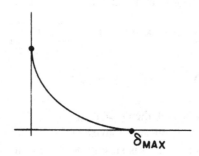

Figure 3.2 Without loss of generality, $\delta_{\min} = 0$.

Example 3.1 (continued). Here the source statistics are $P\{U = 0\} = p$, $P\{U = 1\} = q$, with $p \leqslant \frac{1}{2}$, and the distortion matrix is

$$D = \begin{bmatrix} 0 & 1 \\ 1 & 0 \end{bmatrix}.$$

Clearly $\delta_{\min} = 0$, $\delta_{\max} = \min\{p, q\} = p$, and by the preceding remarks $R(0) = H(U) = H(p)$. To find $R(\delta)$ for $0 < \delta < p$, note that for a test channel achieving $R(\delta)$, $I(U; V) = H(U) - H(U|V) = H(p) - H(U|V)$, and $E(d) = P\{U \neq V\} = \delta$. But by Fano's inequality (corollary to Theorem 1.2) $H(U|V) \leqslant H(\delta)$, and so $R(\delta) \geqslant H(p) - H(\delta)$. This lower bound turns out to be the value of $R(\delta)$ for $0 \leqslant \delta \leqslant p$. To prove this we must produce a test channel with $E(d) = \delta$, $I(U; V) = H(p) - H(\delta)$. The best way to do this is to define a "backwards" test channel, that is, to give the transition probabilities, $p(u|v)$, by $p(u|v) = \delta$ if $u \neq v$, $= 1 - \delta$ if $u = v$. The backwards test channel is shown in Fig. 3.4. Clearly such a test channel will have $E(d) = \delta$, $H(U|V) = H(\delta)$. However, we must make sure that it is possible to choose $\alpha = P\{V = 0\}$ so that $P\{U = 0\} = p$ and $P\{U = 1\} = q$. We must have $p = \alpha(1 - \delta) + (1 - \alpha)\delta$, that is, $\alpha = (p - \delta)/(1 - 2\delta)$. Since $0 < \delta < p \leqslant \frac{1}{2}$, this value of α lies in $[0, 1]$, and so the required backwards test channel exists. Hence finally (see Fig. 3.5)

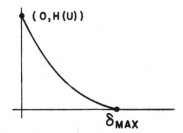

Figure 3.3 $R(0) =$ source entropy, usually.

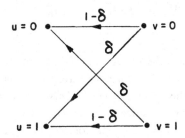

Figure 3.4 A backwards test channel for Example 3.1.

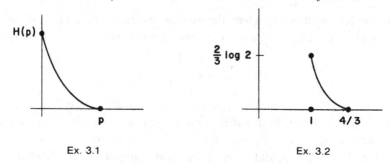

Figure 3.5 $R(\delta)$ for Examples 3.1 and 3.2.

$$R(\delta) = \begin{cases} H(p) - H(\delta), & 0 \leq \delta \leq p, \\ 0, & \delta \geq p. \end{cases} \qquad \square$$

Example 3.2 (continued). Here the source statistics are $p(u) = \frac{1}{3}$ for $u = 0, \pm 1$, and the distortion matrix is

$$D = \begin{bmatrix} 1 & 2 \\ 1 & 1 \\ 2 & 1 \end{bmatrix}.$$

An easy calculation gives $\delta_{\min} = 1$, $\delta_{\max} = \frac{4}{3}$. Since $I(U; V)$ is a convex $\cup$ function in the test channel transition probabilities, symmetric in $P\{V = +\frac{1}{2}|U = 0\}$ and $P\{V = -\frac{1}{2}|U = 0\}$, and $d(0, \frac{1}{2}) = d(0, -\frac{1}{2})$, it follows (see Prob. 3.6) that in a test channel achieving $R(\delta)$ we must have $P\{V = \pm\frac{1}{2}|U = 0\} = \frac{1}{2}$. By similar reasoning, we must have $P\{V = +\frac{1}{2}|U = +1\} = P\{V = -\frac{1}{2}|U = -1\} = 1 - \alpha$ for some $0 \leq \alpha \leq \frac{1}{2}$. For such a test channel $I(U; V) = H(V) - H(V|U) = \log 2 - \frac{2}{3}H(\alpha) - \frac{1}{3}\log 2 = \frac{2}{3}[\log 2 - H(\alpha)]$, and $E(d) = \frac{2}{3}[(1 - \alpha) + 2\alpha] + \frac{1}{3} = 1 + 2\alpha/3$. Hence finally (see Fig. 3.5)

$$R(\delta) = \begin{cases} \frac{2}{3}(\log 2 - H(3(\delta - 1)/2)), & 1 \leq \delta \leq \frac{4}{3}, \\ 0, & \delta \geq \frac{4}{3} \end{cases}.$$

We conclude this section by computing the rate-distortion function in general for an r-ary *symmetric source* with respect to the *Hamming distortion measure*, also called the *error probability distortion*. Here the source and destination alphabets are identical: $A_U = A_V = \{0, 1, \ldots, r - 1\}$, and the source statistics are $P\{U = u\} = 1/r$ for $u = 0, 1, \ldots, r - 1$. The distortions are given by

$$d(u, v) = \begin{cases} 0 & \text{if } u = v \\ 1 & \text{if } u \neq v. \end{cases}$$

The matrix D is as follows for $r = 4$:

$$D = \begin{bmatrix} 0 & 1 & 1 & 1 \\ 1 & 0 & 1 & 1 \\ 1 & 1 & 0 & 1 \\ 1 & 1 & 1 & 0 \end{bmatrix}.$$

Notice that for a (U, V) test channel, $E(d) = \sum\{p(u, v): u \neq v\} = P\{U \neq V\}$, which explains why this particular distortion measure is called the error probability distortion.

From (3.4) $\delta_{\min} = 0$, and from (3.14) $\delta_{\max} = 1 - 1/r$. The value of $R(\delta)$ in the range $0 \leq \delta \leq 1 - 1/r$ is given in the following theorem.

Theorem 3.3[6] *With respect to the error probability (Hamming) distortion measure, the r-ary symmetric source has rate-distortion function*

$$R(\delta) = \begin{cases} \log r - \delta \log(r - 1) - H(\delta), & 0 \leq \delta \leq 1 - 1/r, \\ 0, & \delta \geq 1 - 1/r. \end{cases}$$

Corollary $R(\delta)$ *for a binary symmetric source, same distortion measure, is*

$$R(\delta) = \begin{cases} \log 2 - H(\delta), & 0 \leq \delta \leq \frac{1}{2}, \\ 0, & \delta \geq \frac{1}{2}. \end{cases}$$

(*Note that this corollary, which is the special case $r = 2$ of Theorem 3.3, is the special case $p = \frac{1}{2}$ of Example 3.1; see pp. 81–82.*)

Proof Consider a (U, V) test channel that achieves $R(\delta)$ for a fixed δ, $0 \leq \delta \leq 1 - 1/r$. Then $R(\delta) = I(U; V) = H(U) - H(U|V)$. But $H(U) = \log r$ (cf. Theorem 1.1); and since $\delta = E(d) = P\{U \neq V\}$, by Fano's inequality (corollary to Theorem 1.2) $H(U|V) \leq \delta \log(r - 1) + H(\delta)$. Hence $R(\delta) \geq \log r - \delta \log(r - 1) - H(\delta)$. To show the opposite inequality, assume $0 \leq \delta \leq 1 - 1/r$, and define a test channel by

$$p(v|u) = \begin{cases} 1 - \delta & \text{if } v = u, \\ \dfrac{\delta}{r - 1} & \text{if } v \neq u. \end{cases}$$

Then a simple calculation gives $E(d) = \delta$, $I(U; V) = H(V) - H(V|U) = \log r - H[1 - \delta, \delta/(r - 1), \ldots, \delta/(r - 1)] = \log r - \delta \log(r - 1) - H(\delta)$. □

3.2 The source coding theorem

The rate-distortion function, which was studied from a purely mathematical point of view in Section 3.1, has a beautiful communication-theoretic significance, which is this: $R(\delta)$ *is the number of bits needed to represent a source symbol, if a distortion δ is allowable.* Thus a source symbol can be "compressed" into $R(\delta)$ bits; since $R(\delta)$ decreases as δ increases, more compression is possible as δ increases. For this reason "rate-distortion theory" is sometimes called "data-compression theory."

To see why $R(\delta)$ might have this significance, consider the following situation. Let $(U_1, U_2, \ldots, U_k) = \mathbf{U}$ represent the first k symbols emitted by a certain DMS. Let us suppose that these k symbols are "compressed" into n bits $(X_1, X_2, \ldots, X_n) = \mathbf{X}$, and that it is somehow possible to recover from $\mathbf{X}$ k destination symbols $(V_1, \ldots, V_k) = \mathbf{V}$ such that $\sum_{i=1}^{k} E[d(U_i, V_i)] \leq k\delta$. Under these circumstances it is reasonable to assert that the n bits $X_1, \ldots, X_n$ represent the k source symbols $U_1, \ldots, U_k$ with an average distortion $\leq \delta$. The relationship between $\mathbf{U}, \mathbf{X}, \mathbf{V}$ can be sketched as shown in Fig. 3.6.

Now we know from Eqs. (3.3) and (3.5) that $I(\mathbf{U}; \mathbf{V}) \geq R_k(\delta) \geq kR(\delta)$; from Theorem 1.5 that $I(\mathbf{U}; \mathbf{V}) \leq I(\mathbf{X}; \mathbf{V})$; from the definition of $I(\mathbf{X}; \mathbf{V})$ (see Eq. (1.4)) that $I(\mathbf{X}; \mathbf{V}) \leq H(\mathbf{X})$; and from Theorem 1.1 that $H(\mathbf{X}) \leq n$ bits. Combining these results, we have $kR(\delta) \leq I(\mathbf{U}; \mathbf{V}) \leq I(\mathbf{X}; \mathbf{V}) \leq H(\mathbf{X}) \leq n$, that is (implied logs are base 2),

$$\frac{n}{k} \geq R(\delta). \tag{3.16}$$

The ratio n/k in (3.16) represents the number of bits per source symbol in the above data compression scheme. Thus we see immediately that at least $R(\delta)$ bits are needed to represent a source symbol if the average distortion must be $\leq \delta$. The *source coding theorem*, to be proved below, assets that in a certain sense no more than $R(\delta)$ bits are needed.

First we must define a source code. A source code of length k is a subset of A_V^k, that is, a set $C = \{\mathbf{v}_1, \mathbf{v}_2, \ldots, \mathbf{v}_M\}$ of *destination* sequences of length k. Its *rate* is defined to be $R = k^{-1} \log_2 M$. For each source sequence $\mathbf{u} = (u_1, \ldots, u_k)$ of length k, let $f(\mathbf{u})$ be a codeword $\mathbf{v}_i$ which is "closest" to $\mathbf{u}$ in the sense that

$$d(\mathbf{u}, f(\mathbf{u})) \leq d(\mathbf{u}, \mathbf{v}_j), \qquad j = 1, 2, \ldots, M. \tag{3.17}$$

$$(U_1, \ldots, U_k) \longrightarrow (X_1, \ldots, X_n) \longrightarrow (V_1, \ldots, V_k)$$

Figure 3.6 A general data-compression scheme.

The *average distortion* of the code C is defined to be

$$d(C) = \frac{1}{k} \sum_{\mathbf{u} \in A_U^k} p(\mathbf{u}) d(\mathbf{u}, f(\mathbf{u})), \tag{3.18}$$

where in (3.18) $p(\mathbf{u}) = p(u_1) p(u_2) \dots p(u_k)$ is the probability that the first k symbols emitted by the source will be $u_1, \dots, u_k$.

If $M \leqslant 2^n$, such a source code can be used to devise a data compression scheme of the sort depicted in Fig. 3.6 as follows. To each of the M source codewords $\mathbf{v}_i$ assign a distinct binary n-tuple $\mathbf{x}(\mathbf{v}_i) = (x_1(\mathbf{v}_i), \dots, x_n(\mathbf{v}_i))$. Since $M \leqslant 2^n$, this will be possible. The source sequence $\mathbf{u} = (u_1, \dots, u_k)$ is then represented by the n bits $\mathbf{x} = \mathbf{x}[f(\mathbf{u})]$, and the destination sequence $\mathbf{v}$ is taken to be the codeword $f(\mathbf{u})$. (Since the mapping $\mathbf{v} \to \mathbf{x}(\mathbf{v})$ is one to one, $f(\mathbf{u})$ can be uniquely recovered from $\mathbf{x}[f(\mathbf{u})]$.) It is clear that the average distortion of the scheme is exactly $d(C)$, as defined in (3.18), and the compression ratio n/k is just $\lceil \log_2 M \rceil / k$.

Example 3.1 (continued). Everything is the same as in the previous appearances of Example 3.1 (pp. 75, 81), except that now we specialize the source statistics to $p = q = \frac{1}{2}$. We consider a source code of length 7 with 16 codewords, namely, the 16 words of the $(7, 4)$ Hamming code described in the Introduction (see p. 4 ff.). It was shown there that each of the 128 binary vectors of length 7 differed in at most one position from some codeword. Hence

$$d(C) = \frac{1}{7} \left(\frac{128 - 16}{128} \right) = \frac{1}{8}.$$

The compression ratio is $n/k = 4/7 = 0.5714$. (Since as we have seen $R(\delta) = 1 - H_2(\delta)$ in this case, the inequality (3.16) becomes $4/7 \geqslant 1 - H_2(0.125) = 0.4564$.) $\square$

Example 3.2 (continued). Here the source, distortion, and so on are as in the previous appearances of Example 3.2 (pp. 76, 82), and we consider the following source code of length 2:

$$C = \left\{ \left(+\tfrac{1}{2}, -\tfrac{1}{2} \right), \left(-\tfrac{1}{2}, +\tfrac{1}{2} \right) \right\}.$$

Here a straightforward calculation (see Prob. 3.14) gives $d(C) = \frac{10}{9} = 1.11$. The compression ratio is $n/k = 1/2 = 0.50$. Since $R\left(\frac{10}{9}\right) = \frac{2}{3}\left[1 - H_2\left(\frac{1}{6}\right)\right]$ (see p. 82) $= 0.2333$, again we see that inequality (3.16) is satisfied, as it must be. $\square$

The thrust of the source coding theorem must now be easy to guess; it is, approximately, that there exist source codes for which $d(C) \approx \delta$ and $\log_2 M/k \approx R(\delta)$. More precisely, we have the following theorem.

Theorem 3.4 *(Shannon's source coding theorem). Fix $\delta \geq \delta_{\min}$. For any $\delta' > \delta$ and $R' > R(\delta)$, for sufficiently large k there exists a source code C of length k with M codewords, where:*

(a) $M \leq 2^{\lfloor kR' \rfloor}$,
(b) $d(C) < \delta'$.

Remarks: *Conclusion (a) guarantees that we can take $n = \lfloor kR' \rfloor$ and get a compression ratio $n/k \leq R'$; conclusion (b) guarantees that the resulting distortion will be $< \delta'$. It is annoying, but necessary, to have the two quantities R', δ', rather than the desired $R(\delta), \delta$. (For one thing we could not in general hope to achieve $n/k = R(\delta)$, since $R(\delta)$ is likely to be irrational! But see Prob. 3.15.)*

Proof We begin by selecting numbers R'' and δ'' which satisfy

$$R(\delta) < R'' < R', \qquad \delta < \delta'' < \delta'. \tag{3.19}$$

Now if $C = \{v_1, v_2, \ldots, v_M\}$ is a particular source code of length k, and if $f(\mathbf{u})$ is a source encoding function as defined on p. 84, define the subsets S and T of A_U^k by

$$S = \{\mathbf{u}: d(\mathbf{u}, f(\mathbf{u})) \leq k\delta''\},$$

$$T = \{\mathbf{u}: d(\mathbf{u}, f(\mathbf{u})) > k\delta''\}.$$

Here S is the set of source sequences well represented by C; T is the set poorly represented. Then, by definition (3.18) of $d(C)$,

$$d(C) = \frac{1}{k}\sum_{\mathbf{u}} p(\mathbf{u})d(\mathbf{u}, f(\mathbf{u}))$$

$$= \frac{1}{k}\sum_{\mathbf{u} \in S} p(\mathbf{u})d(\mathbf{u}, f(\mathbf{u})) + \frac{1}{k}\sum_{\mathbf{u} \in T} p(\mathbf{u})d(\mathbf{u}, f(\mathbf{u})). \tag{3.20}$$

The first sum in (3.20) is clearly $\leq \delta''$; hence if we define B as the largest entry in the distortion matrix D, that is, $B = \max\{d(u, v): u \in A_U, v \in A_V\}$, we get

$$d(C) \leqslant \delta'' + B \sum_{\mathbf{u} \in T} p(\mathbf{u}). \tag{3.21}$$

The sum in (3.21) is just the probability that the source sequence will be poorly represented by C, that is, $P\{d(\mathbf{u}, f(\mathbf{u})) > k\delta''\}$.

Now $d(\mathbf{u}, f(\mathbf{u}))$ will be greater than $k\delta''$ iff $d(\mathbf{u}, \mathbf{v}_i) > k\delta''$ for each $i = 1, 2, \ldots, M$ (see (3.17)). Hence if we define a threshold function

$$\Delta(\mathbf{u}, \mathbf{v}) = \begin{cases} 1 & \text{if } d(\mathbf{u}, \mathbf{v}) \leqslant k\delta'', \\ 0 & \text{if } d(\mathbf{u}, \mathbf{v}) > k\delta'', \end{cases} \tag{3.22}$$

the restricted sum in (3.21) becomes the unrestricted sum $\sum_{\mathbf{u}} p(\mathbf{u})$ $[1 - \Delta(\mathbf{u}, \mathbf{v}_1)] \ldots [1 - \Delta(\mathbf{u}, \mathbf{v}_M)]$, and so if we define

$$K(C) = \sum_{\mathbf{u}} p(\mathbf{u}) \prod_{i=1}^{M} [1 - \Delta(\mathbf{u}, \mathbf{v}_i)], \tag{3.23}$$

the estimate (3.21) becomes

$$d(C) \leqslant \delta'' + B \cdot K(C). \tag{3.24}$$

In view of (3.24), our proof will be complete if we can find a source code C of length k with at most $2^{\lfloor kR' \rfloor}$ codewords for which $K(C) < (\delta' - \delta'')/B$. We will not be able to find such a code directly, but we will be able to deduce the existence of such a code indirectly by means of *random coding*. In other words, we will average $K(C)$ with respect to a certain probability distribution over the set of all possible source codes of length k with $2^{\lfloor kR' \rfloor}$ codewords; this average will be shown to approach 0 as $k \to \infty$. Thus for sufficiently large k the average will be $< (\delta - \delta'')/B$, from which it follows that at least one particular source code also has $K(C) < (\delta - \delta'')/B$; this code will satisfy the conclusions of Theorem 3.4.

So now our task is the average $K(C)$ over all source codes of length k with $M = 2^{\lfloor kR' \rfloor}$ codewords. First, of course, we must specify the probability distribution we are averaging with respect to. The right choice turns out to be closely related to the distribution of the random variable V in a test channel that achieves $R(\delta)$. Thus for the rest of the proof let $p(u, v)$ denote a probability distribution on $A_U \times A_V$ achieving $R(\delta)$, that is,

$$I(U; V) = R(\delta), \tag{3.25a}$$

$$E[d(U; V)] \leqslant \delta. \tag{3.25b}$$

The marginal distributions on A_U and A_V are given by

$$p(u) = \sum_v p(u, v) \qquad \text{(source statistics)},$$

$$p(v) = \sum_u p(u, v).$$

We extend this probability distribution to pairs $(\mathbf{u}, \mathbf{v}) = (u_1, \ldots, u_k, v_1, \ldots, v_k)$ from $A_U^k \times A_V^k$ by assuming that the source and test channel are memoryless, that is, by defining

$$p(\mathbf{u}) = \prod_{i=1}^k p(u_i),$$

$$p(\mathbf{v}|\mathbf{u}) = \prod_{i=1}^k p(v_i|u_i), \tag{3.26a}$$

from which follow

$$p(\mathbf{u}, \mathbf{v}) = \prod_{i=1}^k p(u_i, v_i),$$

$$p(\mathbf{v}) = \prod_{i=1}^k p(v_i). \tag{3.26b}$$

The probability assignment on the set of all source codes of length k with M codewords that we want is the one that assigns to the code $C = \{\mathbf{v}_1, \ldots, \mathbf{v}_M\}$ the probability

$$p(C) = \prod_{i=1}^M p(\mathbf{v}_i),$$

where $p(\mathbf{v}_i)$ is given by (3.26b). (This probability assignment is sometimes described by saying the source code is chosen at "random" according to the probability distribution $p(v)$.)

Recalling the definition of $K(C)$ (see (3.23)), we begin to compute its average $E(K)$:

$$E(K) = \sum_{\mathbf{v}_1,\ldots,\mathbf{v}_M} p(\mathbf{v}_1) \ldots p(\mathbf{v}_M) \sum_{\mathbf{u}} p(\mathbf{u}) \prod_{i=1}^{M} [1 - \Delta(\mathbf{u}, \mathbf{v}_i)]$$

$$= \sum_{\mathbf{u}} p(\mathbf{u}) \sum_{\mathbf{v}_1,\ldots,\mathbf{v}_M} \prod_{i=1}^{M} p(\mathbf{v}_i)[1 - \Delta(\mathbf{u}, \mathbf{v}_i)]$$

$$= \sum_{\mathbf{u}} p(\mathbf{u}) \left\{ \sum_{\mathbf{v} \in A_V^k} p(\mathbf{v})[1 - \Delta(\mathbf{u}, \mathbf{v})] \right\}^M. \tag{3.27}$$

(The last step above relies on the fact, which is obvious if not obscured by notation, that if $f(x)$ is a function defined on a finite set A, then

$$\left[\sum_{x \in A} f(x) \right]^M = \sum_{x_1 \in A} \cdots \sum_{x_M \in A} f(x_1) \ldots f(x_M).)$$

The inner sum in (3.27) is

$$\sum_{\mathbf{v}} p(\mathbf{v})[1 - \Delta(\mathbf{u}, \mathbf{v})] = 1 - \sum_{\mathbf{v}} p(\mathbf{v})\Delta(\mathbf{u}, \mathbf{v}),$$

and so (3.27) becomes

$$E(K) = \sum_{\mathbf{u}} p(\mathbf{u}) \left[1 - \sum_{\mathbf{v}} p(\mathbf{v})\Delta(\mathbf{u}, \mathbf{v}) \right]^M. \tag{3.28}$$

(At this point the reader should be able to see directly that (3.28) represents the probability that a source sequence will be "poorly represented" by a source code $\mathbf{v}_1, \ldots, \mathbf{v}_M$ chosen at "random.") The next step in the proof is the estimation of the inner sum in (3.28). To this end define

$$\Delta_0(\mathbf{u}, \mathbf{v}) = \begin{cases} 1 & \text{if } d(\mathbf{u}, \mathbf{v}) \leq k\delta'' \text{ and } I(\mathbf{u}; \mathbf{v}) \leq kR'', \\ 0 & \text{otherwise,} \end{cases}$$

where $I(\mathbf{u}, \mathbf{v}) = \log_2[p(\mathbf{v}|\mathbf{u})/p(\mathbf{v})]$. Then from (3.22) $\Delta_0(\mathbf{u}, \mathbf{v}) \leq \Delta(\mathbf{u}, \mathbf{v})$, and so

$$\sum_{\mathbf{v}} p(\mathbf{v})\Delta_0(\mathbf{u}, \mathbf{v}) \leq \sum_{\mathbf{v}} p(\mathbf{v})\Delta(\mathbf{u}, \mathbf{v}). \tag{3.29}$$

If $\Delta_0(\mathbf{u}, \mathbf{v}) = 1$, then $I(\mathbf{u}, \mathbf{v}) = \log_2[p(\mathbf{v}|\mathbf{u})/p(\mathbf{v})] \leq kR''$, and so $p(\mathbf{v}) \geq 2^{-kR''} p(\mathbf{v}|\mathbf{u})$. Hence

$$\sum_{\mathbf{v}} p(\mathbf{v})\Delta_0(\mathbf{u}, \mathbf{v}) \geq 2^{-kR''} \sum_{\mathbf{v}} p(\mathbf{v}|\mathbf{u})\Delta_0(\mathbf{u}, \mathbf{v}). \tag{3.30}$$

Combining (3.29) and (3.30), we have

$$\left[1 - \sum_{\mathbf{v}} p(\mathbf{v})\Delta(\mathbf{u}, \mathbf{v})\right]^M \le \left[1 - 2^{-kR''} \sum_{\mathbf{v}} p(\mathbf{v}|\mathbf{u})\Delta_0(\mathbf{u}, \mathbf{v})\right]^M. \qquad (3.31)$$

and now (*deus ex machina!*) we invoke the following inequality:

$$(1 - xy)^M \le 1 - x + e^{-yM} \qquad (\text{if } 0 \le x,\ y \le 1,\ M > 0) \qquad (3.32)$$

(proof left as Prob. 3.16), with $x =$ the sum on the right side of (3.31), and $y = 2^{-kR''}$. The result is

$$\left[1 - \sum_{\mathbf{v}} p(\mathbf{v})\Delta(\mathbf{u}, \mathbf{v})\right]^M$$

$$\le 1 - \sum_{\mathbf{v}} p(\mathbf{v}|\mathbf{u})\Delta_0(\mathbf{u}, \mathbf{v}) + \exp(-2^{-kR''} \cdot M). \qquad (3.33)$$

Combining (3.28) and (3.33), we get

$$E(K) \le 1 - \sum_{\mathbf{u}, \mathbf{v}} p(\mathbf{u}, \mathbf{v})\Delta_0(\mathbf{u}, \mathbf{v}) + \exp(-2^{-kR''} \cdot M).$$

$$= \sum_{\mathbf{u}, \mathbf{v}} p(\mathbf{u}, \mathbf{v})[1 - \Delta_0(\mathbf{u}, \mathbf{v})] + \exp(-2^{-kR''} \cdot M). \qquad (3.34)$$

We shall now show that both terms in (3.34) approach 0 as k approaches infinity. First note that since $M = 2^{\lfloor kR' \rfloor}$ and $R' > R''$ (see (3.19)), $\exp(-2^{-kR''} \cdot M) < \exp[-2^{k(R'-R'')-1}]$ approaches 0 very rapidly. Second, notice that $1 - \Delta_0(\mathbf{u}, \mathbf{v})$ is 1 iff either $d(\mathbf{u}, \mathbf{v}) > k\delta''$ or $I(\mathbf{u}; \mathbf{v}) > kR''$, and so

$$\sum_{u, v} p(\mathbf{u}, \mathbf{v})[1 - \Delta_0(\mathbf{u}, \mathbf{v})]$$

$$\le P\{d(\mathbf{U}, \mathbf{V}) > k\delta''\} + P\{I(\mathbf{U}; \mathbf{V}) > kR''\}, \qquad (3.35)$$

the probabilities in (3.35) being taken over the $(\mathbf{U}, \mathbf{V})$ space whose probability distribution is described in (3.26). But

$$d(\mathbf{U}, \mathbf{V}) = \sum_{i=1}^{k} d(U_i, V_i)$$

is the sum of independent, identically distributed random variables, each of which has mean $E[d(U, V)] \le \delta < \delta''$ by (3.25b) and (3.19), and so by the weak law of large numbers (see Appendix A), the first probability in (3.35) approaches 0 as k increases, Similarly,

$$I(\mathbf{U}; \mathbf{V}) = \sum_{i=1}^{k} I(U_i; V_i)$$

(see note on p. 37) is a sum of independent, identically distributed random variables, each with mean $I(\mathbf{U}, \mathbf{V}) = R(\delta) < R''$ (see (3.25a) and (3.19)) and so by another application of the weak law, the second probability in (3.35) approaches 0 as k increases. Combining all these facts, we see that the upper bound on $E(K)$ given by (3.34) approaches 0; in particular, for large enough k it will be less than $(\delta' - \delta'')/B$, and in view of the discussion on p. 87, this completes the proof. $\qquad\square$

Problems

3.1 Compute $R(\delta)$ for the source $\mathbf{p} = (\frac{1}{2}, \frac{1}{2})$ relative to the distortion matrix

$$D = \begin{bmatrix} 0 & 1 \\ 2 & 0 \end{bmatrix}.$$

3.1 Compute $R(\delta)$ for the source $\mathbf{p} = (\frac{1}{2}, \frac{1}{2})$ relative to the distortion matrix

$$D = \begin{bmatrix} 0 & 1 & \frac{1}{4} \\ 1 & 0 & \frac{1}{4} \end{bmatrix}.$$

(The $R(\delta)$ of this problem illustrates the fact that $R(\delta)$ need not be strictly convex.)

3.3 Show that the k-dimensional test channel defined at the end of the proof of Theorem 3.2 has $E[d(\mathbf{U})] \leq k\delta$ and $I(\mathbf{U}; \mathbf{V}) = kR_1(\delta)$.

3.4 The object of this problem is for you to prove that $R(\delta)$ is continuous at $\delta = \delta_{\min}$. Do this by supplying the details in the following argument.

Let $Q_1, Q_2, \ldots$ be a sequence of stochastic matrices describing the transition probabilities of test channels achieving $R(\delta_1), R(\delta_2), \ldots$, where $\lim_{n \to \infty} \delta_n = \delta_{\min}$. Then there is a subsequence $Q_{n_1}, Q_{n_2}, \ldots$ which converges to a stochastic matrix Q. The test channel corresponding to Q has $E[d(U, V)] = \delta_{\min}$ and $I(U; V) = \lim_{k \to \infty} R(\delta_{n_k})$. Thus $R(\delta_{\min}) \leq \lim_{\delta \to \delta_{\min}^+} R(\delta)$, and so R is continuous at $\delta = \delta_{\min}$.

3.5 On p. 80 we showed that a sufficient condition for $R(0) = H$, the source entropy, is for each row of D to have at least one zero entry, and each column to have at most one zero entry. Show that this condition is also necessary.

3.6 In Example 3.2, p. 82, we invoked "symmetry" in finding a test channel for achieving $R(\delta)$. In this problem we will make that notion precise. Let π be a permutation of A_U such that $p(u) = p(\pi(u))$ for all $u \in U$ (if the source statistics are uniform this is no restriction on π), and ρ be a permutation of A_V

such that $D(u, v) = D(\pi(u), \rho(v))$ for all $u \in U$, $v \in V$. This is the "symmetry" required; in Example 3.2,

$$\pi(-1) = +1, \pi(0) = 0, \pi(+1) = -1; \rho\left(-\tfrac{1}{2}\right) = +\tfrac{1}{2}, \rho\left(+\tfrac{1}{2}\right) = -\tfrac{1}{2}.$$

Show that for each $\delta \geqslant \delta_{\min}$ there exists a test channel $p(v|u)$ achieving $R(\delta)$ such that the stochastic matrix $Q(u, v) = p(v|u)$ has the same symmetry as D, i.e., $Q(\pi(u), \rho(v)) = Q(u, v)$ for all $u \in U$, $v \in V$. [*Hint:* If $Q_0(u, v)$ describes a test channel achieving $R(\delta)$, define $Q_i(u, v) = Q_0(\pi^i(u), \rho^i(v))$, and show that Q_i describes a test channel achieving $R(\delta)$ for $i = 1, 2, 3, \ldots$. Then define $Q(u, v) = n^{-1}\sum_{i=0}^{n-1} Q_i(u, v)$, where n is the least common multiple of the orders of π and ρ.]

3.7 In the most general situation, the computation of $R(\delta_{\min})$ is not a simple matter. The following result is often helpful, however. Verify that it is correct.

For each $u \in A_U$, let $B(u)$ denote the set of "best destination representatives of u," that is, $B(u) = \{v \in A_V: d(u, v) \leqslant d(u, v') \text{ for all } v' \in A_V\}$. Then there exists a backward test channel $p(u|v)$ achieving $R(\delta_{\min})$ such that $p(u|v_1) = p(u|v_2)$ whenever $v_1, v_2 \in B(u)$

Apply this result to the problem of finding $R(\delta_{\min})$ for the source $\left(\tfrac{1}{3}, \tfrac{1}{3}, \tfrac{1}{3}\right)$ relative to the distortion matrix

$$D = \begin{bmatrix} 0 & 0 & 1 \\ 1 & 1 & 0 \\ 1 & 1 & 1 \end{bmatrix}.$$

3.8 Consider a fixed source $\mathbf{p} = (p_1, p_2, \ldots, p_r)$ and distortion matrix D, with rate-distortion function $R(\delta)$. Consider a new distortion matrix $\hat{D}$, which is formed from D by adding a constant w_i to the ith row, that is, $\hat{d}(i, j) = d(i, j) + w_i$. Show that the new rate-distortion function is $\hat{R}(\delta) = R(\delta - \overline{w})$, where $\overline{w} = \sum_i w_i p_i$. Use this result to show that there is no essential loss of generality in assuming that $\delta_{\min} = 0$. (Result due to J. Pinkston [25], p. 300.)

3.9 Consider a source $\mathbf{p} = (p_1, \ldots, p_r)$ whose distortion matrix D has all 0's in its first row. Show that $R(\delta) = (1 - p_1)\hat{R}(\delta/(1 - p_1))$, where $\hat{R}$ is the rate-distortion function of the source $(p_2/(1 - p_1), \ldots, p_r/(1 - p_1))$ relative to the distortion matrix $\hat{D}$ obtained from D by deleting its first row. (Result due to J. Pinkston [25], p. 300.)

3.10 (The Shannon lower bound to $R(\delta)$). Suppose the distortion matrix D has the property that each column is a permutation of $(d_1, d_2, \ldots, d_r)$. Define $\Phi(\delta) = \max\{H(\alpha_1, \ldots, \alpha_n): \sum_{i=1}^r \alpha_i d_i = \delta\}$ (see Prob. 1.8). Prove that $R(\delta) \geqslant H(U) - \Phi(\delta)$, where $H(U)$ is the source entropy, by verifying the following steps:

(a) If (U, V) is a test channel achieving $R(\delta)$, $R(\delta) = H(U) - \sum_v p(v)H(U|V = v)$

(b) If $\delta(v) = \sum_u p(u|v)d(u, v)$, then $H(U|V = v) \leqslant \Phi(\delta(v))$

(c) $\sum_v p(v)\Phi(\delta(v)) \leqslant \Phi(\sum_v p(v)\delta(v)) \leqslant \Phi(\delta)$

3.11 (Continuation). If now D has the further property that each *row* is a permutation of every other row, show that $R(\delta) = H(U) - \Phi(\delta)$ for $\delta_{\min} \le \delta \le \delta_{\max}$, if the source is symmetric.

3.12 Consider two DMS's with input alphabets $A_U^{(i)}$, output alphabets $A_V^{(i)}$, and distortion functions $d^i(u, v)$, $i = 1, 2$. Their *product* has source alphabet $A_U^{(1)} \times A_U^{(2)}$, destination alphabet $A_V^{(1)} \times A_V^{(2)}$, and distortion matrix with entries $d[(u_1, u_2), (v_1, v_2)] = d^{(1)}(u_1, v_1) + d^{(2)}(u_2, v_2)$. (Physically this corresponds to two independent, parallel sources.) Show that the rate-distortion function of the product source is $R(\delta) = \min_+\{R^{(1)}(t) + R^{(2)}(\delta - t)\}$, where $R^{(1)}$ and $R^{(2)}$ are the rate-distortion functions of sources 1 and 2, respectively. Thus show that the $R(\delta)$ curve is obtained by adding both coordinates of the curves $R^{(1)}$ and $R^{(2)}$ at points on the two curves having the same slope. (Result due to Shannon.)

3.13 (Perfect codes in the Lee metric[7]). If q is an odd number, and if $A_U = A_V = \{0, 1, \ldots, q - 1\}$, the *Lee distortion* (or Lee metric) is defined by $d(u, v) = \min\{|u - v|, |q - u + v|\}$. If $q = 2t^2 + 2t + 1$, consider this source code of length $k = 2$ with $M = q$ codewords: $C = \{(v_1, v_2): v_2 \equiv (2t + 1)v_1 (\mathrm{mod}\, q)\}$. Show that the spheres of radius t around these code-words completely cover the set of q^2 pairs (u_1, u_2) without overlap. (A sphere of radius t around (v_1, v_2) is the set $\{(u_1, u_2): d(u_1, v_1) + d(u_2, v_2) \le t\}$.) Show that the average distortion of this code is $\frac{1}{3} \cdot (2t^3 + 3t + t)/(2t^2 + 2t + 1)$.

3.14 Verify that $d(C) = \frac{10}{9}$ for the source code of Example 3.2 (p. 85).

3.15 Show that Theorem 3.4 remains true if conclusion (a) is replaced by "$M \le 2^{< \lfloor kR(\delta) \rfloor}$"; and, if $\delta > \delta_{\min}$, that it remains true if (a) is unchanged, but (b) is replaced by "$d(C) \le \delta$."

3.16 Prove that, if $0 \le x, y \le 1$, $M \ge 0$, then $(1 - xy)^M \le 1 - x + e^{-yM}$ (see Eq. (3.32)).

3.17 Describe how you could explicitly achieve what is promised by the source coding theorem (Theorem 3.4) at $\delta = \delta_{\max}$.

Notes

1 (p. 75). Technically $d(u, v)$ is called a *single-letter* distortion measure, to distinguish it from distortion measures that are defined on certain k-tuples $(\mathbf{u}, \mathbf{v})$ $= ((u_1, \ldots, u_k), (v_1, \ldots, v_k))$ directly, rather than by Eq. (3.1).

2 (p. 76). This example is due to Shannon (see [25], pp. 246–247, and Fig. 2 on p. 262). In describing this rather peculiar source, he wrote, "... the [source] alphabet consists of three possible readings, -1, 0, and $+1$. Perhaps, for some reasons of economy, it is desired to work with a reproduced alphabet of two letters, $-\frac{1}{2}$ and $+\frac{1}{2}$. One might then have the matrix that is shown in [Example 3.2]."

3 (p. 76). The definitions that follow (Eqs. (3.3) and (3.5)) can be given for an arbitrary stationary source; see Eqs. (5.3) and (5.4).

4 (p. 77). See Note 3, Chapter 2.

5 (p. 78). In view of Theorem 3.2, the reader may wonder why the functions $R_k(\delta)$ were introduced in the first place. The reason is twofold. First, for a stationary source with memory, definitions (3.3) and (3.5) make sense and typically $R_k(\delta)$ is a decreasing

function of k, for fixed δ. (However we shall not study sources with memory in this book.) Second (and more important for our purposes) the definitions given simplify the proof of the converse of the source-channel coding theorem. (See Chapter 5, especially Eq. (5.10).)

6 (p. 83). Pinkston (see [25], pp. 296 ff.) has generalized this theorem to an arbitrary DMS, with Hamming distortion, as follows. If the source probabilitites are ordered: $p_1 \leqslant p_2 \cdots \leqslant p_r$, define $S_k = \sum_{i=1}^{k} p_i$, $\delta_k = S_{k-1} + (r - k)p_k$, and $H_k = H(p_{k+1}/(1 - S_k), \ldots, p_r/(1 - S_k))$. Then, for $\delta_{k-1} \leqslant \delta \leqslant \delta_k$, $R(\delta) = (1 - S_{k-1})$ $(H_{k-1} - H((\delta - S_{k-1})/(1 - S_k)) - (\delta - S_k)/(1 - S_k)\log(r - k - 1))$, for $k = 0, 1, \ldots, r$.

7 (p. 93). For more on this subject, see Berlekamp [14], pp. 305–309.

4

The Gaussian channel and source

4.1 The Gaussian channel

This channel, whose full name is the "discrete-time memoryless additive Gaussian channel with average power constraint," has channel input alphabet A_X and channel output alphabet A_Y, both equal to the set of all real numbers. If $X_1, X_2, \ldots$ are the inputs to the channel at times $1, 2, \ldots$, then the corresponding outputs $Y_1, Y_2, \ldots$ are given by $Y_i = X_i + Z_i$, where Z_1, $Z_2, \ldots$ are independent, identically distributed normal random variables with mean 0 and variance σ^2. This channel is often depicted as in Fig. 4.1. Furthermore, there is a "cost" associated with each input x; it is $b(x) = x^2$. In this section we shall show that the capacity–cost function of this channel is given by

$$C(\beta) = \tfrac{1}{2}\log(1 + \beta/\sigma^2). \tag{4.1}$$

Just as for the discrete channels of Chapter 2, we shall show that $C(\beta)$ represents the maximum rate at which the channel can transmit error-free information, if the average input cost is restricted to be $\leqslant \beta$.

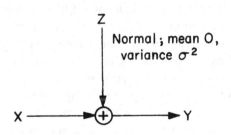

Figure 4.1 The Gaussian channel.

95

Before deriving formula (4.1), however, we pause to give a brief, informal description of how such a channel might arise in practice.

Suppose we wish to transmit a sequence of n real numbers $x_1, x_2, \ldots, x_n$ from one point to another in T seconds, and are required to do this by converting the x_i's into a continuous function of time $x(t)$. (For definiteness, let's assume $x(t)$ represents the voltage across a 1-ohm load.) One way to do this is to find n functions $\phi_i(t)$, $i = 1, 2, \ldots, n$, which are orthonormal on the interval $[0, T]$, that is,

$$\int_0^T \phi_i(t)\phi_j(t)dt = \begin{cases} 1 & \text{if } i = j, \\ 0 & \text{if } i \neq j, \end{cases} \tag{4.2}$$

and to transmit the signal

$$x(t) = \sum_{i=1}^n x_i\phi_i(t). \tag{4.3}$$

The numbers x_i are recoverable from $x(t)$:

$$x_i = \int_0^T x(t)\phi_i(t)dt. \tag{4.4}$$

However, several complications can arise. First, our transmitter may be limited in power, say to P watts. Then the total energy dissipated in T seconds cannot exceed PT joules. This energy is given by the integral $\int_0^T x^2(t)dt$, which because of orthonormality (4.2) is $\sum_{i=1}^n x_i^2$. Hence we must have

$$\frac{1}{n}\sum_{i=1}^n x_i^2 \leq \frac{PT}{n}, \tag{4.5}$$

which is to say that the input vector $\mathbf{x} = (x_1, \ldots, x_n)$ cannot be arbitrary; it must lie within a Euclidean sphere of radius $\sqrt{PT}$.

Another complication is that when $x(t)$ is transmitted the received signal will often be of the form $\hat{x}(t) = x(t) + z(t)$, where $z(t)$ is some kind of noise process. A common type of noise is Johnson (thermal) noise, which is caused by the thermal agitation of the electrons in the receiver. In this case it is reasonable to model $z(t)$ as a white Gaussian noise process; for our purposes this means that there is a number N_0, the noise spectral density, such that the integrals $z_i = \int_0^T z(t)\phi_i(t)dt$ are independent, mean 0, variance $N_0/2$ Gaussian random variables.[1] Thus if the receiver attempts to recover x_i by computing (cf. (4.4)) $\int_0^T \hat{x}(t)\phi_i(t)dt$, he will obtain the estimate $\hat{x}_i = x_i + z_i$.

In summary: we transmit $\mathbf{x} = (x_1, \ldots, x_n)$, where $\mathbf{x}$ must satisfy (4.5), and receive $\hat{\mathbf{x}} = (x_1 + z_1, \ldots, x_n + z_n)$, where $z_1, \ldots, z_n$ are independent, mean

0, variance $N_0/2$ Gaussian random variables. But of course that is just the situation modeled by the Gaussian channel introduced at the beginning of this section, with noise variance $\sigma^2 = N_0/2$ and input constraint $\beta = PT/n$. According to (4.1) the capacity of this channel is $\frac{1}{2}\log_2(1 + 2PT/nN_0)$ bits per symbol. If now we define the transmission "bandwidth"[2] by $W = n/2T$, and observe that we are transmitting $n/T = 2W$ symbols per second, the capacity becomes

$$C = W\log_2(1 + P/N_0 W) \text{ bits per second.} \tag{4.6}$$

Formula (4.6) is Shannon's famous expression for the capacity of a band-limited, power-limited Gaussian channel. Note that if $W \gg P/N_0$, the resulting "wideband" Gaussian channel has capacity

$$C = \frac{1}{\ln 2}\frac{P}{N_0}$$

$$= 1.4427 P/N_0 \text{ bits per second.} \tag{4.7}$$

Let us now return to the main concern of this section, the derivation of formula (4.1).

Analogously with Eq. (2.3), let us define the nth capacity–cost function $C_n(\beta)$ of the Gaussian channel by

$$C_n(\beta) = \sup\left\{ I(\mathbf{X}; \mathbf{Y}) : \sum_{i=1}^{n} E(X_i^2) \leq n\beta \right\}, \tag{4.8}$$

where the supremum is taken over all pairs $\mathbf{X} = (X_1, \ldots, X_n)$, $\mathbf{Y} = (Y_1, \ldots, Y_n)$ of n-dimensional random vectors such that:

$$\mathbf{X} \text{ has a continuous density function } p(\mathbf{x}), \tag{4.9a}$$

$$\sum_{i=1}^{n} E(X_i^2) \leq n\beta, \tag{4.9b}$$

$$Y_i = X_i + Z_i, \qquad i = 1, 2, \ldots, n, \tag{4.9c}$$

where $Z_1, Z_2, \ldots, Z_n$ are independent (of each other and of the X_i's), mean 0, variance σ^2 random variables. The overall capacity–cost function for the Gaussian channel is now defined as

$$C(\beta) = \sup_n \frac{1}{n} C_n(\beta). \tag{4.10}$$

We immediately prove the following theorem.

Theorem 4.1 $C_n(\beta) = (n/2)\log(1 + \beta/\sigma^2)$, *and so* $C(\beta) = \frac{1}{2}\log(1 + \beta/\sigma^2)$.

Proof Let $\mathbf{X} = (X_1, \ldots, X_n)$ be any test source satisfying (4.9a) and (4.9b). Then by (4.9c) the joint density of $\mathbf{X}$ and $\mathbf{Y}$ is given by

$$p(\mathbf{x}, \mathbf{y}) = p(\mathbf{x})g(\mathbf{z}),$$

where $\mathbf{z} = (y_1 - x_1, \ldots, y_n - x_n)$ and $g(\mathbf{z})$ is the joint density of $Z_1, \ldots, Z_n$, that is,

$$g(z_1, \ldots, z_n) = \frac{1}{(2\pi\sigma^2)^{n/2}} \exp\left(-\frac{\sum z_i^2}{2\sigma^2}\right),$$

(cf. Eq. (1.27)).

Let $A_i = E(X_i^2)$. Since X_i and Z_i are independent, $E(Y_i^2) = E(X_i^2) + E(Z_i^2) = A_i + \sigma^2$, and so by Theorem 1.11

$$h(\mathbf{Y}) \leq \frac{n}{2}\log 2\pi e \left[\prod_{i=1}^{n}(A_i + \sigma^2)\right]^{1/n}. \tag{4.11}$$

Now by (4.9b) $\sum_{i=1}^{n}(A_i + \sigma^2) \leq n(\beta + \sigma^2)$, and so, by the arithmetic–geometric mean inequally, the product in (4.11) is not larger than $(\beta + \sigma^2)^n$. Hence

$$h(\mathbf{Y}) \leq \frac{n}{2}\log 2\pi e(\beta + \sigma^2). \tag{4.12}$$

By Theorem 1.10, $I(\mathbf{X}; \mathbf{Y}) = h(\mathbf{Y}) - h(\mathbf{Y}|\mathbf{X})$, but $h(\mathbf{Y}|\mathbf{X}) = h(\mathbf{Z}) = (n/2)\log 2\pi e\sigma^2$ (see Example 1.13 and also Prob. 1.27). Thus

$$I(\mathbf{X}; \mathbf{Y}) = h(\mathbf{Y}) - h(\mathbf{Z})$$

$$\leq \frac{n}{2}\log\left(1 + \frac{\beta}{\sigma^2}\right), \tag{4.13}$$

which proves that $C_n(\beta) \leq (n/2)\log(1 + \beta/\sigma^2)$. To prove the opposite inequality, let $X_1, X_2, \ldots, X_n$ be independent, mean 0, variance β Gaussian random variables. Then (4.9a) and (4.9b) will be satisfied. Also, $Y_1, Y_2, \ldots, Y_n$ will be independent, mean 0, variance $\beta + \sigma^2$ random variables (cf. Prob. 1.29) and $I(\mathbf{X}; \mathbf{Y}) = h(\mathbf{Y}) - h(\mathbf{Z}) = (n/2)\log(1 + \beta/\sigma^2)$. This completes the proof of Theorem 4.1. $\qquad\square$

Note *Another way to define $C_n(\beta)$ is to replace condition (4.9a) with*

$$\mathbf{X} \text{ assumes only finitely many values.} \tag{4.9a'}$$

In Prob. 4.11 it is shown that these two definitions are equivalent.

The final result in this section is the coding theorem for the Gaussian channel. It is completely analogous to the discrete coding Theorem 2.4: A (channel) code of length n is just a set of M n-dimensional vectors $\{\mathbf{x}_1, \ldots, \mathbf{x}_M\}$. If a codeword $\mathbf{x} = (x_1, \ldots, x_n)$ is transmitted over the channel, it is received as $\mathbf{y} = \mathbf{x} + \mathbf{z}$, where the components of $\mathbf{z}$ are independent Gaussian random variables with mean 0 and variance σ^2. A decoding rule for such a code is a mapping f from the set of all n-dimensional vectors $\mathbf{y}$ into the code. The error probabilities $P_E^{(i)}$ represent the probabilities of decoder error, given that the i-th codeword was sent, that is, $P_E^{(i)} = P\{f(y) \neq \mathbf{x}_i | \mathbf{x}_i$ transmitted$\}$, where $\mathbf{y} = \mathbf{x}_i + \mathbf{z}$ as above.

Theorem 4.2 (*coding theorem for Gaussian channels*). *Fix $\beta \geq 0$. Then, for any $\beta' > \beta$, $R < C(\beta) = \frac{1}{2}\log_2(1 + \beta/\sigma^2)$, and $\varepsilon > 0$, there exists a code $C = \{\mathbf{x}_1, \ldots, \mathbf{x}_M\}$ of length n, and a corresponding decoding rule such that*:

(a) $\sum_{j=1}^{n} x_{ij}^2 \leq n\beta'$ *for $i = 1, 2, \ldots, M$; $\mathbf{x}_i = (x_{i1}, \ldots, x_{in})$,*
(b) $M \geq 2^{\lceil Rn \rceil}$,
(c) $P_E^{(i)} < \varepsilon$ *for all $i = 1, 2, \ldots, M$.*

Proof According to the preceding note, it is possible to choose a finitely based random variable $\overline{X}$ with $E(\overline{X}^2) \leq \beta$ such that $I(\overline{X}; Y)$ is arbitrarily close to $C(\beta)$. The corresponding Y will not be discrete, but since $I(\overline{X}; Y) = \lim I(\overline{X}; [Y])$ as the quantization $[Y]$ becomes finer and finer (see Eq. (1.17)); it will be possible to find a discrete random variable $[Y]$ such that $I(\overline{X}; [Y]) > R$. The discrete memoryless channel which has as inputs the values assumed by $\overline{X}$, and as outputs the values assumed by $[Y]$, with inputs constrained to satisfy $E(X^2) \leq \beta$, thus has a capacity greater than R, and so the existence of a code satisfying (a), (b), and (c) follows immediately from Theorem 2.4. $\qquad\square$

4.2 The Gaussian source

This source, whose full name is the "discrete-time memoryless Gaussian source," has as source alphabet A_U the set of all real numbers, and the source output is modeled by a sequence $U_1, U_2, \ldots$ of independent identically distributed normal random variables with mean 0 and variance σ^2. Our object in this section is to compute the rate-distortion function of this source relative to the "squared-error" distortion criterion, in which the destination alphabet A_V is again the set of real numbers, and the distortion between a source symbol u and a destination symbol v is given by

$$d(u, v) = (u - v)^2.$$ (4.14)

This rate-distortion function turns out to be

$$R(\delta) = \begin{cases} \frac{1}{2}\log\dfrac{\sigma^2}{\delta} & \text{if } \delta \leqslant \sigma^2, \\ 0 & \text{if } \delta \geqslant \sigma^2. \end{cases}$$ (4.15)

It is plotted (in units of bits) in Fig. 4.2.

Of course a Gaussian source is likely to be encountered in almost any kind of data-gathering experiment. Since an infinite number of bits are required to represent an arbitrary real number u with perfect fidelity, it is of interest to know the tradeoff between the number of bits used to represent the experimental outcome and the resulting distortion. As defined in (4.15), $R(\delta)$ does in fact represent the minimum possible number of bits sufficient to represent a Gaussian (variance σ^2) random variable if the maximum permissible mean-squared error is δ. This fact will be verified by the source coding theorem, Theorem 4.5, and by the results of Chapter 5.

To establish (4.15), we first define the kth rate-distortion function $R_k(\delta)$ of the Gaussian source with respect to the mean-squared error criterion by (see Eq. (3.3))

$$R_k(\delta) = \inf\{I(\mathbf{U}; \mathbf{V}) : E(\|\mathbf{U} - \mathbf{V}\|^2) \leqslant k\delta\},$$ (4.16)

where the infimum is taken over all pairs of k-dimensional random vectors $\mathbf{U} = (U_1, \ldots, U_k)$ and $\mathbf{V} = (V_1, \ldots, V_k)$ such that:

$U_1, U_2, \ldots, U_k$ are independent, mean 0, variance σ^2 Gaussian random variables, (4.17a)

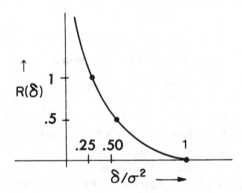

Figure 4.2 The rate-distortion function for a Gaussian source.

$$E(\|\mathbf{U} - \mathbf{V}\|^2) = \sum_{i=1}^{k} E[(U_i - V_i)^2] \leqslant k\delta, \tag{4.17b}$$

the joint distribution of $\mathbf{U}$ and $\mathbf{V}$ is given by a continuous
density function $p(u,v)$. \hfill (4.17c)

The overall rate-distortion function $R(\delta)$ is then defined (cf. Eq. (3.5)) by

$$R(\delta) = \inf_{k} \frac{1}{k} R_k(\delta). \tag{4.18}$$

As happened for discrete memoryless sources in Chapter 3, it will turn out that the infimum in (4.18) is already achieved at $k = 1$. Before proving this fact, however, we give a closed form expression for $R_1(\delta)$, which is of course the value of $R(\delta)$ as well.

Theorem 4.3

$$R_1(\delta) = \begin{cases} \frac{1}{2} \log \dfrac{\sigma^2}{\delta} & \text{if } \delta \leqslant \sigma^2, \\ 0 & \text{if } \delta \geqslant \sigma^2. \end{cases}$$

Proof Pick δ, $\varepsilon > 0$, and a pair (U, V) of random variables such that:

$$I(U; V) < R_1(\delta) + \varepsilon, \tag{4.19a}$$

$$U \text{ is Gaussian, with mean 0, variance } \sigma^2, \tag{4.19b}$$

$$E[(U - V)^2] \leqslant \delta, \tag{4.19c}$$

$$U, V \text{ has a continuous joint density } p(u, v). \tag{4.19d}$$

That this is possible follows from definition (4.16) and the conditions (4.17). By (4.19c) and (4.19d),

$$\delta \geqslant \iint p(u, v)(u - v)^2 \, du \, dv$$

$$= \int p(v) \int p(u|v)(u - v)^2 \, du \, dv, \tag{4.20}$$

where $p(v) = \int p(u, v) du$ is the marginal density of V, and $p(u|v) = p(u, v)/p(v)$ is the conditional density of u, given v. It follows from (4.20) that if we define

$$\delta(v) = \int p(u|v)(u - v)^2 \, du, \tag{4.21}$$

then $\delta(v)$ is finite for almost all v. Hence according to Theorem 1.11, the conditional entropy $h(U|V = v)$ exists for almost all v, and in fact

$$h(U|V = v) = -\int p(u|v)\log p(u|v)\, dv$$

$$\leqslant \tfrac{1}{2}\log 2\pi e\delta(v). \tag{4.22}$$

Now by Theorem 1.10

$$I(U; V) = h(U) - h(U|V). \tag{4.23}$$

But $h(U) = \tfrac{1}{2}\log 2\pi e\sigma^2$ (cf. Example 1.13) and by (4.20), (4.21), and (4.22)

$$h(U|V) = \int p(v)h(U|V = v)\, dv$$

$$\leqslant \tfrac{1}{2}\log 2\pi e\int \delta(v)p(v)\, dv$$

$$\leqslant \tfrac{1}{2}\log 2\pi e\delta.$$

Hence $R_1(\delta) + \varepsilon > I(U; V) \geqslant \tfrac{1}{2}\log \sigma^2/\delta$; but since this is true for all $\varepsilon > 0$, and since in any event $R_1(\delta) \geqslant 0$, it follows that

$$R_1(\delta) \geqslant \max\left(\tfrac{1}{2}\log \frac{\sigma^2}{\delta}, 0\right). \tag{4.24}$$

It remains to show that the lower bound (4.24) is tight. To do this, notice that if $\delta < \sigma^2$, if V is a mean 0, variance $\sigma^2 - \delta$ Gaussian random variable, and if $U = V + G$, where G is a mean 0, variance δ Gaussian random variable independent of V, then (U, V) satisfy (4.17) with $k = 1$, and so $R_1(\delta) \leqslant I(U; V) = h(U) - h(U|V) = h(U) - h(G) = \tfrac{1}{2}\log 2\pi e\sigma^2 - \tfrac{1}{2}\log 2\pi e\delta = \tfrac{1}{2}\log \sigma^2/\delta$ (see Prob. 1.27). On the other hand, if $\delta > \sigma^2$, pick $\varepsilon > 0$ and let V be Gaussian with variance ε, G be Gaussian with variance $\sigma^2 - \varepsilon$. Then again (4.17) is satisfied, and so $R_1(\delta) \leqslant I(U; V) = h(U) - h(G) = \tfrac{1}{2}\log 2\pi e\sigma^2 - \tfrac{1}{2}\log 2\pi e(\sigma^2 - \varepsilon) = \tfrac{1}{2}\log[1 + \varepsilon/(\sigma^2 - \varepsilon)]$. Since this is true for all $\varepsilon > 0$, $R_1(\delta) \leqslant 0$ in this range. Hence the lower bound (4.24) is also an upper bound, and this completes the proof. $\qquad\square$

(The odd relationship between U and V described in the last part of the proof of Theorem 4.3 is often called the "backward" test channel for computing $R(\delta)$, because it can be depicted as in Fig. 4.3, in spite of the fact that U is the source and V is the destination! For a "forward" version of this test channel, see Prob. 4.18.)

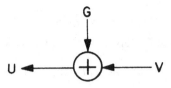

Figure 4.3 The backward test channel for a Gaussian source.

We can now compute $R(\delta)$.

Theorem 4.4 $R_k(\delta) = kR_1(\delta)$ *for all k, and so* $R(\delta) = R_1(\delta) = \max\left(\frac{1}{2}\log \sigma^2/\delta,\, 0\right)$.

Proof Pick $\varepsilon > 0$, and let $\mathbf{U} = (U_1, \ldots, U_k)$ and $\mathbf{V} = (V_1, \ldots, V_k)$ be a pair of random vectors satisfying (4.17) and

$$I(\mathbf{U}; \mathbf{V}) < R_k(\delta) + \varepsilon. \tag{4.25}$$

Then by Theorem 1.8 $I(\mathbf{U}; \mathbf{V}) \geq \sum_{i=1}^{k} I(U_i; V_i)$. If we define $\delta_i = E[(U_i - V_i)^2]$, then, by the definition of $R_1(\delta)$, $I(U_i; V_i) \geq R_1(\delta_i)$, and by (4.17b) $\sum_{i=1}^{k}\delta_i \leq k\delta$. Furthermore, $\sum_{i=1}^{k} R_1(\delta_i) \geq kR_1(\bar{\delta}) \geq kR_1(\delta)$, where $\bar{\delta} = k^{-1}\sum_{i=1}^{k}\delta_i$, since $R_1(\delta)$ is convex $\cup$ and monotonically decreasing by Theorem 4.3. Hence

$$R_k(\delta) + \varepsilon > I(\mathbf{U}; \mathbf{V})$$

$$\geq \sum_{i=1}^{k} I(U_i; V_i)$$

$$\geq \sum_{i=1}^{k} R_1(\delta_i)$$

$$\geq kR_1(\bar{\delta})$$

$$\geq kR_1(\delta).$$

Since this is true for all $\varepsilon > 0$, it follows that $R_k(\delta) \geq kR_1(\delta)$. To prove the opposite inequality, let $(U_1, V_1), \ldots, (U_k, V_k)$ be independent identical copies of the backward test channel achieving $R_1(\delta)$. Then, for $\delta < \sigma^2$, (4.17) is satisfied and $I(U; V) = (k/2)\log \sigma^2/\delta$. (For $\delta \geq \sigma^2$, the proof is similar and is left as Prob. 4.20.) $\qquad\square$

Note *If $R_k(\delta)$ had been defined with condition (4.17c) replaced by*

$$\text{V assumes only finitely many distinct values,} \qquad (4.17c')$$

the value of $R_k(\delta)$ would be the same. This fact is the content of Prob. 4.21.

We conclude this section with the source coding theorem for the Gaussian source. In this context a source code of length k is a set of M k-dimensional Euclidean vectors $C = \{v_1, v_2, \ldots, v_M\}$. If u is another k-dimensional Euclidean vector, define $f(u)$ to be a codeword v_i for which the Euclidean distance $\|u - v_i\|$ is as small as possible. Then the average distortion of this code is

$$d(C) = \frac{1}{k} E(\|u - f(u)\|^2), \qquad (4.26)$$

where the expectation is with respect to a k-dimensional mean 0, variance σ^2 Gaussian distribution on u. Clearly if such a code is used to represent the Gaussian source, each vector u, if represented by its codeword $f(u)$, can be specified with $\lceil \log_2 M \rceil$ bits, and the resulting per-symbol mean-squared distortion is $d(C)$.

Theorem 4.5 *Let $R(\delta) = \max\left(\frac{1}{2} \log \sigma^2/\delta, 0\right)$ denote the rate-distortion function of the Gaussian source with respect to the mean-squared error distortion criterion. Fix $\delta \geqslant 0$. If $R' > R(\delta)$ and $\delta' > \delta$, then for sufficiently large k there exists a source code C with M codewords satisfying:*

(a) $M \leqslant 2^{\lfloor kR' \rfloor}$,
(b) $d(C) < \delta'$.

Proof If $\delta \geqslant \sigma^2$, then, by choosing the single codeword 0 of length 1, we will achieve what is promised. If on the other hand $\delta < \sigma^2$, let (U, V) be the backward test channel achieving $R(\delta)$ described in the proof of Theorem 4.3; see Fig. 4.3. Choose δ_1 and δ_2 so that

$$\delta < \delta_1 < \delta_2 < \delta'. \qquad (4.27)$$

Next pick functions α and β, each assuming only finitely many values such that

$$E\{[\alpha(U) - \beta(V)]^2\} < \delta_1, \qquad (4.28)$$

$$E\{[U - \alpha(U)]^2\} = \varepsilon, \qquad (4.29)$$

where

$$\varepsilon + 2\sqrt{\varepsilon\delta_2} + \delta_2 < \delta'. \tag{4.30}$$

(That this is possible is the content of Prob. 4.22.) We now consider the discrete memoryless source $\alpha(U)$, with destination alphabet equal to the range of β, and distortion function $d(u, v) = (u - v)^2$. Since, by Theorem 1.5, $I[\alpha(U); \beta(V)] \leq I(U; V) = R_1(\delta)$, if we denote the rate-distortion function of this source by $\bar{R}$, it follows from (4.28) that

$$\bar{R}(\delta_1) \leq R(\delta) < R'. \tag{4.31}$$

Thus by the source coding theorem for discrete memoryless sources (Theorem 3.4) there exists a source code of length k for the source $\alpha(U)$ satisfying:

$$M \leq 2^{\lfloor kR' \rfloor}, \tag{4.32}$$

$$d(C) < \delta_2. \tag{4.33}$$

We now consider using this code for the original source U. If $\mathbf{u} = (u_1, \ldots, u_k)$ is an arbitrary vector of length k, and if $\mathbf{v}_i$ is the source codeword closest to $\alpha(\mathbf{u}) = (\alpha(u_1), \ldots, \alpha(u_k))$, then $\|\mathbf{u} - \mathbf{v}_i\| = \|[\mathbf{u} - \alpha(\mathbf{u})] + [\alpha(\mathbf{u}) - \mathbf{v}_i]\|$, and so by Schwarz's inequality[3]

$$E(\|\mathbf{U} - \mathbf{V}_i\|^2) \leq E[\|\mathbf{U} - \alpha(\mathbf{U})\|^2] + E[\|\alpha(\mathbf{U}) - \mathbf{V}_i\|^2]$$
$$+ 2E[\|\mathbf{U} - \alpha(\mathbf{U})\|^2]^{1/2} E[\|\alpha(\mathbf{U}) - \mathbf{V}_i\|^2]^{1/2}, \tag{4.34}$$

where $\mathbf{V}_i$ denotes the codeword closest to the random vector $\mathbf{U}$. But from (4.29) $E[\|\mathbf{U} - \alpha(\mathbf{U})\|^2] = k\varepsilon$, and from (4.33) $E[\|\alpha(\mathbf{U}) - \mathbf{V}_i\|^2) < k\delta_2$. Hence, combining (4.30) and (4.34), we obtain

$$E[\|\mathbf{U} - \mathbf{V}_i\|^2] < k\delta', \tag{4.35}$$

which, combined with (4.32), proves the theorem. $\qquad\square$

Problems

4.1 Consider a channel with $A_X = A_Y =$ the real line, for which the output Y is the sum of the input X and an independent noise random variable Z, which is uniformly distributed on $[-\frac{1}{2}, \frac{1}{2}]$. Suppose the input X is constrained to satisfy $|X| \leq \beta$. If β is an integer, show that the capacity (defined analogously to Eqs. (4.8), (4.9), (4.10)) is $C(\beta) = \log(2\beta + 1)$. What if β is not an integer?

4.2 Consider an additive noise channel such as the one depicted in Fig. 4.1, where Z is not necessarily Gaussian, but is independent of X and has variance σ^2. If $E(X^2) \leq \beta$, show that the corresponding capacity satisfies $\frac{1}{2}\log(1 + \beta/\sigma^2) \leq C(\beta) \leq \frac{1}{2}\log((\beta + \sigma^2)/\sigma_1^2)$, where $\sigma_1^2 = \exp[2h(Z)]/2\pi e$. [*Hint:* see Prob. 1.30.]

In the next seven problems we will sketch a proof of a coding theorem for the Gaussian channel which is in some ways stronger, and in some ways weaker, than Theorem 4.2. It will point out the significance of the quantity

$$R_0 = \log_2\left(\frac{2}{1 + e^{-\beta/2\sigma^2}}\right).$$

The theorem is this:

R_0 *theorem for a Gaussian channel.* For any $R < R_0$ there exists a code $\{\mathbf{x}_1, \ldots, \mathbf{x}_M\}$ with $M = \lfloor 2^{Rn} \rfloor$ codewords of length n, each of which has coordinates $\pm\sqrt{\beta}$, together with a decoding rule, such that, if $P_E = (\sum_{i=1}^{M} P_E^{(i)})/M$ denotes the average decoding error probability, $P_E < 2^{-n(R_0 - R)}$.

(This is stronger than Theorem 4.2 in that it gives an explicit estimate of how small P_E can be made as a function of n. It is weaker in that $R_0 < C(\beta)$, so that for rates $R_0 < R < C(\beta)$ it does not show that $P_E \to 0$ is possible.)

4.3 First show that for a code $\{\mathbf{x}_1, \mathbf{x}_2\}$ with only two codewords, if the decoder chooses the one nearest (Euclidean distance) to the received vector $\mathbf{y}$, the error probability is given by

$$P_2[\mathbf{x}_1, \mathbf{x}_2] = Q(\|\mathbf{x}_1 - \mathbf{x}_2\|/2\sigma),$$

where $Q(\alpha) = (1/\sqrt{2\pi}) \int_\alpha^\infty e^{-s^2/2}\, ds$.

4.4 Next show that in general, if the code $\{\mathbf{x}_1, \ldots, \mathbf{x}_M\}$ is decoded by the nearest codeword strategy, and if $P_E^{(i)}$ denotes the probability of decoding error if $\mathbf{x}_i$ is transmitted, then

$$P_E^{(i)} \leq \sum_{\substack{j=1 \\ j \neq i}}^{M} P_2[\mathbf{x}_i, \mathbf{x}_j].$$

4.5 Next prove the estimate

$$Q(\alpha) < \tfrac{1}{2} e^{-\alpha^2/2}, \qquad \alpha > 0.$$

4.6 Now suppose that all the codewords have coordinates $\pm\sqrt{\beta}$, and that $d_H(\mathbf{x}_i, \mathbf{x}_j)$ (the *Hamming distance* between $\mathbf{x}_i$ and $\mathbf{x}_j$) is defined to be the number of coordinates in which $\mathbf{x}_i$ and $\mathbf{x}_j$ differ. Then show that

$$P_E^{(i)} < \sum_{\substack{j=1 \\ j \neq i}}^{M} \gamma^{d_H(\mathbf{x}_i, \mathbf{x}_j)},$$

where $\gamma = e^{-\beta/2\sigma^2}$.

4.7 Using the results of Prob. 4.4, obtain the following estimate of the average value $E[P_E^{(i)}]$, where the codeword coordinates are selected independently according to the distribution $P\{x = +\sqrt{\beta}\} = P\{x = -\sqrt{\beta}\} = \tfrac{1}{2}$:

$$E[P_E^{(i)}] < M \cdot \left(\frac{1 + \gamma}{2}\right)^n.$$

4.8 Now complete the proof of the R_0 theorem by choosing $M = \lfloor 2^{Rn} \rfloor$.

4.9 Finally, show that:

(a) $\lim\limits_{\beta \to +\infty} R_0(\beta) = \log 2$,

(b) $\lim\limits_{\beta \to 0+} \dfrac{R_0(\beta)}{C(\beta)} = \dfrac{1}{2}$.

In the next 7 problems we shall study the effects of quantization on the Gaussian channel. Denote by $C^{(r,s)}(\beta)$ the capacity of the Gaussian channel if we insist that the input X assumes at most r distinct values, and that the output Y has been quantized to one of at most s different values. Formally, we define, for each n, $C_n^{(r,s)}(\beta) = \sup\{I(X_1, \ldots, X_n; f(Y_1), \ldots, f(Y_n))\}$, where each X_i can assume at most r distinct values, $\sum_{i=1}^{n} E(X_i^2) \leq n\beta$, $Y_i = X_i + Z_i$, where the Z_i's are independent mean 0, variance σ^2 Gaussian random variables, and f is any function which assumes at most s different values. Then $C^{(r,s)}(\beta) = \sup_n n^{-1} C_n^{(r,s)}(\beta)$.

4.10 Show that $C^{(r,s)}(\beta) = C_1^{(r,s)}(\beta)$.

4.11 Show that $\lim\limits_{r,s \to \infty} C^{(r,s)}(\beta) = \frac{1}{2}\log(1 + \beta/\sigma^2)$. [Sketch of proof: First show that for a fixed value of r, $\lim\limits_{s \to \infty} C^{(r,s)}(\beta) = C^{(r,\infty)}(\beta)$ is given by the formula $C^{(r,\infty)}(\beta) = \sup\{I(X; Y)\}$, where the supremum is taken over all random variables X, assuming at most r different values, $E(X^2) \leq \beta$, and $Y = X + Z$ with Z an independent, mean 0, variance σ^2 Gaussian random variable. Using the formula $I(X; Y) = h(Y) - h(Z)$ (cf. Prob. 1.27) show that $C^{(r,\infty)}(\beta) \leq \frac{1}{2}\log(1 + \beta/\sigma^2)$. Next let X be a mean 0, variance β Gaussian random variable, let $I_1, \ldots, I_r$ be a partition of the real line into r intervals, and for each $i = 1, 2, \ldots, r$ choose $x_i \in I_i$ with $|x_i|$ as small as possible. Define the new random variable X' by $P\{X' = x_i\} = P\{X \in I_i\}$, $i = 1, 2, \ldots, r$. Then $C^{(r,\infty)}(\beta) \geq I(X'; X' + Z)$. Now, as $r \to \infty$ and the partition becomes increasingly fine, $h(X' + Z) \to h(X + Z) = \frac{1}{2}\log(\sigma^2 + \beta)$ by Lebesgue's dominated convergence theorem. Hence

$$\lim_{r \to \infty} C^{(r,\infty)}(\beta) = \tfrac{1}{2}\log(1 + \beta/\sigma^2).$$

4.12 For fixed r and s, show that $\lim\limits_{\beta/\sigma^2 \to \infty} C^{(r,s)}(\beta) = \min(\log r, \log s)$.

4.13 Show that, if $r \geq s$, then $C^{(r,s)}(\beta) = C^{(s,s)}(\beta)$. [Result due to H. Rumsey, Jr.]

4.14 If $\beta/\sigma^2 = \alpha^2$, show that $C^{(2,\infty)}(\beta) = \alpha^2 - \int_{-\infty}^{\infty} g(y)\log\cosh(\alpha^2 + \alpha y)\,dy$, where $g(y) = (2\pi)^{-1/2} e^{-y^2/2}$. Show that for small α, $C^{(2,\infty)}(\beta) = \alpha^2/2 - \alpha^4/4 + O(\alpha^6)$, and observe that this is virtually the same as the unquantized capacity $C(\beta) = \frac{1}{2}\log(1 + \alpha^2)$. Hence the folk theorem "For small signal-to-noise ratios, binary input quantization doesn't hurt" (but see Prob. 4.9b). Indeed for $\alpha \leq 1$ the ratio $C^{(2,\infty)}(\beta)/C(\beta)$ is always ≥ 0.97.

4.15 Find an expression for $C^{(2,2)}(\beta)$, and show that $\lim\limits_{\alpha \to 0} C^{(2,2)}(\beta)/C(\beta) = 2/\pi$. Hence the folk theorem "For small signal-to-noise ratios binary output quantization costs you 2 dB." [*Note*: A dimensionless ratio θ is equivalent to $10\log_{10}\theta$ "*dB's*."]

4.16 Using numerical methods, show that, for small α, $C^{(r,3)}(\beta) = 0.405\alpha^2 + O(\alpha^3)$, and $C^{(r,4)}(\beta) = 0.441\alpha^2 + O(\alpha^3)$, all $r \geq 2$.

4.17 [*Note*: The results of this problem can be used to compute the capacity of the "waveform channel" introduced on p. 95 ff. when the noise process $z(t)$ is not white.] Consider n "parallel" Gaussian channels:

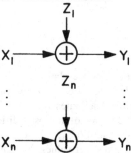

where $Z_1, Z_2, \ldots, Z_n$ are independent mean 0 Gaussian random variables, and var$(Z_i) = \sigma_i^2$. Define the "capacity" by $C(\beta) = \sup\{I(\mathbf{X}; \mathbf{Y})\}$ over all input distributions satisfying $\sum_{i=1}^{n} E(X_i^2) \leq n\beta$. Show that $C(\beta)$ is given parametrically by

$$\beta = \frac{1}{n}\sum_{i=1}^{n}(\sigma^2 - \sigma_i^2)_+,$$

$$C(\beta) = \frac{1}{2}\sum_{i=1}^{n}\left(\log \frac{\sigma^2}{\sigma_i^2}\right)_+$$

where the independent variable σ^2 ranges from $\min_i \sigma_i^2$ to β.

4.18 The object of this exercise is for you to derive a "forward" test channel that achieves $R(\delta)$ for the Gaussian source and mean-squared error distortion criterion. Recall what is needed: U is a mean 0, variance σ^2 Gaussian random variable, $\delta \leq \sigma^2$, and V must be found so that $E[(U - V)^2] = \delta$ and $I(U; V) = \frac{1}{2}\log \sigma^2/\delta$. Show that it is possible to take $V = \alpha U + Z$, where α is a constant and Z is Gaussian with variance σ^2, by choosing δ and σ^2 properly; or to take $V = \beta(U + Z)$, where β is constant and Z is Gaussian. These two "forward" test channels can be depicted like this:

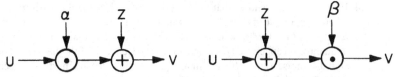

4.19 Consider a source $U_1, U_2, \ldots$ where the U_i's are independent random variables with common density function $p(u)$ and variance σ^2. Show that the rate distortion function of this source (see p. 101) satisfies $R(\delta) \leq \max\left(\frac{1}{2}\log \sigma^2/\delta, 0\right)$. [*Hint*: Use the forward test channel of Prob. 4.18.]

4.20 Show that $R_k(\delta) = 0$ if $\delta \geq \sigma^2$. (See proof of Theorem 4.4. Remember that $(\mathbf{U}, \mathbf{V})$ must have a continuous joint density, so defining $\mathbf{V} \equiv 0$ will not do.)

4.21 The object of this exercise is to show that, if condition (4.17c) were replaced by condition (4.17c'), the value of $R_k(\delta)$ would be unchanged. In the following sketch of a proof, it is assumed that $k = 1$ and $\delta < \sigma^2$. (The generalization to larger k and δ is quite easy and no hints will be given.) Your job is to supply the details. Thus let $R_1'(\delta)$ denote the minimum possible value of $I(U; V)$ subject to (4.17a), (4.17b), (4.17c'). To show that $R_1'(\delta) \leq \frac{1}{2} \log \sigma^2/\delta$, let (U, V) be a test channel achieving $R_1(\delta - \varepsilon)$ as described in the proof of Theorem 4.3. Choose $\varepsilon > 0$, and let V' be a deterministic function of V assuming only finitely many values such that $E[(U - V')^2] < \delta$. Then $I(U; V') \leq I(U; V) = R_1(\delta - \varepsilon)$, and so $R_1'(\delta) \leq \frac{1}{2} \log \sigma^2/(\delta - \varepsilon)$. Since this is true for all $\varepsilon > 0$, $R_1'(\delta) \leq \frac{1}{2} \log \sigma^2/\delta$. To prove $R_1'(\delta) \geq \frac{1}{2} \log \sigma^2/\delta$, let (U, V) be a test channel satisfying (4.17a), (4.17b), (4.17c') for which $I(U; V) < R_1'(\delta) + \varepsilon$. Let $V' = V + G$, where G is a mean 0, variance ε Gaussian random variable independent of V. Then the (U, V) test channel satisfies (4.17a) and (4.17c) and $E[(U - V')^2] \leq \delta + \varepsilon$. Since $I(U; V') \leq I(U; V)$, it follows that $R_1'(\delta) + \varepsilon > \frac{1}{2} \log \sigma^2/(\delta + \varepsilon)$. Since this is true for all ε, $R_1'(\delta) \geq \frac{1}{2} \log \sigma^2/\delta$.

4.22 Show that there are functions $\alpha(u)$, $\beta(v)$ which satisfy Eqs. (4.28), (4.29), and (4.30).

4.23 (The Shannon lower bound). Consider a source $U_1, U_2, \ldots$ where the U_i's are independent, identically distributed random variables with a common density function $p(u)$ and a common differential entropy $h(U)$. Let $f(x)$ be an arbitrary nonnegative even function of a real variable x, and define the rate-distortion function of the source with respect to the distortion measure $d(u, v) = f(u - v)$ as $R(\delta) = \inf\{I(U; V)\}$, the infimum being taken over all pairs of random variables (U, V), where U is distributed according to the density $p(u)$, (U, V) have a joint density function, and $E[f(U - V)] \leq \delta$. (Actually, to be consistent with the presentation of Chapter 4, we should define $R_k(\delta)$ analogously with Eq. (4.16), and $R(\delta) = \inf_k k^{-1} R_k(\delta)$. However, it would turn out that $R_k(\delta) = k R_1(\delta)$, so that the definition of this problem is really the same.) Define $\phi(\delta) = \sup\{h(X) : E[f(X)] \leq \delta\}$. Prove that $R(\delta) \geq R_L(\delta) = h(U) - \phi(\delta)$. *Hint*: First show that $\phi(\delta)$ is convex $\cap$ in δ. Then if $\delta(v) = \int f(u - v) p(u|v) du$, $I(U; V) = h(U) - h(U|V)$,

$$h(U|V) = \int q(v) \int p(u|v) \log \frac{1}{p(u|v)} \, du \, dv$$

$$\leq \int q(v) \phi[\delta(v)] \, dv$$

$$\leq \phi\left[\int q(v) \delta(v) \, dv\right]$$

$$\leq \phi(\delta).$$

4.24 (Continuation). With the aid of Prob. 1.35, evaluate the Shannon lower bound $R_L(\delta)$ in the following cases:

(a) $p(u) = \dfrac{1}{\sqrt{2\pi\sigma^2}} e^{-u^2/2\sigma^2}$, $f(x) = |x|^\alpha$, $\alpha = 1, 2$.

(b) $p(u) = \dfrac{1}{2A} e^{-|u|/A}$, $f(x) = |x|$.

(c) $p(u) = \dfrac{2}{\pi}(1 + u^2)^{-2}$, $f(x) = |x|$.

4.25 By constructing an appropriate test channel, show that $R(\delta) = R_L(\delta)$ for the source of Prob. 4.24b.

4.26 [*Note*: The results of this problem can be used as a starting point in the study of the rate-distortion function of a Gaussian source with memory. See Berger [13], Chapter 4.] Consider k "parallel" Gaussian sources, $U_1, U_2, \ldots, U_k$, that is, the U_i are independent, mean 0, variance σ_i^2 Gaussian random variables. Define $R(\delta) = \inf\{I(\mathbf{U}; \mathbf{V})\}$, where the infimum is over the pairs $(\mathbf{U}; \mathbf{V})$ with a joint density and $\sum_{i=1}^{k} E[(U_i - V_i)^2] \leqslant k\delta$. Show that $R(\delta)$ is given parametrically by

$$R(\delta) = \sum_{i=1}^{k} \frac{1}{2}\left(\log \frac{\sigma_i^2}{\sigma^2}\right)_+,$$

$$\delta = \sum_{i=1}^{k} \min(\sigma^2, \sigma_i^2),$$

where the parameter σ^2 ranges from 0 to $\max_i \sigma_i^2$.

4.27 Verify that the Gaussian channel is memoryless in the sense of the definition on p. 44.

Notes

1 (p. 96). For Johnson noise the spectral density is $N_0 = kT$, where $k =$ Boltzmann's constant $= 1.38 \times 10^{-23}$ joule/°K, and T is the effective noise temperature. Actually the spectral density is dependent on the frequency and is given by $P(f) = hf/(e^{hf/kT} - 1)$ watts/hertz, where f is the frequency in hertz and $h =$ Planck's constant $= 6.6 \times 10^{-34}$ joule-seconds. The assertion that $\int z(t)\phi_i(t)dt$ is a Gaussian random variable of variance $N_0/2$ is true only if the signal $\phi_i(t)$ has most of its energy confined to frequencies where the approximation $P(f) \approx N_0$ is valid (see Note 2). However, this approximation is typically valid up to very high frequencies indeed; for example, if $f < 0.02kT/h = 4 \times 10^8 T$, then $0.99N_0 < P(f) < N_0$. (In one case of practical interest, NASA's 64 meter antenna at Goldstone, California, $T \approx 25°K$, $f \approx 10^9$–10^{10}, and $f \approx 0.01kT/h$.) [Reference: Feynmann, Leighton, and Sands [5], Vol. 1, Chapter 41.]

2 (p. 97). This rather arbitrary definition of bandwidth deserves some elaboration. On physical grounds it is usually necessary to restrict the transmitted signals $x(t)$ to not involve frequencies above some fixed limit, say W hertz. Mathematically, this means that the Fourier transform $X(f) = \int_{-\infty}^{\infty} x(t)e^{-2\pi i f t}\, dt$ vanishes for $|f| > W$. If this is the case, the Fourier integral theorem implies $x(t) = \int_{-W}^{W} X(f)e^{2\pi i f t}\, df$, which in turn implies that x is an analytic function of t. But for any physically realizable signal

$x(t) = 0$ for $t < 0$, and this implies that $x(t)$ is identically 0. So instead of requiring $X(f)$ to vanish for $|f| > W$, we could pick $0 < \varepsilon < 1$ and make the weaker assumption that $x(t)$ has at most a fraction ε of its energy outside the band $[-W, W]$, that is, $\int_{-W}^{W} |X(f)|^2 df / \int_{-\infty}^{\infty} |X(f)|^2 df > 1 - \varepsilon$. This restriction puts a limit on the number $n = n(T, \varepsilon)$ of orthogonal signals that can be chosen on $[0, T]$; indeed, it can be shown that $\lim_{T \to \infty} n(T, \varepsilon)/2T = W$ for all $\varepsilon > 0$. This is why it is reasonable to *define* W as $n/2T$, rather than introduce the necessarily imprecise notion of band-limited functions. [See Gallager [17], Chapter 8.)

3 (p. 105). The probabilistic version of Schwarz's inequality invoked here is that if **X** and **Y** are random vectors defined on the same sample space, $E(\mathbf{X} \cdot \mathbf{Y})^2 \leq E(|\mathbf{X}|^2) E(|\mathbf{Y}|^2)$. (See Feller [4], Vol. II, §v.8.)

5

The source–channel coding theorem

In this chapter, which is the culmination of Part one, we apply the results of Chapters 1–4 to the problem of communicating the output of an information source over a noisy channel. Throughout this chapter, we shall take the block diagram shown in Fig. 5.1 as our paradigm.

Virtually any real communication system can be subsumed under the model of Fig. 5.1. The encoder block represents all the data processing (which may include quantization, modulation, etc., as well as coding for error control) performed on the source output before transmission. It is assumed that there exist integers k and n such that the system processes successive blocks of k source symbols independently and converts them into successive blocks of n channel input symbols. (For example k could be taken as the total number of source symbols emitted during the lifetime of the system, but for most systems k is much smaller than this.) The channel is assumed to accept channel input symbols at discrete intervals, and in response to emit channel output symbols, in one-to-one correspondence to the inputs.[1] The assumption that the channel is "discrete-time" is restrictive, but not very, since most "continuous-time" channels can be viewed as "discrete-time" with very little loss (cf. pp. 96–97). The "decoder" block represents all the data processing

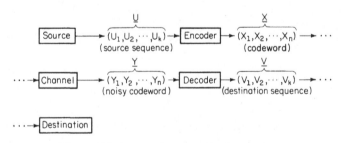

Figure 5.1 A general communication system.

112

performed on the channel output before delivery to the destination. It is assumed that the decoder processes successive blocks of n channel output symbols independently, and converts them into blocks of k destination symbols, the destination sequence $\mathbf{V} = (V_1, V_2, \ldots, V_k)$ being the system's estimate of the source sequence $\mathbf{U} = (U_1, U_2, \ldots, U_k)$.

We now pause to review the basic concepts of Part one, as they apply to the block diagram of Fig. 5.1. In the following discussion we shall refer to very vague and general channel and source models; the reader should be aware, however, that we have proved the required coding theorems only for certain special cases.

First, we consider the channel. We are given the channel's input alphabet A_X and the output alphabet A_Y; with each input symbol $x \in A_X$ there is associated a nonnegative real number $b(x)$, the "cost of sending x". More generally, for each $\mathbf{x} = (x_1, \ldots, x_n) \in A_X^n$, $b(\mathbf{x})$ is the cost of sending $\mathbf{x}$. Also, we are given the channel statistics, that is, for each n we are given a conditional probability distribution[2] that allows us to compute the distribution of the output $\mathbf{Y} = (Y_1, \ldots, Y_n)$ once we know the input $\mathbf{X} = (X_1, \ldots, X_n)$. For each positive integer n and $\beta \geqslant 0$, define $C_n(\beta)$ by

$$C_n(\beta) = \sup\{I(\mathbf{X}; \mathbf{Y}): E[b(\mathbf{X})] \leqslant n\beta\}. \tag{5.1}$$

In (5.1) the supremum is taken over all n-dimensional random vectors $\mathbf{X} = (X_1, \ldots, X_n)$ taking values in A_X^n and satisfying $E[b(\mathbf{X})] \leqslant n\beta$, and $\mathbf{Y}$ is the n-dimensional random vector taking values in A_Y^n which describes the channel's output if $\mathbf{X}$ is used as the input. In this context $\mathbf{X}$ is called an n-dimensional, β-admissable *test source*. The capacity–cost function $C(\beta)$ of the channel is defined by

$$C(\beta) = \sup\left\{\frac{1}{n} C_n(\beta): n = 1, 2, \ldots\right\}. \tag{5.2}$$

In general $C(\beta)$ is a continuous, convex $\cap$, and monotonically increasing function of $\beta \geqslant \beta_{\min} = \inf\{b(x): x \in A_X\}$. The crucial fact about $C(\beta)$ is the following. (As usual, all implied logs are base 2.)

Channel Coding Theorem: Fix $\beta_0 \geqslant \beta_{\min}$. Then for any triple (β, C', ε) with $\beta > \beta_0$, $C' < C(\beta_0)$, $\varepsilon > 0$, there exists for all sufficiently large n, a code $\{\mathbf{x}_1, \mathbf{x}_2, \ldots, \mathbf{x}_M\}$ of length n and a decoding rule such that:

(a) each codeword satisfies $b(\mathbf{x}_i) \leqslant n\beta$,
(b) $M \geqslant 2^{\lceil C'n \rceil}$,
(c) $P_E^{(i)} < \varepsilon$, $i = 1, 2, \ldots, M$.

($P_E^{(i)}$ is the probability of decoder failure, given that $\mathbf{x}_i$ is transmitted.)

Next we turn to the source. We are given a source alphabet A_U, a destination alphabet A_V, and for each pair $(u, v) \in A_U \times A_V$ a nonnegative real number $d(u, v)$, the "distortion" that results if the source symbol u arrives at the destination as v. More generally, for each positive integer k and each pair $(\mathbf{u}, \mathbf{v}) \in A_U^k \times A_V^k$, $d(\mathbf{u}, \mathbf{v})$ measures the total distortion if the source sequence $\mathbf{u} = (u_1, \ldots, u_k)$ arrives as the destination sequence $\mathbf{v} = (v_1, \ldots, v_k)$. Also, we are given the source statistics, that is, for each k we are given the distribution of $\mathbf{U} = (U_1, \ldots, U_k)$, the random vector that describes k successive source outputs.[2] For each positive integer k and $\delta \geqslant \delta_{\min}$ define $R_k(\delta)$ by

$$R_k(\delta) = \inf\{I(\mathbf{U}, \mathbf{V}): E[d(\mathbf{U}, \mathbf{V})] \leqslant k\delta\}, \tag{5.3}$$

where the infimum is taken over all pairs $(\mathbf{U}; \mathbf{V}) = (U_1, \ldots, U_k; V_1, \ldots, V_k)$ of k-dimensional random vectors taking values in $A_U^k \times A_V^k$ such that the marginal distribution of $\mathbf{U}$ is that given by the source statistics and $E[d(\mathbf{U}; \mathbf{V})] \leqslant k\delta$. Thus what we are really varying in (5.3) is the conditional distribution of $\mathbf{V}$, given $\mathbf{U}$, and this conditional distribution is called a k-dimensional *test channel*. The infimum in (5.3) is said to take place over all k-dimensional, δ-admissible test channels. The *rate-distortion function* of the source is defined by

$$R(\delta) = \inf\left\{\frac{1}{k}R_k(\delta): k = 1, 2, \ldots\right\}. \tag{5.4}$$

In general $R(\delta)$ is a continuous, convex $\cup$, and monotonically decreasing function of $\delta > \delta_{\min}$. The crucial fact about $R(\delta)$ is the following theorem.

Source Coding Theorem: Fix $\delta > \delta_{\min}$. Then for any pair (δ', R') with $\delta' > \delta$ and $R' > R(\delta)$, for all sufficiently large k there exists a source code $C = \{\mathbf{v}_1, \mathbf{v}_2, \ldots, \mathbf{v}_M\}$ of length k such that:

(a) $M \leqslant 2^{\lfloor R'k \rfloor}$,
(b) $d(C) < \delta'$.

(Recall that $d(C)$, the average distortion of the source code C, is defined by $d(C) = k^{-1}E[d_{\min}(\mathbf{U})]$, where, for each $\mathbf{u} \in A_U^k$, $d_{\min}(\mathbf{u}) = \min\{d(\mathbf{u}, \mathbf{v}_j): j = 1, 2, \ldots, M\}$.)

We now return to the block diagram of Fig. 5.1. Observe that once the encoder and the decoder have been specified, we can view $\mathbf{U}$, $\mathbf{X}$, $\mathbf{Y}$, and $\mathbf{V}$ as random vectors, for $\mathbf{U}$ is defined to be a random vector in the first place, and the conditional probabilities of **X** given **U**, of **Y** given **X**, and of **V** given **Y**, are respectively given by the encoder's design, the channel statistics, and the decoder's design. We now define three parameters that will tell us much about how well the system performs.

The average cost:

$$\overline{\beta} = \frac{1}{n} E[b(\mathbf{X})]. \tag{5.5}$$

The average distortion:

$$\overline{\delta} = \frac{1}{k} E[d(\mathbf{U}; \mathbf{V})]. \tag{5.6}$$

The rate of transmission:

$$\overline{r} = \frac{k}{n}. \tag{5.7}$$

The physical significance of these parameters should be obvious: $\overline{\beta}$ tells us how much the system costs to run (on a per-channel input basis); $\overline{\delta}$ indicates how reliably the system transmits the source output; and $\overline{r}$ measures how fast (in units of source symbols per channel symbol) the system is transmitting information. For a given source and channel, we would like to design a system with small $\overline{\beta}$, small $\overline{\delta}$, and large $\overline{r}$, but of course these are conflicting goals. The following theorem, which is the central result of information theory, tells us exactly what is and is not possible.

Theorem 5.1 *(the source–channel coding theorem). For a given source and channel:*

(a) *The parameters $\overline{\beta}$, $\overline{\delta}$, and $\overline{r}$ must satisfy*

$$\overline{r} \leq \frac{C(\overline{\beta})}{R(\overline{\delta})}.$$

(b) *Conversely, given numbers $\beta > \beta_{\min}$, $\delta > \delta_{\min}$, and $r < C(\beta)/R(\delta)$, it is possible to design a system of the type depicted in Fig. 5.1 such that $\overline{\beta} \leq \beta, \overline{\delta} \leq \delta$, and $\overline{r} \geq r$.*

Proof We shall first prove (a), and begin by remarking that the proof will depend only on definitions (5.2) and (5.4) and not on the coding theorems. Thus the first part of the theorem (sometimes called the "converse" or "weak converse" of the coding theorem) will be seen to be true in extremely general circumstances.

Thus suppose we are given a system of the type depicted in Fig. 5.1. Our first observation is that the sequence $(\mathbf{U}, \mathbf{X}, \mathbf{Y}, \mathbf{V})$ of random vectors is a Markov chain ($\mathbf{Y}$ depends on $\mathbf{U}$ only through $\mathbf{X}$, $\mathbf{V}$ depends on $\mathbf{X}$ only through $\mathbf{Y}$), and so by the data-processing theorem (Eq. (1.15))

$$I(\mathbf{U}; \mathbf{V}) \leq I(\mathbf{X}; \mathbf{Y}). \tag{5.8}$$

Next observe that since $E[b(\mathbf{X})] = n\bar{\beta}$ (see (5.5)), it follows from (5.1) that $I(\mathbf{X}; \mathbf{Y}) \leq C_n(\bar{\beta})$. On the other hand from (5.2), $C_n(\bar{\beta}) \leq nC(\bar{\beta})$, and so

$$I(\mathbf{X}; \mathbf{Y}) \leq nC(\bar{\beta}). \tag{5.9}$$

Also, since $E[d(\mathbf{U}; \mathbf{V})] = k\bar{\delta}$ (see (5.6)), it follows from (5.3) that $I(\mathbf{U}; \mathbf{V}) \geq R_k(\bar{\delta})$, and from (5.4) that $R_k(\bar{\delta}) \geq kR(\bar{\delta})$. Hence

$$I(\mathbf{U}; \mathbf{V}) \geq kR(\bar{\delta}). \tag{5.10}$$

Combining (5.8), (5.9), and (5.10), we arrive at conclusion (a) of the source–channel coding theorem.[3]

Of course as we have arranged things (a) is not a deep result; it is merely a corollary of the definitions of $C(\beta)$ and $R(\delta)$! The astonishing thing is that the converse (b) holds; but with the help of the channel coding theorem and the source coding theorem, it is not hard to prove.

So from now on we assume that $\beta > \beta_{min}$, $\delta > \delta_{min}$, and $r < C(\beta)/R(\delta)$ are given. Our job is to design an encoder and decoder so that the resulting $\bar{\beta}$, $\bar{\delta}$, and $\bar{r}$ satisfy $\bar{\beta} \leq \beta, \bar{\delta} \leq \delta, \bar{r} \geq r$.

First the technical preliminaries. Select numbers β_0, δ_0, δ_1, C', and R' satisfying:

$$\beta_{min} \leq \beta_0 < \beta, \tag{5.11}$$

$$\delta_{min} \leq \delta_0 < \delta_1 < \delta, \tag{5.12}$$

$$C' < C(\beta_0), \tag{5.13}$$

$$R' > R(\delta_0), \tag{5.14}$$

$$r < C'/R'. \tag{5.15}$$

(That this can always be done is the content of Prob. 5.2.)

Our encoder will be designed in two stages, a source encoder and a channel encoder, as shown in Fig. 5.2. We now describe the source encoder of Fig. 5.2. According to the source coding theorem, for sufficiently large k_0 there exists a source code C of length k_0 with M_1 codewords, where

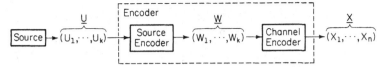

Figure 5.2 The general design of the encoder.[4]

$$M_1 \leq 2^{\lfloor R' k_0 \rfloor} \tag{5.16}$$

and

$$d(C) < \delta_1. \tag{5.17}$$

For a certain integer m (to be specified later), we take $k = k_0 m$. The source encoder of Fig. 5.2 partitions the source sequence $\mathbf{U}$ of length k into m blocks of length k_0, and outputs the m source codewords corresponding to this sequence of source blocks. Thus the intermediate vector $\mathbf{W} = (W_1, \ldots, W_k)$ will always be a sequence of m codewords from the given source code C. In particular, $\mathbf{W}$ assumes at most $M_1^m \leq 2^{m k_0 R'}$ distinct values. That completes the description of the source encoder (except for the specification of m).

To describe the channel encoder, we first need to define the *worst-case distortion* of the source code C. For each source sequence $\mathbf{u} \in A_U^{k_0}$, let $d_{\max}(\mathbf{u}) = \max\{d(\mathbf{u}, \mathbf{v}_i): \mathbf{v}_i \in C\}$. Thus, if the source emits $\mathbf{u}$ and the source encoding process somehow goes haywire, $d_{\max}(\mathbf{U})$ represents the worst possible resulting distortion. The worst-case distortion for the code C is defined to be the average value of this, normalized to a per-symbol basis:

$$D(C) = \frac{1}{k_0} E[d_{\max}(\mathbf{U})], \tag{5.18}$$

where the expectation is computed with respect to the source statistics, which describe the distribution of $\mathbf{u}$. (If the source alphabet A_U is infinite, it is not immediately obvious that $D(C)$ is finite; see Prob. 5.6.) Now define

$$\varepsilon = (\delta - \delta_1)/D(C) \qquad (\textit{see } (5.12)). \tag{5.19}$$

We can now describe the channel encoder of Fig. 5.2. For each $m = 1, 2, \ldots,$ define the integer n_m by

$$n_m = \lceil m k_0 R'/C' \rceil. \tag{5.20}$$

Then, according to the channel coding theorem, for all sufficiently large m there exists a channel code $\{\mathbf{x}_1, \ldots, \mathbf{x}_{M_2}\}$ of length n_m, together with a decoding rule, such that:

$$b(\mathbf{x}_i) \leqslant n_m \beta, \qquad i = 1, 2, \ldots, M_2, \qquad (5.21)$$

$$M_2 \geqslant 2^{\lceil C' n_m \rceil} \geqslant 2^{m k_0 R'}, \qquad (5.22)$$

$$P_E^{(i)} < \varepsilon, \qquad i = 1, 2, \ldots, M_2. \qquad (5.23)$$

In addition we assume m is so large that

$$k_0 m / n_m \geqslant r. \qquad (5.24)$$

(This is possible because of (5.15).) The channel encoder of Fig. 5.2 maps each of the sequences $\mathbf{W} = (W_1, \ldots, W_k)$ of m source codewords onto a distinct channel codeword of length $n = n_m$. This is possible since, as we observed earlier, there are at most $2^{m k_0 R'}$ distinct $\mathbf{W}$'s and by (5.22) at least this many distinct codewords.

Having completely described the encoder in Fig. 5.1, we have immediately from (5.24) that $\bar{r} = k/n = k_0 m / n_m \geqslant r$, and from (5.21) that $n\bar{\beta} = E[b(\mathbf{X})] \leqslant n\beta$. It remains to show that the average distortion $\bar{\delta}$ is $\leqslant \delta$, and to do this we must of course design the decoder in Fig. 5.1. Its general design is depicted in Fig. 5.3. The channel decoder illustrated is the decoder whose existence is promised by the channel coding theorem; it maps the noisy codeword $\mathbf{Y} = (Y_1, \ldots, Y_n)$ into one of the channel codewords, say $\mathbf{Z} = (Z_1, \ldots, Z_n)$. Finally, the source decoder in Fig. 5.3 accepts as input the channel codeword $\mathbf{Z}$ and outputs the sequence $\mathbf{V} = (V_1, \ldots, V_k)$ of m source codewords corresponding to $\mathbf{Z}$, if there is one. (This correspondence is the one given by the source encoder of Fig. 5.2.) If no sequence of m source codewords could have given rise to $\mathbf{Z}$, let us assume the source decoder outputs a fixed "null sequence" $\mathbf{v}_0 = (v_1^0, \ldots, v_k^0)$ of k symbols from the destination alphabet A_V.

Having completely described the communication system of Figs. 5.1–5.3, our remaining task is to compute the average distortion $\bar{\delta} = k^{-1} E[d(\mathbf{U}; \mathbf{V})]$. To do this we introduce a new random variable B, which tells us whether or not the channel decoder in Fig. 5.3 succeeds. Thus

$$B = \begin{cases} 0 & \text{if decoder succeeds;} \quad \mathbf{Z} = \mathbf{X} \\ 1 & \text{if decoder fails;} \quad \mathbf{Z} \neq \mathbf{X}. \end{cases}$$

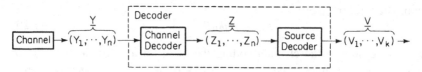

Figure 5.3 The general design of the decoder.[4]

Then:

$$E[d(\mathbf{U}; \mathbf{V})] = E[d(\mathbf{U}; \mathbf{V})|B = 0]P\{B = 0\} + E[d(\mathbf{U}; \mathbf{V})|B = 1]P\{B = 1\}. \tag{5.25}$$

Now if $B = 0$, the decoder has succeeded, and so the destination sequence $\mathbf{V}$ is identical to the output of the source encoder. Hence if $\mathbf{U} = [\mathbf{U}^{(1)}, \mathbf{U}^{(2)}, \ldots, \mathbf{U}^{(m)}]$ is the decomposition of $\mathbf{U}$ into m blocks of k_0 source symbols, then, if $B = 0$, $d(\mathbf{U}; \mathbf{V}) = \sum_{l=0}^{m} d_{\min}[\mathbf{U}^{(l)}]$, and so $E[d(\mathbf{U}; \mathbf{V})|B = 0] = mE[d_{\min}(\mathbf{U})] < k_0 m\delta_1$ by (5.17). Since $P\{B = 0\} \le 1$, we have

$$E[d(\mathbf{U}; \mathbf{V})|B = 0]P\{B = 0\} < k\delta_1. \tag{5.26}$$

If, on the other hand, the decoder fails ($B = 1$), then in any event $d(\mathbf{U}; \mathbf{V}) \le \sum_{l=0}^{m} d_{\max}[\mathbf{U}^{(l)}]$ (recall that $d_{\max}(\mathbf{u}) = \max\{d(\mathbf{u}, \mathbf{v}_j): j = 1, \ldots, M_1\}$), and so $E[d(\mathbf{U}; \mathbf{V})|B = 1] \le mE[d_{\max}(\mathbf{U})|B = 1]$. But $E[d_{\max}(\mathbf{U})|B = 1] = \sum_{i=1}^{M_2} E[d_{\max}(\mathbf{U})|B = 1, \mathbf{X} = \mathbf{x}_i] \cdot P\{\mathbf{X} = \mathbf{x}_i|B = 1\}$; and since the random variables $\mathbf{U}, \mathbf{X}, B$ form a Markov chain (because the decoder's success or failure depends only on the transmitted codeword $\mathbf{X}$ and not on the particular source sequence $\mathbf{U}$ that afforded $\mathbf{X}$), this implies that $E[d_{\max}(\mathbf{U})|B = 1] = \sum_{i=1}^{M_2} E[d_{\max}(\mathbf{U})|\mathbf{X} = \mathbf{x}_i] \cdot P\{\mathbf{X} = \mathbf{x}_i|B = 1\}$. Now $P\{\mathbf{X} = \mathbf{x}_i|B = 1\} = P\{B = 1|\mathbf{X} = \mathbf{x}_i\} \cdot P\{\mathbf{X} = \mathbf{x}_i\}/P\{B = 1\}$. But, from (5.23), $P\{B = 1|\mathbf{X} = \mathbf{x}_i\} = P_E^{(i)} < \varepsilon$ for all i, and so:

$$P\{B = 1\}E[d(\mathbf{U}; \mathbf{V})|B = 1]$$

$$\le m\varepsilon \sum_{i=1}^{M_2} E[d_{\max}(\mathbf{U})|\mathbf{X} = \mathbf{x}_i]P\{\mathbf{X} = \mathbf{x}_i\}$$

$$= m\varepsilon E[d_{\max}(\mathbf{U})]$$

$$= k(\delta - \delta_1), \quad \text{from (5.19).} \tag{5.27}$$

Combining (5.25), (5.26), and (5.27), we obtain $\overline{\delta} = k^{-1}E[d(\mathbf{U}; \mathbf{V})] < \delta$, as promised. $\qquad\square$

Discussion. We assume the average cost $\overline{\beta}$ is fixed (or that the cost function $b(x)$ is identically zero, in which case the capacity function $C(\beta)$ is a constant $C_{\max}$, the channel capacity), and study the tradeoff between the rate $\overline{r}$ and the average distortion $\overline{\delta}$ of the communication system of Fig. 5.1. The assertions of the source–channel coding theorem are summarized in Fig. 5.4. Assertion (a) of the source–channel coding theorem is that for any realizable system the

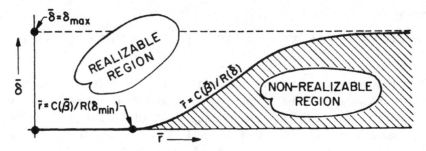

Figure 5.4 A graph of the source–channel coding theorem.

pair $(\bar{r}, \bar{\delta})$ must lie somewhere in the "realizable region", including the heavy black boundary. Assertion (b) is that any point in the realizable region, excluding the boundary, can be achieved. In general it is not known which points on the boundary can be achieved, and this is a question of considerable theoretical interest (see Prob. 5.4). However, from a practical standpoint it is unimportant, since a communications engineer desiring a system whose parameters $(\bar{r}, \bar{\delta})$ lie on the boundary will surely be satisfied with one having parameters $(\bar{r}, \bar{\delta}(1 + 10^{-50}))$!

Problems

5.1 An experimenter wishes to design a system for communicating observations of a Gaussian random process via a binary symmetric channel which: accepts 100,000 bits per second, has a raw bit error probability of $\frac{1}{10^5}$; and for which transmitting "0" costs nothing but transmitting "1" costs 10^{-6} dollars. He plans to do this by sampling the process at a rate of R samples per second (the samples are mean 0, variance 1 Gaussian random variables), and encoding the samples before transmission. Suppose he can tolerate an average mean-squared error of at most δ and can spend an average of at most B dollars per day on the channel. Which of the following three sets of (B, δ, R) are in principle realizable?

B	δ	R
\$ 864	0.1	12,500
2592	0.2	150,000
4320	0.001	11,000

5.2 Show that it is possible to select numbers β_0, δ_0, δ_1, C', and R' satisfying Eqs. (5.11)–(5.15).

5.3 Consider communicating the output of a binary symmetric source that produces R bits per second over a "wideband" Gaussian channel (see p. 97). Denote the ratio P/R by E_b (its dimensions are joules per bit, and the ratio E_b/N_0 is called the bit signal-to-noise ratio). Show that if the communication is done via a scheme of the form depicted in Fig. 5.1, an overall bit error probability of P_e is achievable as long as $E_b/N_0 > [1 - H_2(P_e)] \log 2$, but not otherwise. However, if the code rate $\bar{r} = k/n$ must also satisfy $\bar{r} \geq r$, show that the minimum value of E_b/N_0 is $\{2^{2r[1-H_2(P_e)]} - 1\}/2r$.

5.4 Consider communicating the output of a binary symmetric source over a binary symmetric channel with raw bit error probability p. Referring to Fig. 5.4, show that every point on the boundary satisfying $\bar{\delta} = p$ is realizable. (Note that if $p = 0$ this part of the boundary is a continuum.)

5.5 Show that the condition $\beta > \beta_{\min}$ in part (b) of the source–channel coding theorem can be removed.

5.6 Show that the quantity $D(C)$ defined for a source code in Eq. (5.18) is finite, provided that for each $\mathbf{v} \in A_V^{k_0}$ the expectation $E[d(\mathbf{U}; \mathbf{v})]$ is finite. Then show that this condition is satisfied for the Gaussian source of Chapter 4.

5.7 (E. C. Posner). This problem illustrates the possible pitfalls involved in building a "factored" communication system to achieve what is promised by the source–channel coding theorem. Suppose we want to communicate the output of a binary symmetric source over a binary symmetric channel whose raw error probability is $\frac{1}{10}$. (Assume there is no cost associated with channel use.) Further assume that it is required to have $\bar{\delta} \leq 0.10$, $\bar{r} \geq 1.0$. According to our proof of Theorem 5.1, this can be done by first designing a source code for the BSS with $d(C) \approx 0.10$, and then designing an appropriate channel code for the BSC with very small error probability. Show, however, that there is a simpler system that achieves the desired $(\bar{\delta}, \bar{r})$.

5.8 The object of this problem is to show that, if it is desired to communicate the output of a Gaussian source over a Gaussian channel, what is promised by the source–channel coding theorem can sometimes be achieved without any coding. To be precise, consider the following figure:

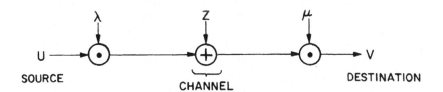

Here U is a mean 0, variance σ_U^2 Gaussian random variable which passes through a device (amplifier) which multiplies it by a constant λ. Then a mean 0, variance σ_Z^2 Gaussian random variable Z, independent from U, is added to it, and finally the result is passed through a device (attenuator) which multiplies by the constant μ. The input X to the channel must satisfy $E(X^2) \leq \beta$.

Show that there exists a distortion δ and constants λ, μ such that:

(a) $R(\delta) = C(\beta)$,

(b) $E[(U - V)^2] = \delta$,

where $R(\delta)$ is the rate-distortion function of the source, and $C(\beta)$ is the capacity–cost function of the channel. Hence show that the inequality $k/n \leq C(\beta)/R(\delta)$ can be achieved with equality with $k = n = 1$.

Notes

1 (p. 112). Notice that channels which mischievously insert or delete symbols into the transmitted stream are thus beyond our paradigm.

2 (pp. 113, 114). In order for the given definitions of $C(\beta)$ and $R(\delta)$ to be unambiguous, we must assume that the channel and source are *stationary*, that is, the statistics do not depend on when we begin to transmit the source's output or when we begin to use the channel.

3 (p. 116). Actually, it is not necessary to assume that the encoder and the decoder are deterministic devices in order to prove the inequality $k/n \leq C(\overline{\beta})/R(\overline{\delta})$. All that is required is that the encoder and decoder be such that the sequence $(\mathbf{U}, \mathbf{X}, \mathbf{Y}, \mathbf{V})$ forms a Markov chain; for example, an encoder such that $\mathbf{X}$ depends not only on $\mathbf{U}$, but on some locally generated random variables as well, would be acceptable.

4 (pp. 117, 118). The peculiar construction we used in the proof of the source–channel coding theorem, in which the encoders and decoders were factored into source and channel encoders and decoders, deserves comment. It can reasonably be argued that by doing the source and channel coding separately, we may end up with a system more complex than necessary (see Probs. 5.7 and 5.8, e.g.). However, from the standpoint of $(\overline{\beta}, \overline{\delta}, \overline{r})$ nothing is lost since every such triple achievable by any means is also achievable by a factored system.

6

Survey of advanced topics for Part one

6.1 Introduction

In this chapter we briefly summarize some of the important results in information theory which we have not been able to treat in detail. We shall give no proofs, but instead refer the interested reader elsewhere, usually to a textbook, sometimes to an original paper, for details.

We choose to restrict our attention solely to generalizations and extensions of the twin pearls of information theory, Shannon's *channel coding theorem* (Theorem 2.4 and its corollary) and his *source coding theorem* (Theorem 3.4). We treat each in a separate section.

6.2 The channel coding theorem

We restate the theorem for reference (see Corollary to Theorem 2.4).

Associated with each discrete memoryless channel, there is a nonnegative number C (called channel capacity) with the following property. For any $\varepsilon > 0$ and $R < C$, for large enough n, there exists a code of length n and rate $\geq R$ (i.e., with at least 2^{Rn} distinct codewords), and an appropriate decoding algorithm, such that, when the code is used on the given channel, the probability of decoder error is $< \varepsilon$.

We shall now conduct a guided tour through the theorem, pointing out as we go places where the hypotheses can be weakened or the conclusions strengthened. The points of interest will be the phrases *discrete memoryless channel, a nonnegative number C, for large enough n* and *there exists a code ... and ... decoding algorithm*. We shall also briefly discuss various converses to the coding theorem.

• *Discrete memoryless channel.* The theorem has also been proved for many other channel models. Perhaps the simplest example of such a channel is one for which the input alphabet A_X and the output alphabet A_Y are both equal to a finite abelian group A, and where the ith channel output Y_i is related to the ith channel input X_i by the equation

$$Y_i = X_i + Z_i,$$

where $Z_1, Z_2, \ldots$ form an ergodic random process taking values in the group A. The channel capacity can be defined by

$$C = \sup_n \left(\frac{1}{n} \max_{\mathbf{X}} I(\mathbf{X}; \mathbf{Y}) \right), \tag{6.1}$$

where the inner maximum is taken over all n-dimensional random vectors $\mathbf{X}$ taking values in A^n, and $\mathbf{Y} = (Y_1, \ldots, Y_n)$ is the random vector that describes the channel output if $\mathbf{X}$ is the input, that is, $\mathbf{Y} = \mathbf{X} + \mathbf{Z}$, where $\mathbf{Z} = (Z_1, \ldots, Z_n)$ consists of the first n components of the noise process. It is easy to see that the inner maximum is achieved when $\mathbf{X}$ is uniformly distributed on A^n, and so, since $I(\mathbf{X}; \mathbf{Y}) = H(\mathbf{Y}) - H(\mathbf{Z})$ (see Prob. 1.13),

$$C = \log q - \inf_n \frac{1}{n} H(Z_1, Z_2, \ldots, Z_n),$$

where q is the number of elements in A. It turns out that

$$\inf_n \frac{1}{n} H(Z_1, \ldots, Z_n) = \lim_{n \to \infty} \frac{1}{n} H(Z_1, \ldots, Z_n) = H_0.$$

This quantity is called the *entropy* of the process $Z_1, Z_2, \ldots$.

A proof of the coding theorem for this additive ergodic noise channel can be given along the lines of the proof we gave for Theorem 2.4. The only difficulty arises when we try to assert that for each $\varepsilon > 0$, the probability

$$P\left\{ \left| \frac{1}{n} I(\mathbf{x}; \mathbf{y}) - C \right| > \varepsilon \right\}$$

approaches 0 if n is large. Fortunately there is a theorem, McMillan's *asymptotic equipartition property*, which asserts that for each $\varepsilon > 0$, the probability

$$P\left\{ \left| \frac{1}{n} H(z_1, \ldots, z_n) - H_0 \right| > \varepsilon \right\}$$

approaches 0 if n is large. Since $I(\mathbf{x}; \mathbf{y}) = n \log q - H(z_1, \ldots, z_n)$, this is just what is needed.

There are many other discrete channels with memory for which the coding theorem has been proved; consult Gallager [17] or Wolfowitz [27] for details.

The Gaussian channel of Chapter 4 can be generalized as follows. Let $A_X = A_Y =$ the real numbers, and let $Y_i = X_i + Z_i$, where now $Z_1, Z_2, \ldots$ is a real-valued random process, and where there may also be some kind of input constraints. The most successful theorems of this kind occur when the Z_i's are independent (i.e., the channel is memoryless but not discrete), or when the Z_i's form a Gaussian process; see Gallager [17], Chapters 7 and 8.

All the channels we have discussed so far have been the same in the sense that a single stream of information flows in one direction only. Some of the most interesting extensions of the coding theorem, however, have been to *multiterminal* channels. The earliest such result, due to Shannon himself, concerns channels with feedback, that is, channels in which the transmitter can find out the output y_i corresponding to his input x_i. Shannon (see [25], pp. 119–120) proved the surprising result that adding feedback to a discrete memoryless channel does not change its capacity. What it does do is to make communication at rates less than capacity much easier (see the eight papers on pp. 373–436 in [25] and also Prob. 2.20).

Channels with feedback still have only one sender and one receiver. More recently, impressive results have been obtained for channels with several receivers or senders. For example consider the following example of what is called a *multiaccess channel*. Here there are two senders and one receiver. The channel is memoryless, and every unit of time each of the two senders is permitted to transmit one symbol from $\{0, 1\}$ over the channel. If the two inputs are labeled x_1, x_2, the corresponding output $y \in \{0, 1, ?\}$ is given by the following table:

x_1	x_2	y
0	0	0
0	1	?
1	0	?
1	1	1

(Here "?" is a special erasure symbol.) As a four-input, three-output channel, this channel is noiseless and has capacity $\log 3$. However if the two senders attempt to transmit independent information streams simultaneously over the channel, they will interfere with each other. Thus an important question is: For a given pair (R_1, R_2) of rates, is it possible for sender 1 to find a code of rate R_1, for sender 2 to find a code of rate R_2, and for the receiver to find a decoding rule such that he can correctly decode both messages with as small an error probability as desired? The answer turns out to be "Yes," iff the point (R_1, R_2) lies in the *capacity region* for the channel, which in this case is shown in Fig. 6.1.

For more general memoryless two-input, one-output channels, the situation is essentially the same. There is a convex region in the first quadrant, again called the capacity region, such that arbitrarily reliable communication is attainable for any pair of rates in the region. Similarly, a *broadcast channel* in one with one input and more than one, usually two, outputs. Again it is possible to prove the existence of a capacity region. (Consult Wyner [49] for a more detailed account of these multiterminal channels.)

• *A nonnegative number C.* Even for a discrete memoryless channel we saw in Chapter 2 that the calculation of channel capacity can be very difficult, unless the channel enjoys a high degree of symmetry. Recall that C is given by

$$C = \max_X I(X; Y),$$

the maximum being taken over all (one-dimensional) test sources X, that is, over all possible probability distributions on the channel input alphabet A_X. This variational problem is not trivial but is somewhat easier than might at first appear, since $I(X; Y)$ is a convex $\cap$ function of X, and hence any local maximum is also a global maximum. Using this fact as a starting point, Gallager [17] devotes two sections to the problem of finding channel capacity for a DMC. More recently, Arimoto [29] and Blahut [32] have given efficient numerical procedures for computing C.

For channels other than DMC's, the calculation of capacity is usually much more difficult, for the definition, Eq. (5.2), is in general not even effective. For channels with additive ergodic noise, the calculation of C, as we have seen, is equivalent to the calculation of the noise entropy. If the noise process is

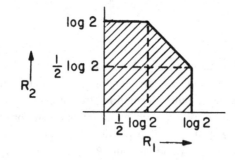

Figure 6.1 The capacity region for a simple multiple-access channel.

Markov, it is easy to compute the entropy (see Prob. 1.20), but in general very little is known. For multiterminal channels the corresponding problem, that is, the calculation of the capacity region, is at this writing virtually untouched.

• *For large enough n.* For discrete memoryless channels, the problem of determining the relationship between ε, the desired decoder error probability, R, the desired rate, and the smallest n such that there exists a code of length n that does the job has been the subject of some extermely intensive research.

Roughly speaking it has been discovered that for each $0 \leqslant R < C$, the best code of length n and rate R has a decoder error probability given approximately by

$$P_E \approx 2^{-nE(R)},$$

where $E(R)$, the channel's *reliability exponent*, is a convex $\cup$ function of R that looks something like the curve shown in Fig. 6.2. To be more precise, denote by $P_E(R, n)$ the error probability of the best code of length n and rate $\geqslant R$, and define

$$E(r) = \lim_{n \to \infty} -\frac{1}{n} \log P_E(R, n).$$

(Actually, nobody has ever proved that this limit exists, but in the following discussion we will ignore this technicality, it being understood that upper bounds on $E(r)$ are really upper bounds on $\overline{E}(R) = \lim \sup -n^{-1} \log P_E(R, n)$, etc.) Although $E(R)$ is not known exactly for any nontrivial DMC, there exist good upper and lower bounds to $E(R)$, which in general look as shown in Fig. 6.3. Thus there is a "critical rate," R_{crit}, above which the upper and lower bounds on $E(R)$ agree, but for $0 < R < R_{crit}$ there is a gap. For example, consider a BSC with raw error probability $p = .01$. Then the best known upper and lower bounds on $E(R)$ are plotted in Fig. 6.4. (Note the straight line supporting the lower bound in Fig. 6.4. This corresponds to the R_0 coding theorem, Probs. 2.21–2.26.)

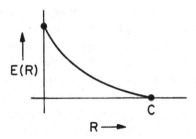

Figure 6.2 The approximate shape of the reliability exponent for a DMC.

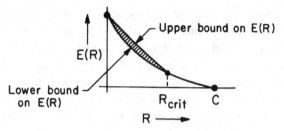

Figure 6.3 The approximate shape of the known bounds on $E(R)$.

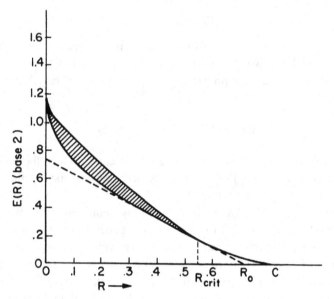

Figure 6.4 The best known bound on $E(R)$ for a BSC with $p = .01$. (Taken from McEliece and Omura [40].)

• *There exist a code ... and ... decoding algorithm.* The most unsatisfactory aspect of the channel coding theorem is its nonconstructive nature. The theorem asserts the existence of good codes, but tells neither how to find nor how to implement them. In a sense the subject we have called "coding theory" (and devoted the second part of this book to) can be viewed as a search for a constructive and practical channel coding theorem. However, none of the explicit coding schemes now known can be used to achieve very low error probabilities at rates very close to channel capacity. The coding theorist, in other words, has been willing to sacrifice the "$P_E \rightarrow 0$ for all

$R < C$" conclusion of the coding theorem for the sake of constructiveness. To an information theorist, however, that conclusion is sacred, and so a considerable amount of effort has been devoted to the problem of retaining it while at the same time strengthening the "there exist a code and decoding algorithm" part of the conclusion. Let us descirbe some of those efforts.

In the first place, it is possible to restrict attention to a relatively small class of codes and still prove the coding theorem. For example, the class of *linear codes* (see Chapter 7) contains codes that are good in the sense of the coding theorem (see e.g. Gallager [17], Section 6.2). Since linear codes are very easy to encode, this shows that encoding algorithms for good codes need not be impractically complex. The coding theorem has also been proved for *time-varying convolutional codes*; see Massey [21], Sections 6 and 7. Neither of these classes of codes, however, has a practical decoding algorithm.

What exactly do we mean by a "practical" algorithm? Let us adopt the view, widely accepted by computer scientists (see e.g. Aho, Hopcroft, and Ullman [1], Chapter 10) that an algorithm with n imputs is "practical" if its running time and storage requirements are both bounded by a polynomial in n. If we accept this viewpoint, the following result, due to Forney (see [25], pp. 90–95), is indisputably a "practical" coding theorem:

For any $R < C$, there exists a sequence of codes C_1, C_2, ... (code C_n has length n), each having rate $\geqslant R$, such that:

(a) $P_{E,n} < 2^{-nE_c(R)}$ *(here $P_{E,n}$ is the error probability of the nth code and $E_c(R)$ is a monotonically decreasing function of R which is positive for all $R < C$),*

(b) *The complexity of encoding and decoding C_n is $0(n^4)$.*

The key idea in Forney's proof is that of *concatenation*, in which the channel encoder and decoder are factored in a certain way. Consider an ordinary encoder and decoder for a given channel:

$$\underline{U} \rightarrow \boxed{\text{Encoder}} \rightarrow \underline{X} \rightarrow \boxed{\text{Channel}} \rightarrow \underline{Y} \rightarrow \boxed{\text{Decoder}} \rightarrow \underline{V}$$

Here $\mathbf{V} = (V_1, \ldots, V_k)$ is an estimate of $\mathbf{U} = (U_1, \ldots, U_k)$, which is usually, but not always, equal to $\mathbf{U}$. In other words, $\mathbf{V}$ is a noisy version of $\mathbf{U}$, and so the boxes "encoder–channel–decoder" can be viewed as a kind of "super-channel," or outer channel. We could then design a code for this outer channel and arrive at a block diagram like the one shown in Fig. 6.5. By choosing an inner code randomly and taking a particular kind of outer code (a Reed–

Figure 6.5 A concatenated coding system.

Solomon code; see Chapter 9), Forney was able to prove the low-complexity coding theorem quoted above.

Forney's results greatly diminished the gap between information theory (which shows what it should be possible to do) and coding theory (which shows what it is practical to do). Unfortunately, while Forney's theorem is "practical," it is still not "constructive" in the sense that it does not tell us how to find the appropriate inner code. Thus the problem of finding an explicitly constructible, practical sequence of codes for which P_E approaches 0 exponentially for all $R < C$ remains open. If someone ever produces such a theorem, it will unite the subjects of information theory and coding theory (and probably put both out of business).

- *Converses to the coding theorem.* Traditionally, converses to the coding theorem have been classified as "weak converses" or "strong converses," but this nomenclature is misleading, since the strong converses do not imply the weak ones.

We have already seen (p. 115) the weak converse. It says that if we attempt to communicate the output of, say, a binary symmetric source over a channel at a rate exceeding the channel's capacity, the resulting bit error probability is bounded away from 0. For the source–channel coding Theorem 5.1 (see also discussion on p. 84) tells us that in a communication system that fits the model of Fig. 5.1 the inequality $k/n \leqslant C/R(\delta)$ must hold. (We have specialized to a channel with no input costs, and a binary symmetric source for which $R(\delta) = 1 - H_2(\delta)$ bits.) The ratio k/n measures the rate of transmission; clearly if k/n exceeds C, then $R(\delta) < 1$, which forces δ, the bit error probability, away from 0. The proof of this result depends only on the definition of channel capacity in terms of mutual information (see Eq. (6.1)), and so it holds for very general channels.

The *strong converse* is more properly a converse to the coding theorem stated on p. 123, which says that for $R < C$ the error probability of the best code of length n and rate R approaches 0 as $n \to \infty$. The strong converse says that for $R > C$ the error probability of the best code approaches 1 as $n \to \infty$. (See Viterbi and Omura [26], Chapter 3.) The current strongest form shows that P_E approaches 1 exponentially, that is,

$$P_E \geq 1 - 2^{-nE_A(R)}, \quad R > C,$$

where $E_A(R)$ is *Arimoto's error exponent*, which has the general shape shown in Fig. 6.6.

For the special case of a BSC with raw error probability $p = 0.1$, Fig. 6.7 gives an accurate picture of $E_A(R)$.

6.3 The source coding theorem

This section will follow the same general outline as Section 6.2. However, the reader is warned from the outset that despite the strong similarity (or rather duality) between channel coding and source coding that we have emphasized, the subject of source coding or *rate-distortion theory* is by a wide margin more difficult. Hence the subject is considerably less advanced at this time.

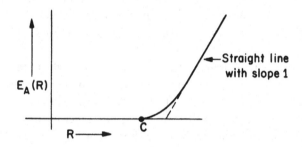

Figure 6.6 The general shape of Arimoto's exponent $E_A(R)$.

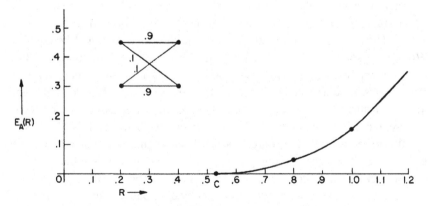

Figure 6.7 $E_A(R)$ for a BSC with $p = 0.1$.

We now restate the source coding Theorem 3.4 for reference.

Associated with each discrete memoryless source and single-letter distortion measure, there is a nonnegative, convex $\cup$, decreasing function $R(\delta)$, defined for all $\delta \geqslant \delta_{\min}$, with the following property. For any $\varepsilon > 0$ and $\delta \geqslant \delta_{\min}$, for large enough k there exists a source code of length k with at most $2^{k[R(\delta)+\varepsilon]}$ codewords, and average distortion $< \delta + \varepsilon$.

Again we conduct a tour, stopping at the phrases *discrete memoryless source, a function $R(\delta)$, for large enough k*, and *there exists a source code*. We shall also briefly discuss *converses* to the source coding theorem.

• *Discrete memoryless source.* The most natural way to extend the source coding theorem is to prove it for a wider class of sources than the discrete memoryless ones. The theorem has in fact been extended in this way to a vast number of sources. Indeed, practically any source which can be modeled as a stationary random process ..., x_{-1}, x_0, x_1, x_2, ... has been studied. Unfortunately, however, the rate-distortion function $R(\delta)$ can be calculated effectively for relatively few of these source models. (See Berger [13], and also p. 34.)

We saw in Section 6.2 that quite a lot is known about the capacity region of certain multiterminal channels. Much less is known about source coding for multiple users, but one result, the Slepian–Wolf theorem on noiseless coding of correlated sources, is worth mentioning. To describe the Slepian–Wolf theorem, we must pause to point out an important implication of the source coding theorem (to which incidentally, Chapter 11 is entirely devoted!).

Let X_1, X_2, ... be an independent, identically distributed sequence of random variables taking values in a set A. Think of this sequence as an information source relative to the Hamming distortion function, that is, take the destination alphabet as A also, and define $d(x, y) = 1$ if $x \neq y$, $= 0$ if $x = y$. Then as we saw on p. 80, in this case the value $R(0)$ will be equal to $H(X)$, the source entropy. Thus for this special case the source coding theorem implies that, for any $\varepsilon > 0$ and $R > H(X)$, for sufficiently large k there exists a source code of length k with at most 2^{kR} codewords, and average distortion $< \varepsilon$. In other words, the source can be essentially perfectly represented using only about $H(X)$ bits per sample. This much is of course a simple consequence of the source coding theorem, and in Chapter 11 we will give explicit techniques for doing this *noiseless coding*, as it is called.

Now let us turn to the results of Slepian and Wolf. Let (X_i, Y_i), $i = 1, 2, 3, \ldots$, be an indpendent, identically distributed sequence of pairs of discrete random variables with common distribution function $p(x, y)$, common entropies $H(X)$, $H(X, Y)$, $H(X|Y)$, etc. The problem is to find independent source codes for the X and Y sequences, such that the resulting distortion is negligible. Specifically, we want a configuration such as the one depicted in Fig. 6.8. The vector $\mathbf{X}$ is encoded into the codeword $f(\mathbf{X})$, which can assume any one of M_X values. Similarly $g(\mathbf{Y})$ can assume any one of M_Y values. The *rates* are defined as

$$R_X = \frac{1}{k} \log M_X,$$

and

$$R_Y = \frac{1}{k} \log M_Y,$$

and the *error probability* is

$$P_E = P\{\hat{\mathbf{X}} \neq \mathbf{X} \text{ or } \hat{\mathbf{Y}} \neq \mathbf{Y}\}.$$

A pair of rates (R_X, R_Y) is said to be *admissible* iff for every $\varepsilon > 0$ there exists a configuration of the sort depicted in Fig. 6.8 with $P_E < \varepsilon$. Slepian and Wolf (see [25], pp. 450–457) showed that the set of admissible rates is as depicted in Fig. 6.9.

The significance of the Slepian–Wolf theorem is perhaps not immediately obvious. But consider: if the encoders in Fig. 6.8 were combined into one, the problem would reduce to the error-free encoding of a sequence of independent random vectors with entropy $H(X, Y)$, and so the admissible region would be the half-plane bounded by the line $R_X + R_Y = H(X, Y)$. If the encoders are then disconnected, the size of the admissible region cannot increase, and so the S.–W. admissible region must lie entirely above this line. Also note that if, for example, Y is known perfectly at the decoder, the remaining uncertainty about X is $H(X|Y)$, and so every point in the admissible region must satisfy $R_X \geq H(X|Y)$, and similarly $R_Y \geq H(Y|X)$. Thus the admissible region must be a subset of the region indicated in Fig. 6.9. The remarkable fact is that it is this large.

More recently, Wyner and Ziv [50] have investigated the best possible

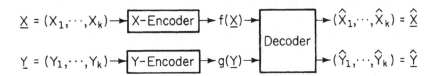

Figure 6.8 Block diagram for the Slepian–Wolf theorem.

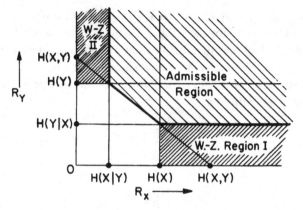

Figure 6.9 The set of admissible rates.

performance of the configuration in Fig. 6.8 when the rates R_X and R_Y lie in the region marked "W.–Z. region" in Fig. 6.9. In W.–Z. region I X can be reproduced exactly, since $R_X \geqslant H(Z)$. But since $R_Y < H(Y|X)$, inevitably there will be distortion in the reception of Y. Wyner and Ziv have determined the minimum possible distortion.

The Slepian–Wolf and Wyner–Ziv results adumbrate a deep multiuser generalization of rate-distortion theory.

• *A functon $R(\delta)$*. We have already seen in Chapter 3 that even for a discrete memoryless source, unless a very large amount of symmetry is present, it is quite difficult to compute the rate-distortion function $R(\delta)$. However, Blahut [32] (see also Viterbi and Omura, [26], Appendix 7A) has discovered a rapidly convergent algorithm for computing $R(\delta)$ numerically, and so the problem of computing $R(\delta)$ in this case is widely regarded as solved.

For more general sources and distortion measures, however, the problem of computing $R(\delta)$ is much more difficult. Perhaps the most impressive general class of sources for which a tractable formula for $R(\delta)$ is known is the class of *disccrete-time stationary Gaussian* sources, relative to the mean-squared error criterion. Here the source is modeled as a stationary Gaussian sequence of random variables $\ldots, X_{-1}, X_0, X_1, X_2, \ldots$. The rate distortion function $R(\delta)$ is given parametrically by

$$\delta(\theta) = \frac{1}{2\pi} \int_{-\pi}^{\pi} \min(\theta, S(\omega))\, d\omega,$$

$$R(\theta) = \frac{1}{4\pi} \int_{-\pi}^{\pi} \max\left(0, \log \frac{S(\omega)}{\theta}\right) d\omega,$$

where $S(\omega)$ is the spectral density of the process, and θ ranges through the interval $[0, \text{ess. sup } S(\omega)]$. In the special case that the X_i's are independent, with common variance σ^2, $S(\omega) = \sigma^2$ for all ω, and this theorem reduces to $\delta(\theta) = \theta$, $R(\theta) = \frac{1}{2}\log\sigma^2/\theta$, $0 \le \theta \le \sigma^2$, that is, to Theorem 4.4. For a proof and several examples see Berger [13], Chapter 4.

Unfortunately, for most sources with memory there is no known effective formula for $R(\delta)$. (There is always a noneffective formula, which is essentially the one given by Eq. (5.4).) For example, there is apparently no known discrete source with memory for which $R(\delta)$ can be computed exactly. However, there exist many techniques for bounding $R(\delta)$, and we again refer the interested reader to Berger [13].

• *For large enough k.* We saw in Section 2 that for a DMC the channel error probability P_E can be made to approach 0 exponentially with increasing block length. The analogous question for the source coding theorem is, How quickly does the distortion of an optimum source code of length k and rate $\le R(\delta)$ approach δ? In this direction Pilc [43] (see also Gallager [17], Section 9.3) has shown for a DMS that, if $d_{\min}(k)$ denotes the minimum possible average distortion of a source code of length k and rate $\le R(\delta)$, then

$$d_{\min}(k) = \delta + 0\left(\frac{\log k}{k}\right).$$

Thus the convergence of $\delta_{\min}(k)$ to δ is much less rapid than the convergence of P_E to 0.

• *There exists a source code.* The source coding theorem suffers from the same difficulty as the channel coding theorem, namely, it is nonconstructive. Thus a considerable amount of research has been devoted to the problem of finding a class of easily implementable source codes for which the source coding theorem can be proved. This research has led to source coding theorems of the form "For any $\varepsilon > 0$ and $\delta \ge \delta_{\min}$, there exists a source code of type T with rate $\le R(\delta) + \varepsilon$ and average distortion $\le \delta + \varepsilon$." Such theorems have been proved where type T is the set of *linear codes* (see Chapter 7) or *tree codes* (see Chapter 10). For details, consult Berger [13], chapter 6, or Viterbi and Omura [26], chapter 7. We will have more to say about the performance of certain explicit linear and tree codes as source codes in Chapter 12.

• *Converses.* There is really only one kind of converse of the source coding theorem, and we have already seen it in disguise (p. 84). We did not state it explicitly previously because our main concern has been the application of the source coding theorem to the problem of communicating the source output over a given channel. The following result, which we state for completeness, is really just a special case of conclusion (a) of the source–channel coding Theorem 5.1 where the channel is noiseless. The converse to the source coding theorem is as follows:

If the average distortion of a source code is $\leq \delta$, its rate must be $\geq R(\delta)$.

To prove this theorem, let the given code be of length k, and denote it by $\{v_1, v_2, \ldots, v_M\}$. Consider the block diagram of Fig. 6.10, where the source encoder maps each soruce sequence $\mathbf{u} = (u_1, \ldots, u_k)$ onto a codeword $\mathbf{v}_i$ for which $d(\mathbf{u}, \mathbf{v}_i)$ is as small as possible. Now since the code's average distortion is $\leq \delta$, we have that $E[d(\mathbf{U}; \mathbf{V})] \leq k\delta$, and so (see Eqs. (3.3) and (5.10)) $I(\mathbf{U}; \mathbf{V}) \geq R_k(\delta) \geq kR(\delta)$. But also $I(\mathbf{U}; \mathbf{V}) \leq H(\mathbf{V}) \leq \log M$, and so

$$\frac{1}{k} \log M \geq R(\delta).$$

However, $(1/k)\log M$ is by definition the rate of the code, and this proves the theorem.

Note that the proof of this result depends only on the definition of $R(\delta)$ as the infimum of the $(1/k)R_k(\delta)$'s, together with some very general facts about mutual information and entropy. Thus the converse to the source coding theorem holds for virtually every source and distortion measure. The remarkable fact is that the source coding theorem can also be proved in very general circumstances.

$$\underline{U} = (U_1, \cdots, U_k) \rightarrow \boxed{\begin{array}{c} \text{Source} \\ \text{Encoder} \end{array}} \rightarrow (V_1, \cdots, V_k) = \underline{V}$$

Figure 6.10 Proof of the converse to the source coding theorem.

Part two
Coding theory

7

Linear codes

7.1 Introduction: The generator and parity-check matrices

We have already noted that the channel coding Theorem 2.4 is unsatisfactory from a practical standpoint. This is because the codes whose existence is proved there suffer from at least three distinct defects:[1]

(a) *They are hard to find* (although the proof of Theorem 2.4 suggests that a code chosen "at random" is likely to be pretty good, provided its length is large enough).

(b) *They are hard to analyze.* (Given a code, how are we to know how good it is? The impossibility of computing the error probability for a fixed code is what led us to the random coding artifice in the first place!)

(c) *They are hard to implement.* (In particular, they are hard to decode: the decoding algorithm sugggested in the proof of Theorem 2.4—search the region $S(\mathbf{y})$ for codewords, and so on—is hopelessly complex unless the code is trivially small.)

In fact, virtually the only coding scheme we have encountered so far which suffers from none of these defects is the (7, 4) Hamming code of the Introduction. In this chapter we show that the Hamming code is a member of a very large class of codes, the *linear codes*,[2] and in Chapters 7–9 we show that there are some very good linear codes which are free from the three defects cited above.

In an attempt to find codes which are simultaneously good in the sense of the coding theorem, and reasonably easy to implement, it is natural to impose some kind of structure on the codes. Our first step in this direction will be to impose structure on the channel input alphabet A_X: from now on we assume that $A_X =$ a finite field F_q with q elements.[3] (The basic facts about finite fields are summarized in Appendix C.) Having made this assumption, we note

139

that a codeword $\mathbf{x} = (x_1, \ldots, x_n)$ of length n can be viewed as an n-dimensional vector over F_q.

Definition[4]   *An (n, k) linear code over F_q is a k-dimensional subspace of the n-dimensional vector space $V_n(F_q) = \{(x_1, \ldots, x_n): x_i \in F_q\}$; n is called the length of the code, k the dimension. The code's rate is the ratio k/n.*

One immediate advantage of linear codes over nonlinear ones is that they are much easier to specify. An (n, k) linear code C can be completely described by any set of k linearly independent codewords $\mathbf{x}_1, \mathbf{x}_2, \ldots, \mathbf{x}_k$, for every codeword is one of the q^k linear combinations $\sum_{i=1}^{k} a_i \mathbf{x}_i$, $a_i \in F_q$. If we arrange the codewords into a $k \times n$ matrix G, G is called a *generator matrix* for C. More generally:

Definition *Let C be an (n, k) linear code over F_q. A matrix G whose rowspace equals C is called a generator matrix for C. Conversely, if G is a matrix with entries from F_q, its rowspace is called the code generated by G.*

By specifying a generator matrix it is already possible to give compact descriptions of some interesting codes. The three examples that follow (all with $q = 2$; they are *binary* codes) will be used for illustrative purposes throughout this chapter.

Example 7.1 A $(5, 1)$ linear code C_1 with generator matrix

$$G_1 = [1 \ 1 \ 1 \ 1 \ 1].$$ □

Example 7.2 A $(5, 3)$ linear code C_2 with generator matrix

$$G_2 = \begin{bmatrix} 1 & 1 & 1 & 0 & 0 \\ 0 & 0 & 1 & 1 & 0 \\ 1 & 1 & 1 & 1 & 1 \end{bmatrix}.$$ □

Example 7.3 A $(7, 4)$ linear code C_3 with generator matrix

$$G_3 = \begin{bmatrix} 1 & 0 & 0 & 0 & 0 & 1 & 1 \\ 0 & 1 & 0 & 0 & 1 & 0 & 1 \\ 0 & 0 & 1 & 0 & 1 & 1 & 0 \\ 0 & 0 & 0 & 1 & 1 & 1 & 1 \end{bmatrix}.$$ □

(C_1 contains only the two codewords 00000 and 11111; it is the rate $1/5$

repetition code discussed in the Introduction. C_3 is the (7, 4) Hamming code of the Introduction. C_2 is new.)

Another advantage enjoyed by linear codes is *ease of encoding*. An (n, k) linear code has q^k codewords, and so can be used to communicate any one of q^k distinct messages; if we assume that these messages are (indexed by) the q^k k-tuples $\mathbf{u} = (u_1, u_2, \ldots, u_k) \in V_k(F_q)$, and that the rows of G are linearly independent, then one very simple encoding rule which maps messages $\mathbf{u}$ into codewords $\mathbf{x}$ is

$$\mathbf{u} \to \mathbf{u}G, \tag{7.1}$$

where in (7.1) $\mathbf{u}G$ denotes multiplication of the $1 \times k$ vector $\mathbf{u}$ by the $k \times n$ matrix G. For example, using the generator matrix G_2 for C_2, the mapping (7.1) becomes

$$(u_1, u_2, u_3) \to (u_1 + u_3, u_1 + u_3, u_1 + u_2 + u_3, u_2 + u_3, u_3).$$

The mapping (7.1) can usually be simplified still further by using the fact that, if G is a generator matrix for C, then so is any matrix row-equivalent to G. Now it is well known that any matrix is row-equivalent to a row reduced echelon (RRE) matrix, and so *every linear code has a unique RRE generator matrix*. (To refresh the reader's memory, an RRE matrix over a field F has the following three properties:[5]

(a) The leftmost nonzero entry in each row is 1.
(b) Every column containing such a leftmost 1 has all its other entries 0.
(c) If the leftmost nonzero entry in row i occurs in column t_i, then $t_1 < t_2 \cdots < t_r$.

Thus the generator matrices G_1 and G_3 are already in RRE form, but G_2 is not. The unique RRE generator matrix for C_2 is in fact

$$G_2' = \begin{bmatrix} ① & 1 & 0 & 0 & 1 \\ 0 & 0 & ① & 0 & 1 \\ 0 & 0 & 0 & ① & 1 \end{bmatrix},$$

where for clarity we have circled the leftmost 1 in each row. Using G_2' rather than G_2, the encoding (7.1) for C_2 becomes:

$$(u_1, u_2, u_3) \to (\widehat{u_1}, u_1, \widehat{u_2}, \widehat{u_3}, u_1 + u_2 + u_3).$$

This encoding for C_2 has the desirable feature that the information symbols u_1, u_2, u_3 appear in the clear in the codeword; in general, the symbol u_i will appear as the t_ith component of the codeword $\mathbf{x} = \mathbf{u}G$, if the leftmost entry of the ith row of G occurs in column t_i. Any code, linear or not, which has the property that there exists an encoding rule such that the information symbols appear in the clear is said to be *systematic*. We have thus demonstrated that *every linear code is systematic*.

Notice next that the RRE generator matrices for C_1 and C_3 have the form $G = [I_k A]$, where I_k is the $k \times k$ identity matrix. Clearly not every linear code enjoys this property (e.g., C_2); but if the code is for use on a memoryless channel, then performing a column permutation on G will not alter the code's performance, so we can always assume $G = [I_k A]$ in this case. For example by reordering the columns of G_2' into the order (13425), we obtain

$$G_2'' = \begin{bmatrix} 1 & 0 & 0 & 1 & 1 \\ 0 & 1 & 0 & 0 & 1 \\ 0 & 0 & 1 & 0 & 1 \end{bmatrix},$$

which generates a code whose performance is the same as that of C_2 on any memoryless channel.

So much, for a while, for the generator matrix. There is another perhaps even more useful matrix associated with every linear code, called the *parity-check matrix*, which we now describe.

If C is an (n, k) linear code over F_q, a *parity check*[6] for C is an equation of the form

$$a_1 x_1 + a_2 x_2 + \cdots + a_n x_n = 0, \tag{7.2}$$

which is satisfied for all $\mathbf{x} = (x_1, x_2, \ldots, x_n) \in C$. The set of all vectors $\mathbf{a} = (a_1, \ldots, a_n)$ for which (7.2) is satisfied for all $\mathbf{x} \in C$ is itself a subspace of $V_n(F_q)$. It is denoted by $C^\perp$ (C "perp") and called the *dual code* of C. By a standard result in linear algebra $C^\perp$ has dimension $n - \dim(C)$, that is, $C^\perp$ is an $(n, n - k)$ *linear code over* F_q. A parity-check matrix for C is now defined as a generator matrix for $C^\perp$. More directly:

Definition *Let C be an (n, k) linear code over F_q A matrix H with the property that $H\mathbf{x}^T = \mathbf{0}$ iff $\mathbf{x} \in C$ is called a parity-check matrix for C.*[7]

Of course by the arguments above it follows that every (n, k) linear code has a unique $(n - k) \times n$ RRE parity-check matrix. But it is more common to use a slightly different canonical form for H. For example if $G = [I_k A]$ it follows that H can be taken as

$$H = [-A^T I_{n-k}]. \tag{7.3}$$

If G is not of this form, H can be obtained by first performing a column permutation to put G into the form $[I_k A]$, and then performing the inverse permutation on $[-A^T I_{n-k}]$.[8]

For example, the parity-check matrices formed in this way from G_1, G_2, and G_3 (p. 134) are

$$H_1 = \begin{bmatrix} 1 & 1 & 0 & 0 & 0 \\ 1 & 0 & 1 & 0 & 0 \\ 1 & 0 & 0 & 1 & 0 \\ 1 & 0 & 0 & 0 & 1 \end{bmatrix}, \quad H_2 = \begin{bmatrix} 1 & 1 & 0 & 0 & 0 \\ 1 & 0 & 1 & 1 & 1 \end{bmatrix},$$

$$H_3 = \begin{bmatrix} 0 & 1 & 1 & 1 & 1 & 0 & 0 \\ 1 & 0 & 1 & 1 & 0 & 1 & 0 \\ 1 & 1 & 0 & 1 & 0 & 0 & 1 \end{bmatrix}.$$

Notice that when H is in this form, the defining equations $H\mathbf{x}^T = \mathbf{0}$ for the codewords of C yield an explicit functional dependence between the information symbols and the remaining, or parity-check, symbols. For example in C_1, x_1 is the information symbol and $x_2 = x_3 = x_4 = x_5 = x_1$. similarly, in C_2, x_1, x_3, x_4 are information symbols and $x_2 = x_1$, $x_5 = x_1 + x_3 + x_4$. Finally, in C_3, the parity-check bits x_5, x_6, x_7 are related to the information bits x_1, x_2, x_3, x_4 by $x_5 = x_2 + x_3 + x_4$, $x_6 = x_1 + x_3 + x_4$, $x_7 = x_1 + x_2 + x_4$ (see Eq. (0.6)). Of course the location of the information symbols is not unique: see Prob. 7.13.

Let us summarize the results of this section in a theorem.

Theorem 7.1 *Let C be an (n, k) linear code over F_q. Then there is a unique $k \times n$ RRE matrix H with the property that $\mathbf{x} \in C$ iff $\mathbf{x}$ is in the rowspace of G. Furthermore, there is an $(n - k) \times n$ matrix H with the property that $\mathbf{x} \in C$ iff $H\mathbf{x}^T = \mathbf{0}$. If C is to be used on a memoryless channel, there is no loss in assuming that a $k \times (n - k)$ matrix A exists such that*

$$G = [I_k A], \quad H = [-A^T I_{n-k}],$$

in which case the encoding of a vector $\mathbf{u} \in V_k(F_q)$ is given by $\mathbf{u} \rightarrow (\mathbf{u}, \mathbf{u}A)$.

7.2 Syndrome decoding on q-ary symmetric channels

So far our discussion of linear codes has not involved the channel output alphabet A_Y or the channel statistics. From now on, however, we shall assume $A_Y = F_q$, that is, the input and output alphabets are identical (but cf. Probs. 7.11 and 7.20.) Thus if $\mathbf{x} = (x_1, \ldots, x_n) \in V_n(F_q)$ is transmitted, the received vector $\mathbf{y} = (y_1, \ldots, y_n)$ will also lie in $V_n(F_q)$; the difference $\mathbf{z} = \mathbf{y} - \mathbf{x}$ is called the *error pattern*. If $\mathbf{z}_i \neq 0$, an error is said to have occurred in the ith coordinate.

The parity-check matrix H introduced in Section 7.1 proves to be a useful tool for decoding in this situation. If $\mathbf{x}$ is transmitted, $\mathbf{x}$ is a codeword, and so $H\mathbf{x}^T = \mathbf{0}$. If the channel has caused some errors, that is, if $\mathbf{z} \neq \mathbf{0}$, then it is very likely that $H\mathbf{y}^T \neq \mathbf{0}$. The vector $\mathbf{s} = H\mathbf{y}^T$ is called the *syndrome*.[9] The vital fact about the syndrome is that it depends only on the error pattern $\mathbf{z}$ and not on the transmitted codeword:

$$\mathbf{s} = H\mathbf{y}^T$$

$$= H(\mathbf{x} + \mathbf{z})^T$$

$$= H\mathbf{x}^T + H\mathbf{z}^T$$

$$= H\mathbf{z}^T,$$

since $H\mathbf{x}^T = \mathbf{0}$. Of course the receiver is not directly interested in $\mathbf{z}$; he wants to know $\mathbf{x}$. But since he knows $\mathbf{y}$ and $\mathbf{x} = \mathbf{y} - \mathbf{z}$, he can safely focus on the problem of finding $\mathbf{z}$.

The syndrome provides some information about $\mathbf{z}$, but not enough. This is because for a fixed $\mathbf{s} \in V_{n-k}(F_q)$, the set of solutions to $H\mathbf{z}^T = \mathbf{s}$ forms a *coset*[10] of the code C, that is, a subset of $V_n(F_q)$ of the form

$$C + \mathbf{z}_0 = \{\mathbf{x} + \mathbf{z}_0 \colon \mathbf{x} \in C\}. \tag{7.4}$$

There are q^{n-k} cosets of C, corresponding to the q^{n-k} possible syndromes $\mathbf{s}$; each coset contains exactly q^k elements. Thus once the receiver computes $\mathbf{s}$, he has reduced his search for $\mathbf{z}$ from q^n to q^k possibilities, namely, the elements of the coset corresponding to $\mathbf{s}$.

In order to distinguish between these q^k candidates for $\mathbf{z}$, however, it is necessary to know somehting more about the channel. In this section we shall assume that the channel is a *q-ary symmetric channel* (qSC), that is, if $\mathbf{X}$ is a random vector modeling the channel input, and $\mathbf{Y}$ is a random vector modeling the channel output, then $\mathbf{Y} = \mathbf{X} + \mathbf{Z}$, where $\mathbf{Z} = (Z_1, Z_2, \ldots, Z_n)$ is a random vector whose components are independent, identically distributed random variables, with common distribution

$$P\{Z = 0\} = 1 - (q - 1)\varepsilon,$$

$$P\{Z = z\} = \varepsilon \quad \text{if } z \neq 0. \tag{7.5}$$

For this channel it is quite easy to distinguish between competing error patterns, since if $\mathbf{z} \in V_n(F_q)$, then

$$P\{\mathbf{Z} = \mathbf{z}\} = [1 - (q - 1)\varepsilon]^{n - w_H(\mathbf{z})} \varepsilon^{w_H(\mathbf{z})}, \tag{7.6}$$

where $w_H(\mathbf{z})$, the *Hamming weight* of $\mathbf{z}$, is defined to be the number of nonzero components of $\mathbf{z}$. Alternatively, $w_H(\mathbf{z})$ is the number of errors occurring in $\mathbf{z}$. If $\varepsilon \leqslant 1/q$, the right side of (7.6) is a decreasing function of $w_H(\mathbf{z})$, and so the most probable $\mathbf{z}$ is the one of smallest weight. (For the case $1/q < \varepsilon < 1/(q-1)$, see Prob. 7.3.)

Figure 7.1 shows how the syndrome decoder works, at least in principle. Of course step 2 in this algorithm represents a formidable amount of work; indeed the main object of Chapters 8 and 9 is the design of some linear codes for the qSC for which step 2 is tractable. However, if k and $n-k$ are both relatively small, it is possible to implement step 2 via a "table lookup" procedure, which we now describe.

Consider again code C_2. Its parity-check matrix as given on p. 143 is

$$H = \begin{bmatrix} 1 & 1 & 0 & 0 & 0 \\ 1 & 0 & 1 & 1 & 1 \end{bmatrix}.$$

In this simple case there are only four possible syndromes: 00, 01, 10, 11. We can classify the 32 vectors $\mathbf{z} = (z_1, z_2, z_3, z_4, z_5)$ according to their syndromes, and arrange them, as shown in Fig. 7.2, in a 4×8 matrix with entries from $V_5(F_2)$, called the *standard array*. The rows of the standard array are the cosets of C; for example, the first row is the code itself. Within each coset a vector of least weight has been listed first and called the *coset leader*. In general, an entry in a coset other than the code itself is equal to its coset

---

1. Compute the syndrome $\mathbf{s} = H\mathbf{y}^T$.
2. Find a minimum-weight vector in the coset corresponding to $\mathbf{s}$. Call it $\mathbf{z}_0$.
3. Output the codeword $\hat{\mathbf{x}} = \mathbf{y} - \mathbf{z}_0$.

Figure 7.1 The syndrome decoding algorithm for a qSC.

Syndrome	Coset leader							
00	*00000*	00011	00101	00110	11001	11010	11100	11111
01	*00100*	00111	*00001*	*00010*	11101	11110	11000	11011
10	*01000*	01011	01101	01110	10001	10010	10100	10111
11	*10000*	10011	10101	10110	01001	01010	01100	01111

Figure 7.2 The standard array for C_2.

leader plus the codeword above it. For example the entry 01101 in the third row equals its coset leader 01000 plus the codeword 00101. Notice that in this particular example the 01 syndrome has three candidates for coset leader (the three italicized vectors of weight 1), while in the other three cosets the leader is elected without opposition.

Given the standard array, step 2 in the decoding algorithm of Fig. 7.1 is trivial: before transmission we form a table consisting of all pairs $(\mathbf{s}, \mathbf{z}(\mathbf{s}))$, where $\mathbf{s}$ is one of the q^{n-k} possible syndromes and $\mathbf{z}(\mathbf{s})$ is the leader for the coset with syndrom $\mathbf{s}$. Then step 2 becomes simply

$$2'.\text{Set } \mathbf{z}_0 = \mathbf{z}(\mathbf{s}).$$

If this technique is feasible (i.e., if it is possible to precompute and store the coset leaders corresponding to each of the q^{n-k} syndromes), the resulting decoding algorithm is the fastest known.

7.3 Hamming geometry and code performance

The vector space $V_n(F_q)$ can be made into a metric space if we define the *Hamming distance* between two vectors $\mathbf{x}$ and $\mathbf{y}$ as

$$d_H(\mathbf{x}, \mathbf{y}) = \text{the number of components for which } x_i \neq y_i$$

$$= w_H(\mathbf{y} - \mathbf{x})(cf. \text{ p. 145}).$$

(This distance satisfies the formal properties required of a metric; see Prob. 7.4.) There is an interesting relationship between the Hamming geometry of a code and its ability to correct errors on a qSC, which we now investigate.

Thus let $C = \{\mathbf{x}_1, \mathbf{x}_2, \ldots, \mathbf{x}_M\}$ be a code of length n, not necessarily linear, for use on a qSC. Suppose that we want C to be capable of correcting all error patterns of Hamming weight $\leqslant e$; that is, if $\mathbf{x}_i$ is sent, $\mathbf{y} = \mathbf{x}_i + \mathbf{z}$ is received, and $w_H(\mathbf{z}) \leqslant e$, we want our decoder's output to be $\hat{\mathbf{x}} = \mathbf{x}_i$. It is easy to see that if each codeword is sent with probability $1/M$, then the receiver's best strategy for guessing which codeword was sent is to pick the codeword closest to $\mathbf{y}$, that is, the one for which $d_H(\mathbf{x}_i, \mathbf{y})$ is smallest (see Prob. 2.13). [*Note*: Since $d_H(\mathbf{x}_i, \mathbf{y}) = w_H(\mathbf{z})$, it follows that the syndrome decoding for linear codes described in Fig. 7.1 is equivalent to "find the closest codeword" decoding.] It is clear that if this geometric decoding strategy is used, the code will be capable of correcting all patterns of weight $\leqslant e$ iff the distance between each pair of codewords is $\geqslant 2e + 1$. For (Fig. 7.3a) if $d_H(\mathbf{x}_i, \mathbf{x}_j) \geqslant 2e + 1$, that is, if the Hamming spheres of radius e around $\mathbf{x}_i$ and $\mathbf{x}_j$ are disjoint, then if $\mathbf{x}_i$ is sent and $d_H(\mathbf{x}_i, \mathbf{y}) \leqslant e$, $\mathbf{y}$ cannot be closer to

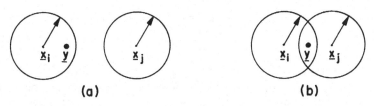

Figure 7.3 Hamming spheres of radius e around adjacent codewords.

$\mathbf{x}_j$ than it is to $\mathbf{x}_i$, and so a geometric decoder will not prefer $\mathbf{x}_j$ to $\mathbf{x}_i$. Conversely, if $d_H(\mathbf{x}_i, \mathbf{x}_j) \leqslant 2e$, that is, if the Hamming spheres of radius e intersect (Fig. 7.3b), then it is clear that, if $\mathbf{x}_i$ is sent, there exists a $\mathbf{y}$ that has $d_H(\mathbf{x}_i, \mathbf{y}) \leqslant e$, but is at least as close to $\mathbf{x}_j$ as it is to $\mathbf{x}_i$. So we are led to define the *minimum distance* of the code C as

$$d_{\min}(C) = \min\{d_H(\mathbf{x}, \mathbf{x}'): \mathbf{x}, \mathbf{x}' \in C, \mathbf{x} \neq \mathbf{x}'\},$$

and we have proved the following theorem.

Theorem 7.2 *A code* $C = \{\mathbf{x}_1, \mathbf{x}_2, \ldots, \mathbf{x}_M\}$ *is capable of correcting all error patterns of weight* $\leqslant e$ *iff* $d_{\min}(C) \geqslant 2e + 1$.

For example, a code with $d_{\min} = 7$ can correct all error patterns of weight $\leqslant 3$; if $d_{\min} = 22$, all patterns of weight $\leqslant 10$; and so forth.

Now let us apply these general remarks, which apply to any code, to the special case of linear codes. Our first observation is that since $d_H(\mathbf{x}, \mathbf{x}') = w_H(\mathbf{x} - \mathbf{x}')$, and $\mathbf{x} - \mathbf{x}'$ must be a (nonzero) codeword if C is linear (and $\mathbf{x} \neq \mathbf{x}'$), the minimum distance of a linear code is the same as its *minimum weight* $w_{\min}(C)$, where

$$w_{\min}(C) = \min\{w_H(\mathbf{x}): \mathbf{x} \in C, \mathbf{x} \neq 0\}.$$

Thus in order to compute $d_{\min}$ for an (n, k) linear code it is not necessary to compute the $(q^{2k} - q^k)/2$ distances $d_H(\mathbf{x}, \mathbf{x}')$ for $\mathbf{x} \neq \mathbf{x}'$; it suffices to compute the $q^k - 1$ weights $w_H(\mathbf{x})$ for $\mathbf{x} \neq 0$. The next theorem gives an alternative procedure for computing $d_{\min}$ for a linear code that is sometimes even simpler.

Theorem 7.3 *If C is an (n, k) linear code over F_q with parity-check matrix H, $d_{\min}(C) = $ the smallest number of columns of H that are linearly dependent. Hence if every subset of $2t$ or fewer columns of H is linearly indpendent, the code is capable of correcting all error patterns of weight* $\leqslant t$.

Note *If $q = 2$, the words "are linearly dependent" can be replaced with "sum to 0."*

Proof The codewords of C are those vectors $\mathbf{x} \in V_n(F_q)$ such that $H\mathbf{x}^T = \mathbf{0}$ (Theorem 7.1). The product $H\mathbf{x}^T$ is however a linear combination of the columns of H; Indeed, if $[\mathbf{c}_1, \mathbf{c}_2, \ldots, \mathbf{c}_n]$ are the columns of H, $H\mathbf{x}^T = x_1\mathbf{c}_1 + \cdots + x_n\mathbf{c}_n$. Hence a nonzero codeword of weight w yields a nontrivial linear dependence among w columns of H; and conversely. This proves the theorem.   □

Corollary *If $q = 2$ and all possible linear combinations of $\leqslant e$ columns of H are distinct, then $d_{\min}(C) \geqslant 2e + 1$, and so C can correct all patterns of weight $\leqslant e$.*

Proof Left as Prob. 7.7.   □

To illustrate Theorem 7.3, consider the three parity-check matrices on p. 143. For H_1 it is clear that no subset of four or fewer columns can be dependent, but the sum of all columns is $\mathbf{0}$; and so $d_{\min}(C_1) = 5$. For H_2 it is clear that $d_{\min}(C_2) = 2$, since, for example the third and fourth columns of H_2 are identical. The study of H_3 is so important we shall devote our next section to it.

7.4 Hamming codes

For reference, we again display the parity-check matrix for C_3:

$$H_3 = \begin{bmatrix} 0 & 1 & 1 & 1 & 1 & 0 & 0 \\ 1 & 0 & 1 & 1 & 0 & 1 & 0 \\ 1 & 1 & 0 & 1 & 0 & 0 & 1 \end{bmatrix}.$$

We now apply Theorem 7.3 to the problem of determining $d_{\min}$. By the note following the statement of the theorem, $d_{\min}$ is the smallest number of columns of H which sum to $\mathbf{0}$. Clearly $d_{\min} \neq 1, 2$, since the columns of H_3 are nonzero and distinct. However, there are many subsets of three columns of H_3 summing to $\mathbf{0}$, for example, columns 1, 2, and 3. Thus $d_{\min} = 3$, and so C_3 is a *single-error-correcting* code, that is, it is capable of correcting all error patterns of weight 0 or 1. Finally, observe that if C is any $(n, n - 3)$ single-error-correcting code, then $n \leqslant 7$ since if the $3 \times n$ parity-check matrix H had $n \geqslant 8$, it would either have a zero column ($d_{\min} = 1$) or a pair of identical columns ($d_{\min} = 2$). The general definition of a binary Hamming code follows.

Definition *Let H be an* $m \times (2^m - 1)$ *binary matrix such that the columns of H are the* $2^m - 1$ *nonzero vectors from* $V_m(F_2)$ *in some order. Then the* $n = 2^m - 1$, $k = 2^m - 1 - m$ *linear code over* F_2 *whose parity-check matrix is H is called a (binary) Hamming code of length* $2^m - 1$.[11]

We make two remarks about Hamming codes. First, notice that syndrome decoding is especially easy. If the error pattern is $\mathbf{z} = 0$, then the syndrome is $\mathbf{s} = \mathbf{0}$; but if $w_H(\mathbf{z}) = 1$, say $z_i = 1$, then $\mathbf{s} = \mathbf{c}_i$, the ith column of H. Hence the syndrome directly identifies the error location, and the general decoding algorithm of Fig. 7.1 becomes the special algorithm of Fig. 7.4. Second, recall that a code C can correct all patterns of weight ≤ 1 iff the Hamming spheres of radius 1 around the codewords are disjoint. But a Hamming sphere of radius 1 in $V_n(F_2)$ contains $n + 1$ vectors, so that a single-error-correcting code can have at most $2^n/(n + 1)$ codewords. In particular if $n = 2^m - 1$, there can be at most $2^{2^m-1}/2^m = 2^{2^m-1-m}$ codewords, the exact number in the Hamming code! Thus the Hamming codes have the beautiful geometric property that the spheres of radius 1 around the codewords exactly fill $V_n(F_2)$ without overlap. This means that the Hamming codes belong to an extremely exclusive class of codes, the *perfect codes*. The only other binary linear perfect codes are the repetition codes (see Prob. 7.18) and the (23, 12) Golay code (see Section 9.8). (Also see the discussion of perfect codes in Chapter 12.)

Several variations of Hamming codes, including nonbinary Hamming codes, are given in Probs. 7.17 and 7.19. However, the mose interesting and useful generalization of Hamming codes is to a class of codes which correct e errors for $e > 1$; this generalization is the main topic treated in Chapter 9.

7.5 Syndrome decoding on general q-ary channels

In Section 7.2 we described a good decoding algorithm for linear codes being used on a q-ary symmetric channel. However, few q-input, q-output channels are well modeled by qSC, and so in this section we shall briefly consider the problem of decoding linear codes on more complicated channels.

---

1. Compute the syndrome $\mathbf{s} = H\mathbf{y}^T$.
2. If $\mathbf{s} = \mathbf{0}$, output $\hat{\mathbf{x}} = \mathbf{y}$.
3. Otherwise $\mathbf{s}$ will be equal to a unique column of H, say $\mathbf{s} = \mathbf{c}_i$. Add 1 (mod 2) to the ith coordinate of $\mathbf{y}$, and output the result as $\hat{\mathbf{x}}$.

Figure 7.4 Syndrome decoding for a Hamming code.

Thus let C be an (n, k) linear code over F_q which is going to be used on a channel whose input and output alphabets are both equal to F_q. We assume that the channel noise is additive, that is, if $\mathbf{x} = (x_1, \ldots, x_n)$ is the transmitted codeword, the received vector $\mathbf{y}$ is given by $\mathbf{y} = \mathbf{x} + \mathbf{Z}$, where $\mathbf{Z} = (Z_1, \ldots, Z_n)$ is a random noise vector with distribution given by $P\{\mathbf{Z} = z\} = p(\mathbf{z})$, $\mathbf{z} \in V_n(F_q)$. (Note that the qSC of Section 7.2 fits this description, with $p(\mathbf{z})$ given by Eq. (7.5).)

How do we decode C on this channel? As before, the first step is to compute the syndrome $\mathbf{s} = H\mathbf{y}^T$; this will identify the coset that the error pattern $\mathbf{z}$ belongs to. But now the most likely error pattern within the coset characterized by the syndrome $\mathbf{s}$ is the one with the largest value $p(\mathbf{z})$, rather than the smallest weight. Hence the decoding algorithm in this general situation (cf. Fig. 7.1) is given in Fig. 7.5.

For example, consider using code C_2 on a channel for which the length 5 error patterns have probabilities given by the following table:

$\mathbf{z}$	$p(\mathbf{z})$
00000	.80
00100	.06
01010	.05
10001	.04
01110	.03
10101	.02
Anything else	0

By examining the standard array for C_2 (Fig. 7.2), we find that the coset with syndrome 00 (the code itself) contains only one of the six possible error patterns, namely, 00000, so 00000 remains the leader (the error pattern with largest $p(\mathbf{z})$) of its coset. Similarly 00100 remains the leader of coset 01. Coset 10 contains both 10001 and 01110, but since 10001 is more probable it is the coset leader. Finally, in coset 11, 01010 is chosen leader over 10101. So for this particular code and channel the table of syndromes and coset leaders should be as follows:

---

1. Compute the syndrome $\mathbf{s} = H\mathbf{y}^T$.
2. Find a vector in the coset corresponding to $\mathbf{s}$ with largest possible $p(\mathbf{z})$. Call it $\mathbf{z}_0$.
3. Output the codeword $\hat{\mathbf{x}} = \mathbf{y} - \mathbf{z}_0$.

Figure 7.5 Syndrome decoding on an arbitrary additive q-ary channel.

Syndrome	Coset Leader
00	00000
01	00100
10	10001
11	01010

and the probability of decoder error (the probability that the true error patten is not a coset leader) is $p(01110) + p(10101) = 0.05$.

Unfortunately, there are two difficulties with the decoding algorithm of Fig. 7.5. First, as we saw in Section 7.2, the implementation of step 2 may be very difficult unless the code is small enough so that a table of coset leaders can be precomputed. The second difficulty is that for most "real" channels the probabilities $p(\mathbf{z})$ will be known only empirically via tests; indeed, if n is even moderately large, most of the possible $\mathbf{z}$'s will never have been observed at all, and so only crude estimates of the form $p(\mathbf{z}) < \alpha$ for some α will be available for them. We now describe a general technique which is sometimes helpful in this situation.

Let F be a subset of $V_n(F_q)$; think of F as the set of error patterns that have "moderate" probability of occurring in the channel Let E be a subset of F; think of E as the set of error patterns with "high" probability of occurring. Given a linear code C, we wish to design, if possible, a decoder that will detect the error patterns in F and correct the error patterns in E. By this we mean the following. The decoder will be allowed to output either a codeword $\hat{\mathbf{x}}$ or a special erasure symbol "?" (cf. p. 60). If now $\mathbf{x}$ is transmitted, and $\mathbf{y} = \mathbf{x} + \mathbf{z}$ is received, there are three possibilities:

(a) The decoder outputs a codeword $\hat{\mathbf{x}} = \mathbf{x}$.
(b) The decoder outputs a codeword $\hat{\mathbf{x}} \neq \mathbf{x}$.
(c) The decoder outputs "?".

In case (a), we say the error pattern $\mathbf{z}$ has been *corrected*; in case (b), the decoder has made an *error*; in case (c), the decoder has *detected* an error. We now make the following definition.

Definition *The code C is called E-correcting, F-detecting if it is possible to design a decoder for C such that, if the error pattern $\mathbf{z}$ lies in E, it will be corrected, and if $\mathbf{z} \in F$ it will be either corrected or detected.*

Theorem 7.4 *Let C be an (n, k) linear code over F_q with parity-check* *matrix H, and let $E \subseteq F$ be subsets of $V_n(F_q)$. Then c is E-correcting, F-detecting iff it has the following properties:*

(a) $\mathbf{z}_1, \mathbf{z}_2 \in E, \mathbf{z}_1 \neq \mathbf{z}_2$ *implies* $H\mathbf{z}_1^T \neq H\mathbf{z}_2^T$.
(b) $\mathbf{z}_1 \in E, \mathbf{z}_2 \in F - E$ *implies* $H\mathbf{z}_1^T \neq H\mathbf{z}_2^T$.

Proof First suppose (a) and (b) are satisfied. We design an appropriate decoder as follows. Start with a table consisting of q^{n-k} pairs $(\mathbf{s}, f(\mathbf{s}))$, one for each possible syndrome $\mathbf{s}$, $f(\mathbf{s}) = ?$ for all $\mathbf{s}$. Then modify the table by setting $f(H\mathbf{z}^T) = \mathbf{z}$ for all $\mathbf{z} \in E$. The promised decoding algorithm is then as follows:

1. Compute $\mathbf{s} = H\mathbf{y}^T$.
2. If $f(\mathbf{s}) \in V_n(F_q)$, output $\hat{\mathbf{x}} = \mathbf{y} - f(\mathbf{s})$.
3. Otherwise output $f(\mathbf{s}) = ?$.

We leave the verification that this decoding algorithm works, as well as the proof of the converse, as Prob. 7.9.   □

Example 7.4 Let C be a $(7, 3)$ code with

$$
H = \begin{bmatrix}
0 & 1 & 1 & 1 & 1 & 0 & 0 \\
1 & 0 & 1 & 1 & 0 & 1 & 0 \\
1 & 1 & 0 & 1 & 0 & 0 & 1 \\
1 & 1 & 1 & 1 & 1 & 1 & 1
\end{bmatrix}.
$$

(This is just the matrix H_3 of p. 143 with an extra row of 1's attached.) Let $E = \{\mathbf{z}: w_H(\mathbf{z}) = 0 \text{ or } 1\}$, $F = \{\mathbf{z}; w_H(\mathbf{z}) = 0, 1, \text{ or } 2\}$. Then it is easy to show that C is E-correcting, F-detecting, or, as it is usually expressed in this special case, *single-error-correcting, double-error-detecting* (see Prob. 7.17 for proof and generalization).   □

Example 7.5 Let C be an arbitrary (n, k) linear code; we design a *detection-only* decoder for C by setting $f(\mathbf{0}) = \mathbf{0}$, $f(\mathbf{s}) = ?$ for all syndromes $\mathbf{s} \neq \mathbf{0}$. This decoder cannot correct any nonzero error pattern, but it will detect any error pattern $\mathbf{z}$ unless $\mathbf{z}$ is a nonzero codeword. The decoder's error probability is then just the probability that $\mathbf{z}$ is a nonzero codeword from C. If we assume the channel is a qSC, then the probability that $\mathbf{z}$ equals a particular codeword of weight w is $[(1 - (q-1)\varepsilon]^{n-w}\varepsilon^w$ (see Eq. (7.6)). Hence if we denote by A_i the number of codewords in C of weight i, the error probability for this detection-only scheme is

$$P_E = \sum_{i=1}^{n} A_i \varepsilon^i [1 - (q-1)\varepsilon]^{n-i}$$

$$= [1 - (q-1)\varepsilon]^n [A(\delta) - 1], \tag{7.7}$$

where $\delta = \varepsilon / [1 - (q-1)\varepsilon]$ and $A(z) = A_0 + A_1 z + \cdots + A_n z^n$ (note that $A_0 = 1$ for any code). The generating function $A(z)$ is called the *weight enumerator* of C. We shall study it further in the next section. □

7.6 Weight enumerators and the MacWilliams identities

To repeat: if C is an (n, k) linear code, its weight enumerator is the polynomial

$$A(z) = A_0 + A_1 z + \cdots + A_n z^n,$$

where A_i denotes the number of codewords in C of Hamming weight i. Clearly, $A_0 = 1$ and $A(1) = q^k$. As we have seen (Eq. (7.7)), $A(z)$ can be used to calculate the error probability when C is used for error detection only.

The following theorem shows that $A(z)$ can be used to bound the error probability in a much more interesting situation, namely, when the code is being used on a discrete memoryless channel with a maximum likelihood decoding rule. [Note. Syndrome decoding is maximum likelihood; see p. 146] For simplicity we shall treat only binary codes, but see Prob. 7.10.

Theorem 7.5 *Let C be a binary linear code which is to be used on a DMC with input alphabet $A_X = \{0, 1\}$ and output alphabet A_Y; a maximum likelihood decoding rule is being used. Then the resulting error probability is bounded by*

$$P_E \leqslant A(\gamma) - 1,$$

where

$$\gamma = \sum_{y \in A_Y} \sqrt{p(y|0)p(y|1)}. \tag{7.8}$$

(*In particular, for a* BSC *with raw bit error probability* ε, $\gamma = 2\sqrt{\varepsilon(1 - \varepsilon)}$.)

Proof Let $C = \{\mathbf{x}_0, \mathbf{x}_1, \ldots, \mathbf{x}_{M-1}\}$ with $\mathbf{x}_0 = (00 \ldots 0)$, and let $P_E^{(i)}$ denote the probability of decoder error, given that $\mathbf{x}_i$ is transmitted. If $\mathbf{y}$ is received, an ML decoder will output a codeword for which $p(\mathbf{y}|\mathbf{x}_i)$ is as large as possible. Now suppose $\mathbf{x}_0$ is transmitted. The decoder will definitely not output $\mathbf{x}_i$ if $p(\mathbf{y}|\mathbf{x}_0) > p(\mathbf{y}|\mathbf{x}_i)$, and so if $Y_i = \{\mathbf{y}: p(\mathbf{y}|\mathbf{x}_i) \geqslant p(\mathbf{y}|\mathbf{x}_0)\}$, it follows that

$$P_E^{(0)} \leqslant \sum_{i=1}^{M-1} Q_i, \tag{7.9}$$

where

$$Q_i = \sum_{\mathbf{y} \in Y_i} p(\mathbf{y}|\mathbf{x}_0). \tag{7.10}$$

Since $\sqrt{p(\mathbf{y}|\mathbf{x}_i)/p(\mathbf{y}|\mathbf{x}_0)} \geqslant 1$ for all $\mathbf{y} \in Y_i$, we can multiply each term in sum (7.10) by this factor and get

$$Q_i \leqslant \sum_{\mathbf{y} \in Y_i} \sqrt{p(\mathbf{y}|\mathbf{x}_0) p(\mathbf{y}|\mathbf{x}_i)}.$$

We can now get a somewhat weaker upper bound on Q_i by extending the summation to all $\mathbf{y} \in A_Y^n$:

$$Q_i \leqslant \sum_{\mathbf{y} \in A_Y^n} \sqrt{p(\mathbf{y}|\mathbf{x}_0) p(\mathbf{y}|\mathbf{x}_i)}. \tag{7.11}$$

We now use the fact that $p(\mathbf{y}|\mathbf{x}) = p(y_1|x_1) \dots p(y_n|x_n)$, where $\mathbf{y} = (y_1, \dots, y_n)$ and $\mathbf{x} = (x_1, \dots, x_n)$. Interchanging the order of product and summation in (7.11), we obtain

$$Q_i \leqslant \prod_{k=1}^{n} \sum_{y \in A_Y} \sqrt{p(y|x_{0k}) p(y|x_{ik})}, \tag{7.12}$$

where $\mathbf{x}_0 = (x_{01}, \dots, x_{0n})$ and $\mathbf{x}_i = (x_{i1}, \dots, x_{in})$. The inner sum in (7.12) is clearly 1 if $x_{0k} = x_{ik}$, and γ if $x_{0k} \neq x_{ik}$ (see Eq. (7.8)). Hence (7.12) reduces to

$$Q_i \leqslant \gamma^{d_{\mathrm{H}}(\mathbf{x}_0, \mathbf{x}_i)}, \tag{7.13}$$

where d_{H} denotes the Hamming distance. Combining Eqs. (7.9) and (7.13), we obtain

$$P_E^{(0)} \leqslant \sum_{i=1}^{n} A_i^{(0)} \gamma^i, \tag{7.14}$$

where $A_i^{(0)} =$ the number of codewords $\neq \mathbf{x}_0$ with Hamming distance i from $\mathbf{x}_0$. But this is just the number of words of Hamming weight i, that is, A_i. Thus the decoder error probability is $\leqslant A(\gamma) - 1$, as promised, if $\mathbf{x}_0$ is transmitted. If some other codeword, say $\mathbf{x}_j$, is transmitted, reasoning identical to that just given shows that the resulting error probability $P_E^{(j)} \leq \sum_{i=1}^{n} A_i^{(j)} \gamma^i$, where $A_i^{(j)}$ = the number of codewords $\neq \mathbf{x}_j$ with Hamming distance i from $\mathbf{x}_j$. But, since the code is linear, $A_i^{(j)} = A_i^{(0)} = A_i$, (see Prob. 7.12), and the theorem is proved. $\qquad\square$

Example 7.6 Code C_1 of this chapter has only two codewords, 00000 and 11111. Its weight enumerator is clearly $A(z) = 1 + z^5$. On a BSC the bound of Theorem 7.5 is $P_E \leq 32[\varepsilon(1-\varepsilon)]^{5/2}$, whereas on p. 3 we saw that $P_E = 10\varepsilon^3(1-\varepsilon)^2 + 5\varepsilon^4(1-\varepsilon) + \varepsilon^5 = 10\varepsilon^3 - 15\varepsilon^4 + 6\varepsilon^5$. Hence for very small values of ε the bound of Theorem 7.5 is not too bad ($32\varepsilon^{5/2}$ vs. $10\varepsilon^3$). But see Prob. 7.26 for an improvement. $\qquad\square$

We could easily at this point compute the weight enumerators of codes C_2 and C_3, since they contain only 8 and 16 codewords, respectively. However, we shall postpone the computation until we have proved the following remarkable result, which says that the weight enumerator for a code C can be obtained via a simple linear transformation of the weight enumerator of the dual code $C^\perp$ (see p. 142).

Theorem 7.6 (*the MacWilliams identities*). *Let $A(z)$ be the weight enumerator of an (n, k) linear code C, and let $B(z)$ be the weight enumerator of the dual code $C^\perp$, that is,*

$$B(z) = \sum_{j=0}^{n} B_j z^j,$$

where B_j = the number of words of weight j in $C^\perp$. Then $A(z)$ and $B(z)$ are related by the formula

$$B(z) = \frac{1}{q^k} \sum_{i=0}^{n} A_i (1-z)^i (1 + (q-1)z)^{n-i}.$$

Proof We shall prove the theorem in detail only for binary codes, and leave the generalization to the reader (see Prob. 7.27).

First we establish some conventions. If $\mathbf{x} = (x_1, \ldots, x_m)$ is a vector of any length, even $m = 1$, over a field F, $w(\mathbf{x})$ denotes its Hamming weight, that is, the number of nonzero components. In particular, if x is a scalar (i.e., $x \in F$), then

$$w(x) = \begin{cases} 0 & \text{if } x = 0, \\ 1 & \text{if } x \neq 0. \end{cases} \qquad (7.15)$$

Now we let $F = F_2$, the field with two elements. If $\mathbf{x} = (x_1, \ldots, x_m)$ and $\mathbf{y} = (y_1, \ldots, y_m)$ are two vectors with components from F, define

$$\langle \mathbf{x}, \mathbf{y} \rangle = (-1)^{x_1 y_1 + \cdots + x_m y_m}, \tag{7.16}$$

where $(-1)^0 = 1, (-1)^1 = -1$.

Lemma 1 *Let C be an (n, k) linear code over F_2, that is, a subspace of $V_n(F_2) = V$. Then, if $\mathbf{y}$ is any element of V,*

$$\sum_{\mathbf{x} \in C} \langle \mathbf{x}, \mathbf{y} \rangle = \begin{cases} 2^k & \text{if } \mathbf{y} \in C^{\perp}, \\ 0 & \text{if } \mathbf{y} \notin C^{\perp}. \end{cases}$$

Proof For a fixed $\mathbf{y}$ the mapping $\mathbf{x} \to \langle \mathbf{x}, \mathbf{y} \rangle$ is a homomorphism of C into the multiplicative group $\{1, -1\}$. If $\mathbf{y} \in C^{\perp}$, then $\langle \mathbf{x}, \mathbf{y} \rangle = 1$ for all $\mathbf{x} \in C$, and clearly the first alternative of the lemma holds. If however $\mathbf{y} \notin C^{\perp}$, the homomorphism is nontrivial, and so $+1$ and -1 are both covered equally often; thus the sum is 0. $\square$

Lemma 2 *Let $\mathbf{x}$ be a fixed vector in $V = V_n(F_2)$ and let z be an indeterminate. Then*

$$\sum_{\mathbf{y} \in V} z^{w(\mathbf{y})} \langle \mathbf{x}, \mathbf{y} \rangle = (1 - z)^{w(\mathbf{x})} (1 + z)^{n - w(\mathbf{x})}.$$

Proof Merely compute:

$$\sum_{\mathbf{y} \in V} z^{w(\mathbf{y})} \langle \mathbf{x}, \mathbf{y} \rangle = \sum_{y_1 \in F} \cdots \sum_{y_n \in F} z^{w(y_1) + \cdots + w(y_n)} \langle x_1, y_1 \rangle \cdots \langle x_n, y_n \rangle$$

(see Eqs. (7.15) and (7.16))

$$= \sum_{y_1 \in F} \cdots \sum_{y_n \in F} \prod_{i=1}^{n} z^{w(y_i)} \langle x_i, y_i \rangle.$$

Interchanging sum and product, we obtain

$$\sum_{\mathbf{y} \in V} z^{w(\mathbf{y})} \langle \mathbf{x}, \mathbf{y} \rangle = \prod_{i=1}^{n} \sum_{y \in F} z^{w(y)} \langle x_i, y \rangle.$$

The last sum is obviously $1 + z$ if $x_i = 0$ and $1 - z$ if $x_i = 1$, and Lemma 2 follows, since $w(\mathbf{x})$ is by definition the number of components x_i of $\mathbf{x}$ equal to 1. $\square$

We now proceed to prove Theorem 7.6. The idea is to compute the sum

$$\sum_{\mathbf{x},\mathbf{x}'\in C}\sum_{\mathbf{y}\in V}z^{w(\mathbf{y})}\langle \mathbf{y}, \mathbf{x}-\mathbf{x}'\rangle \tag{7.17}$$

in two ways. If we calculate the inner sum in (7.17) first, we obtain, by Lemma 2,

$$\sum_{\mathbf{x},\mathbf{x}'\in C}(1-z)^{w(\mathbf{x}-\mathbf{x}')}(1+z)^{n-w(\mathbf{x}-\mathbf{x}')}=\sum_{i=0}^{n}\overline{A}_i(1-z)^i(1+z)^{n-i},$$

where $\overline{A}_i = $ the number of pairs $(\mathbf{x}, \mathbf{x}')$ of codewords with $w(\mathbf{x}-\mathbf{x}')=i$. But it is easily verified that this number is $2^k\cdot A_i$, A_i being the number of codewords of weight i (see Prob. 7.12), so (7.17) is equal to

$$2^k\sum_{i=0}^{n}A_i(1-z)^i(1+z)^{n-i}. \tag{7.18}$$

If now we interchange the order of summation in (7.17), we get

$$\sum_{\mathbf{y}\in V}z^{w(\mathbf{y})}\sum_{\mathbf{x},\mathbf{x}'\in C}\langle \mathbf{y}, \mathbf{x}-\mathbf{x}'\rangle,$$

the inner sum of which is

$$\sum_{\mathbf{x}\in C}\langle \mathbf{y}, \mathbf{x}\rangle\sum_{\mathbf{x}'\in C}\langle \mathbf{y}, -\mathbf{x}'\rangle = \left|\sum_{\mathbf{x}\in C}\langle \mathbf{y}, \mathbf{x}\rangle\right|^2.$$

By Lemma 1, this is 2^{2k} if $\mathbf{y}\in C^\perp$ and 0 if $\mathbf{y}\notin C^\perp$, so that (7.17) has become

$$2^{2k}\sum_{\mathbf{y}\in C^\perp}z^{w(\mathbf{y})}=2^{2k}\sum_{j=0}^{n}B_j z^j. \tag{7.19}$$

Equating (7.18) and (7.19), we have the conclusion of Theorem 7.6. □

The reason the MacWilliams identities are so useful is this. In order to compute the weight enumerator for an (n, k) linear code C, it is in general necessary to find the weight of all q^k codewords. This is clearly a formidable task unless k is relatively small. However, if k is so large that $n - k$ is small, it will be possible to calculate the weight enumerator of the dual code $C^\perp$ and so also the weight enumerator of C. For example, the general binary Hamming code (see Section 7.4) has $n = 2^m - 1$, $k = 2^m - 1 - m$, and the task of enumerating all 2^{2^m-1-m} codewords by weight might seem hopeless. But the dual code has only 2^m codewords and has a very simple weight enumerator, and so by Theorem 7.6 $A(z)$ for the Hamming codes can be computed in general (see the following example, and also Probs. 7.29 and 7.30).

Example 7.7 We are now prepared to compute $A(z)$ for codes C_2 and C_3 of this chapter. C_2 is a $(5, 3)$ code; it has 8 codewords. But its dual code $C_2^\perp$ is a $(5, 2)$ code with only 4 codewords, so we compute its weight enumerator first. By the results of Section 7.1, a parity-check matrix for C_2 will be a generator matrix for $C_2^\perp$, and on p. 143 we found such a matrix:

$$\begin{bmatrix} 1 & 1 & 0 & 0 & 0 \\ 1 & 0 & 1 & 1 & 1 \end{bmatrix}.$$

The four codewords of $C_2^\perp$ are then 00000, 11000, 10111, and 01111; thus $A(z) = 1 + z^2 + 2z^4$ for this code. Hence by Theorem 7.6, the weight enumerator for C_2 itself is

$$\tfrac{1}{4}(1 + z)^5 A\left(\frac{1 - z}{1 + z}\right) = \tfrac{1}{4}\left[(1 + z)^5 + (1 + z)^3(1 - z)^2 + 2(1 + z)(1 - z)^4\right]$$

$$= 1 + 3z^2 + 3z^3 + z^5.$$

Thus C_2 contains 1 word of weight 0, three words of weight 2, three words of weight 3, and one word of weight 5 (a result that could easily have been verified directly, of course). $\qquad \square$

Example 7.8 Similarly, the dual code $C_3^\perp$ of the $(7, 4)$ Hamming code C_3 has generator matrix (see p. 143)

$$\begin{bmatrix} 0 & 1 & 1 & 1 & 1 & 0 & 0 \\ 1 & 0 & 1 & 1 & 0 & 1 & 0 \\ 1 & 1 & 0 & 1 & 0 & 0 & 1 \end{bmatrix}.$$

When we compute its weight enumerator, we get a surprise! Each of the 7 nonzero codewords has weight 4! Thus $A(z) = 1 + 7z^4$, and so the weight enumerator for C_3 itself is, by Theorem 7.6,

$$\tfrac{1}{8}\left[(1 + z)^7 + 7(1 - z)^4(1 + z)^3\right] = 1 + 7z^3 + 7z^4 + z^7. \qquad \square$$

Problems

7.1 Let C be a linear code over the field $F_3 = \{0, 1, 2\}$ of integers (mod 3), with generator matrix

$$G = \begin{bmatrix} 1 & 1 & 1 & 0 \\ 2 & 0 & 1 & 1 \end{bmatrix}.$$

(a) Use syndrome decoding to decode the vectors 2121, 1201, 2222 (assume the channel is symmetric).

(b) Compute the code's weight enumerator.

7.2 Let

$$G = \begin{bmatrix} 1 & 0 & 1 & 0 & 1 & 1 \\ 0 & 1 & 1 & 1 & 1 & 0 \\ 0 & 0 & 0 & 1 & 1 & 1 \end{bmatrix}$$

be a generator matrix for a (6, 3) binary linear code C.

(a) Find a row-reduced echelon generator matrix for C.

(b) Find a parity-check matrix H for C.

(c) Find minimum-weight coset leaders for each of the 8 cosets for C.

(d) Let A_i ($i = 0, 1, \ldots, 6$) denote the number of vectors of weight i in C. Find the A_i.

(e) Decode the following received vectors: 111010, 000011, 101010 (assume a BSC with $p < 1/2$).

7.3 Suppose the error probability ε of the qSC satisfies $\varepsilon > 1/q$. How should the decoding algorithm described in Section 7.2 be modified?

7.4 Show that the Hamming distance d_H defined on p. 146 satisfies the following properties required of a bona fide metric:

(a) $d(\mathbf{x}, \mathbf{x}) = 0$.

(b) $d(\mathbf{x}, \mathbf{y}) > 0$ if $\mathbf{x} \neq \mathbf{y}$.

(c) $d(\mathbf{x}, \mathbf{y}) = d(\mathbf{y}, \mathbf{x})$.

(d) $d(\mathbf{x}, \mathbf{y}) \leq d(\mathbf{x}, \mathbf{z}) + d(\mathbf{z}, \mathbf{y})$.

7.5 Show that the total number of distinct Hamming codes of length $n = 2^m - 1$ is $(2^m - 1)!/\prod_{i=0}^{m-1}(2^m - 2^i)$.

7.6 How does syndrome decoding of code C_1 compare in complexity to majority-vote decoding (see p. 2)?

7.7 Prove the corollary to Theorem 7.3.

7.8 For each n, describe a code C of largest possible rate with $d_{\min}(C) = 2$. Is it unique?

7.9 Complete the proof of Theorem 7.4.

7.10 Let C be a q-ary linear code, being used on a qSC with maximum likelihood decoding. Show that the resulting error probability is bounded by $A(\gamma) - 1$, where A is the code's weight enumerator and $\gamma = 2\sqrt{\varepsilon[1 - (q - 1)\varepsilon]} + \varepsilon(q - 2)$. (For the definition of a qSC, see Eq. (7.5).)

7.11 A q-ary *erasure channel* is a discrete memoryless channel with input alphabet $A_X = F_q$, the finite field with q elements, and $A_Y = F_q \cup \{?\}$, where "?" is a special erasure symbol. If $\mathbf{x} = (x_1, \ldots, x_n)$ is transmitted and $\mathbf{y} = (y_1, \ldots, y_n)$ is received, an *error* is said to occur in the ith position if $y_i \in F_q$ but $y_i \neq x_i$, an *erasure* if $y_i = ?$.

(a) Let C be an (n, k) linear code over F_q with minimum distance $d_{\min}$. Show that C can correct all combinations of e errors and f erasures iff $d_{\min} \geq 2e + f + 1$.

(b) Suppose the channel is a *pure erasure channel*, that is, $p(y|x) = 0$ unless $y = x$ or $y = ?$. In this case show that the condition $H\mathbf{x}^T = \mathbf{0}$ for $\mathbf{x}$ to be a codeword can be used to obtain $n - k$ linear equations in the unknown erased codeword components.

(c) Apply the technique suggested in part (b) to decode the following code-words from the $(7, 4)$ Hamming code (parity-check matrix H_3 on p. 146), which have suffered erasures but no errors: 10?0?01, ???0000, ?0??01?.

7.12 Let C be an (n, k) linear code over F_q, and for any $\mathbf{y} \in V_n(q)$ define $C - \mathbf{y} = \{\mathbf{x} - \mathbf{y}: \mathbf{x} \in C\}$. Clearly $C - \mathbf{y}$ is a coset of C; see Eq. (7.4). Show that $C - \mathbf{y} = C$ iff $\mathbf{y} \in C$. Hence show that:
(a) If $\mathbf{x}_j$ is a fixed codeword of C, the number of codewords at Hamming distance i from $\mathbf{x}_j$ is A_i, the number of codewords of Hamming weight i.
(b) The number of pairs of codewords $(\mathbf{x}, \mathbf{x}')$ with $d_H(\mathbf{x}, \mathbf{x}') = i$ is exactly $q^k A_i$.

7.13 Let C be an (n, k) linear code over F_q. An *information set* for C is a subset $I = \{i_1, i_2, \ldots, i_k\} \subseteq \{1, 2, \ldots, n\}$ such that, for any choice of $\alpha_1, \alpha_2, \ldots, \alpha_k$ from F_q, there is exactly one codeword $(x_1, \ldots, x_n)$ from C with $x_{i_1} = \alpha_1, \ldots, x_{i_k} = \alpha_k$. (The terminology comes from the fact that the positions $i_1, \ldots, i_k$ of the codewords can be used to carry the information, that is, the message $\mathbf{u} = (u_1, \ldots, u_k)$; see p. 141.)
(a) Show that every linear code has at least one information set.
(b) Show that I is an information set for C iff columns $i_1, \ldots, i_k$ of the generator matrix are linearly independent.
(c) How many information sets do codes C_1, C_2, C_3 of this chapter have?
(d) Let C be a *binary* linear code such that every k-element subset of $\{1, 2, \ldots, n\}$ is an information set. Show that $k = 0, 1, n - 1$, or n. [Note: There are some nontrivial, nonbinary codes with this property (e.g., the Reed–Solomon codes of Section 9.6).]

7.14 (Continuation). The notion of information set, introduced in Prob. 7.13, can be used to design a decoding algorithm for linear codes that differs markedly from syndrome decoding. It is sometimes called *error trapping*, or *decoding with multipliers*.[12] Suppose that C is an (n, k) linear code, and that we wish to correct all patterns of e or fewer errors with C. Let $I_1, I_2, \ldots, I_r$ be informa-tion sets with the additional property that for any e-element subset $J \subseteq \{1, 2, \ldots, n\}$, there exists at least one set I_i such that $I_i \cap J = \phi$.
(a) Suppose that $\mathbf{y}$ is the received codeword, and that it contains $\leqslant e$ errors. For each $i = 1, 2, \ldots, r$, let $\mathbf{x}_i =$ the unique codeword from C that agrees with $\mathbf{y}$ at the positions specified by I_i. Show that at least one of the $\mathbf{x}_i$'s so generated is the actual transmitted codeword.
(b) Assuming that $d_{\min}(C) \geqslant 2e + 1$, show that by comparing the $\mathbf{x}_i$'s to $\mathbf{y}$, one by one, the transmitted codeword can be uniquely identified.
(c) Apply this technique to the $(7, 4)$ Hamming code. What is the smallest number of information sets that will suffice to correct 1 error?

7.15 Suppose you were approached by a communications engineer who told you that his (binary) channel accepts words of length n and that the only kind of error pattern ever observed is one of the $n + 1$ patterns (000000, 000001, 000011, 000111, 001111, 011111, 111111, illustrated for $n = 6$). Design a linear (n, k)

code that will correct all such patterns with as large a rate as possible. Illustrate
your construction for $n = 7$.

7.16 Let C be an (n, k) code with minimum distance d_{min}, and let $e \leqslant f$ be
nonnegative integers. Define $E = \{\mathbf{z}: w_H(\mathbf{z}) \leqslant e\}$, $F = \{\mathbf{z}: w_H(\mathbf{z}) \leqslant f\}$. Show
that, if $e + f < d_{min}$, then C is E-correcting, F-detecting in the sense of
Theorem 7.4. (In this special case the code is usually said to be "e-error-
correcting, f-error-detecting.")

7.17 Although the binary Hamming code is ideally suited to the problem of
correcting single errors, if the channel makes more than one error its performance is dismal. This problem will expose this flaw and suggest possible ways of
remedying it.

(a) Show that if a Hamming code is used but the channel makes 2 or more
errors, the decoder is always wrong.

(b) The *extended* Hamming code of length 2^m is defined to be the code
obtained from the original Hamming code by adjoining an overall parity-
check bit; that is, if H is the original parity-check matrix, the new one is

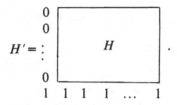

Similarly, the *expurgated*[13] Hamming code has length $2^m - 1$ and

$$H'' = \boxed{\quad H \quad}.$$
$$1\ 1\ \ \ldots\ \ 1$$

Calculate the dimensions of these two codes, and show that $d_{min} = 4$ in
both cases.

(c) Show that both the extended and the expurgated Hamming codes can be
decoded in such a way that single errors are corrected and double errors are
at least detected (see Prob. 7.16).

7.18 Show that the n-fold *repetition code*, that is, the $(n, 1)$ linear code with
generator matrix $G = [1111111]$ and n *odd*, is a perfect code in the sense that
the Hamming spheres of radius $(n - 1)/2$ completely fill $V_n(F_2)$, without
overlap.

7.19 This problem will show you how to construct *nonbinary* Hamming codes. Let
H be an $m \times n$ matrix with entries in F_q, a parity-check matrix for a certain
linear code C.

(a) Show that C will be a single-error-correcting code iff no column of H is all
0, and no column is a scalar multiple of any other column.

(b) Show that for a fixed value of m, if H has the properties in part (a), then

$n \leqslant (q^m - 1)/(q - 1)$, and that this bound is sharp. If $n = (q^m - 1)/(q - 1)$, the resulting code is called a *q-ary Hamming code*. Clearly, it has $n = (q^m - 1)/(q - 1)$, $k = n - m$.

(c) Construct parity-check matrices for the following Hamming codes:

q	n	k
3	4	2
3	13	10
5	6	4

7.20 The object of this problem is to study the performance of a binary linear code on a Gaussian channel (see Section 4.1), with energy constraint $E(X^2) \leqslant \beta$ and noise variance σ^2. In particular, we shall obtain an analogue of Theorem 7.5 for this channel. So let C be an (n, k) binary linear code; we modify it for use on the Gaussian channel by mapping each "0" component of a codeword into $+\sqrt{\beta}$, and each "1" into $-\sqrt{\beta}$. We assume that the codewords are $\mathbf{x}_0, \mathbf{x}_1, \ldots, \mathbf{x}_{M-1}$ (with each codeword now having Euclidean length $\sqrt{n\beta}$), that each codeword is sent with probability $1/M$, and that $\mathbf{y}$ is the received vector.

(a) The minimum error probability decoding rule (see Prob. 2.13) is to set $\hat{\mathbf{x}}$ equal to the $\mathbf{x}_i$ for which $P\{\mathbf{x}_i \text{ transmitted}|\mathbf{y} \text{ received}\}$ is largest. Show that this is equivalent to setting $\hat{\mathbf{x}}$ equal to the $\mathbf{x}_i$ for which the Euclidean distance $\|\mathbf{y} - \mathbf{x}_i\|$ is smallest. Show that this in turn is equivalent to finding the codeword $\mathbf{x}_i$ for which the inner product $\mathbf{x}_i \cdot \mathbf{y}$ is largest.

(b) Show that if there exists a codeword $\mathbf{x}_i$ for which $\mathbf{x}_i \cdot \mathbf{y} > \|\mathbf{y}\| \cdot \sqrt{\beta(n - d_{\min})}$, where $d_{\min}$ is the minimum distance of the code, then $\mathbf{x}_i$ must be the output of the maximum likelihood decoder described in part (a). (This observation is due to L. D. Baumert.)

(c) As in the proof of Theorem 7.5, suppose $\mathbf{x}_0 = (0, 0, \ldots, 0)$ $(= (\sqrt{\beta}, \sqrt{\beta}, \ldots, \sqrt{\beta})$ after modification) is sent, and let $Q_i = P$ {decoder's output is $\mathbf{x}_i|\mathbf{x}_0$ is sent}. Show that $Q_i \leqslant Q(\|\mathbf{x}_0 - \mathbf{x}_i\|/2\sigma)$, where $Q(\alpha) = \int_\alpha^\infty (2\pi)^{-1/2} e^{-t^2/2} \, dt$.

(d) Using the bound of Prob. 4.5 and the results of part (c), show that $Q_i \leqslant \gamma^{w(\mathbf{x}_i)}$, where $\gamma = e^{-\beta/2\sigma^2}$ and $w(\mathbf{x}_i)$ denotes the Hamming weight of $\mathbf{x}_i$.

(e) Finally show that the overall decoder error probability is bounded by $P_E \leqslant A(\gamma) - 1$, where $\gamma = e^{-\beta/2\sigma^2}$.

We saw in Section 7.3 that a code is capable of correcting all patterns of e or fewer errors iff the spheres of radius e around the codewords are disjoint iff no subset of e or fewer columns of H is linearly dependent. This of course implies that the more errors the code corrects, the smaller its rate will be. In the next 5 problems we shall derive some bounds on the relationship between $d_{\min}$ and k for the best codes. For simplicity we shall restrict our attention to *binary linear codes*. For fixed n and d, let $M_L(n, d)$ denote the maximum possible number of codewords in a linear code with length n and minimum distance $\geqslant d$.

7.21 (The Gilbert–Varshamov bound). Show that $M_L(n, d) \geq 2^n/[1 + \binom{n}{1} + \cdots + \binom{n}{d-1}]$. [*Hint:* If the best linear code had fewer than this many words, at least one of its cosets would have all words of weight $\geq d$. But the union of a linear code with one of its cosets is also a linear code.]

7.22 (The Hamming bound). Show that if a binary code of length n, linear or not, is capable of correcting e errors, it contains at most $2^n/[1 + \binom{n}{1} + \cdots + \binom{n}{e}]$ codewords.

7.23 (Continuation). Show that the Hamming codes achieve the bound of Prob. 7.22.

7.24 (The Plotkin bound). Let $\pi_i: (x_1, \ldots, x_n) \to x_i$ be the mapping that projects a binary vector onto its ith component. Show that, if $\mathbf{x}$ ranges through an (n, k) code C, $\pi_i(\mathbf{x}) = 0$ has either 2^k or 2^{k-1} solutions. Use this fact to prove that

$$\sum_{\mathbf{x} \in C} w_H(\mathbf{x}) \leq n \cdot 2^{k-1}.$$

Conclude that $d_{\min}(C) \leq n \cdot 2^{k-1}/(2^k - 1)$.

7.25 (Continuation). Suppose $n \geq 2d - 2$, and let C_0 be the code obtained by selecting all codewords from C that are 0 in the first $n - 2d + 2$ coordinates and deleting these coordinates. Applying the result of Prob. 7.24 to C_0, conclude that

$$M_L(n, d) \leq 2^{n-2d+2} d.$$

7.26 The following result (de to Van de Meeberg) shows how to strengthen Theorem 7.5 for a BSC.

    (a) Show that bound (7.13) can be improved to $Q_i \leq \gamma^{d_H(\mathbf{x}_0, \mathbf{x}_i)+1}$ if $d_H(\mathbf{x}_0, \mathbf{x}_i)$ is odd. [*Hint:* in Prob. 0.2 it was shown that the error probability for a repetition code of length $2n$ is the same as for one of length $2n - 1$.]

    (b) Hence show that Theorem 7.5 can be improved to $P_E \leq \frac{1}{2}[(1 + \gamma) A(\gamma) + (1 - \gamma)A(-\gamma)] - 1$.

    (c) Apply the bound obtained in part (b) to code C_1 of this chapter, and compare the result to that obtained in Example 7.6.

The next two problems deal with generalizations of Theorem 7.6, the MacWilliams identities. The first result merely supplies the details of the proof of Theorem 7.6 when $q \neq 2$. The second deals with nonlinear codes.

7.27 Prove Theorem 7.6 for general q. [*Hint:* The proof given in the text will work once the quantity $\langle \mathbf{x}, \mathbf{y} \rangle$ (see Eq. (7.16)) is suitably generalized. To do this, let $\langle \mathbf{x}, \mathbf{y} \rangle = \lambda(\mathbf{x} \cdot \mathbf{y})$, where λ is any nontrivial homomorphism of the additive group of F_q into the complex pth roots of unity, where $q = p^j$ and p is prime.]

7.28 Prove the nonlinear MacWilliams identities: If $C = \{\mathbf{x}_1, \ldots, \mathbf{x}_M\}$ is a subset of $V_n(F_q)$, define $A_i = M^{-1}\cdot$ the number of pairs $(\mathbf{x}, \mathbf{x}')$ from C with $d_H(\mathbf{x}, \mathbf{x}') = i$. If the B_j's are defined by the formula

$$\frac{1}{M} \sum_{i=0}^{n} A_i (1 - z)^i [1 + (q - 1)z]^{n-i} = \sum_{j=0}^{n} B_j z^j,$$

then show that $B_j \geq 0$ for all j. (The B_j's apparently do not have a natural combinatorial significance, in general.)

7.29 The object of this problem is the computation of the weight enumerator for the general Hamming code.

(a) Show that every nonzero codeword in the dual code of a binary Hamming code of length $n = 2^m - 1$ has weight 2^{m-1}.

(b) Hence show that for a binary Hamming code,

$$A(z) = \frac{1}{2^m}\left[(1+z)^n + n(1 - z^2)^{(n-1)/2}(1+z)\right].$$

(c) Generalize to nonbinary Hamming codes (see Prob. 7.19).

7.30 In this problem we will define and investigate the important *Reed–Muller codes*. Let $P(m, d)$ denote the set of all polynomials of degree $\leq d$ in m variables over F_2. Let $(\mathbf{v}_0, \mathbf{v}_1, \ldots, \mathbf{v}_{M-1})$, $M = 2^m$, denote a list of all the 2^m binary vectors $(x_1, x_2, \ldots, x_m)$ in some order. Then, for each $f \in P(m, d)$, we get a binary vector of length 2^m via the mapping $f \rightarrow (f(\mathbf{v}_0), f(\mathbf{v}_1), \ldots, f(\mathbf{v}_{M-1}))$. The set of all vectors obtained in this way from polynomials in $P(m, d)$ is called the dth *order Reed–Muller code of length* 2^m, or RM(m, d) for short.

(a) Show that RM(m, d) is a binary (n, k) linear code with $n = 2^m$ and $k = 1 + \binom{m}{1} + \cdots + \binom{m}{d}$.

(b) Show that the minimum distance of RM(m, d) is 2^{m-d}. [*Hint:* To show that $d_{\min} \geq 2^{m-d}$, let $f = f(x_1, x_2, \ldots, x_m) \in P(m, d)$, and consider $f(0, x_2, \ldots, x_m)$ and $f(1, x_2, \ldots, x_m)$ as polynomials in $P(m - 1, d)$.]

(c) Show that RM(m, d)$^{\perp}$ = RM($m, m - d - 1$).

(d) Show that RM($m, m - 2$) = the extended Hamming code of Prob. 7.17b.

(e) Compute the weight enumerators of RM(m, d) for $d = 0, 1, m - 2, m - 1$, m, and all m.

7.31 (Shortening a linear code). Let C be an (n, k) linear code with parity-check matrix of the form $H = [BI_{n-k}]$, where B is $(n - k) \times k$ and I_{n-k} is an $(n - k) \times (n - k)$ identity matrix. If $1 \leq t \leq k$, the code C_t whose parity-check matrix is $H_t = [B_t I_{n-k}]$, where B_t is the $(n - k) \times (k - t)$ matrix obtained by deleting the first t columns of B, is called *a shortened version of C*. (See also Prob. 7.25.)

(a) Show that C_t consists of all codewords of C whose first t coordinates are equal to 0, with these coordinates deleted.

(b) Show that C_t is an $(n - t, k - t)$ linear code.

(c) Show that $d_{\min}(C_t) \geq d_{\min}(C)$.

7.32 (Interleaved codes for burst-error protection). A binary error pattern $\mathbf{z}$ of length n is called a *burst of length b* if the 1's in $\mathbf{z}$ are confined to b consecutive components of $\mathbf{z}$. For example, $\mathbf{z} = (00 \ldots 00100101100 \ldots 00)$ is a burst of length b for all $b \geq 7$. Similarly, if the 1's in $\mathbf{z}$ can be confined to e such bursts, $\mathbf{z}$ is called a pattern with e *bursts of length b*. For example, the $\mathbf{z}$ given above can be thought of as 2 bursts of length 4.

Now let H be an $m \times n$ parity-check matrix for an (n, k) binary linear code C. The b-*fold interleaving* of C is the (nb, kb) code $C^{(b)}$ with $mb \times nb$ parity-check matrix

$$H^{(b)} = \begin{bmatrix} \mathbf{c}_1 & & & \mathbf{c}_2 & & & \mathbf{c}_n & \\ & \mathbf{c}_1 & & & & & & \\ & & \ddots & & & \ddots & & \ddots \\ & & & \mathbf{c}_1 & & \mathbf{c}_2 & & \mathbf{c}_n \end{bmatrix},$$

where $[\mathbf{c}_1, \ldots, \mathbf{c}_n]$ are the columns of H, and blanks denote 0's.

(a) Show that each codeword in $C^{(b)}$ is an "interleaving" of b codewords from C.

(b) Assume the original code C corrects all patterns of weight $\le e$, and detects all patterns of weight $\le f$ (cf. Prob. 7.16). Then show that $C^{(b)}$ corrects all patterns of e bursts of length b, and detects all patterns of f bursts of length b.

(c) Construct parity-check matrices for (i) a (21, 12) code that corrects all single bursts of length 3; (ii) a (25, 20) code that detects all bursts of length 5. Given an efficient decoding algorithm for both codes.

Notes

1 (p. 139). We pause here to bid adieu to Shannon; for on the subject of explicit codes that do what his coding theorems promise he has been silent.

2 (p. 139). More precisely, linear *block* codes, to distinguish them from the linear *convolutional* codes of Chapter 10.

3 (p. 139). Since this requires that q be a prime power, we cannot consider input alphabets with 6, 10, 12, 14, ... elements.

4 (p. 140). The discussion that follows assumes the reader is familiar with certain elementary facts about linear algebra, such as can be found, for example, in Birkhoff and MacLane [3], Chapter 7.

5 (p. 141). Usually an RRE matrix is permitted to have some rows that are 0, so that the number of rows in the matrix does not change in the transition to RRE form. We shall assume, however, that if zero rows appear in the reduction they are deleted.

6 (p. 142). The name comes from the fact that in the binary case a parity-check says that each codeword has an even number of 1's in a certain fixed subset of the coordinate positions.

7 (p. 142). Here and elsewhere T is the transpose operator.

8 (p. 143). The simplicity of this transition (from G to H) is of course no accident; the RRE form is designed to do this. For example, to solve the system of linear homogeneous equations represented by G_2 (p. 140):

$$x_1 + x_2 + x_3 \qquad\qquad = 0,$$
$$x_3 + x_4 \quad = 0,$$
$$x_1 + x_2 + x_3 + x_4 + x_5 = 0,$$

one first puts G into RRE form; the equations become (cf. G_2' on p. 141)

$$x_1 + x_2 \qquad\quad + x_5 = 0$$

$$x_3 \quad\; + x_5 = 0$$

$$x_4 + x_5 = 0$$

at which point it becomes obvious that x_2 and x_5 can be chosen independently, and that $x_1 = x_2 + x_5$, $x_3 = x_5$, $x_4 = x_5$, that is, every solution to the original system is a linear combination of the rows of the following matrix:

$$\begin{bmatrix} 1 & 1 & 0 & 0 & 0 \\ 1 & 0 & 1 & 1 & 1 \end{bmatrix},$$

which is just the matrix H_2 on p. 143.

9 (p. 144). See Note 5 of the Introduction.

10 (p. 144). If we forget about the multiplicative structure of the field F_q and view it merely as an additive group with q elements, the code C becomes a subgroup of the direct product $F_q \times \cdots \times F_q$, and the subset $C + z_0$ defined in Eq. (7.4) is a coset of this subgroup. For this reason you will occasionally see linear codes referred to as "group codes,"

11 (p. 149). Usually the codes for $m = 1, 2$ are not considered to be Hamming codes, since they are so simple.

12 (p. 160). This terminology, which is due to G. Solomon, arises as follows. If I is an information set for C, let $\mathbf{m}_I = (m_1, m_2, \ldots, m_n)$ denote the indicator of I, that is, $m_i = 1$ if $i \in I$, $= 0$ otherwise. Then, if $\mathbf{y}$ is the received word, we can multiply $\mathbf{y}$ by $\mathbf{m}_I$, and recover $\mathbf{x}$, provided that no errors have occurred in the information set.

13 (p. 161). So called because the codewords of odd weight have been *expurgated* from C.

8
Cyclic codes

8.1 Introduction

At the beginning of Chapter 7, we said that by restricting our attention to linear codes (rather than arbitrary, unstructured codes), we could hope to find some good codes which are reasonably easy to implement. And it is true that (via syndrome decoding, for example) a "small" linear code, say with dimension or redundancy at most 20, can be implemented in hardware without much difficulty. However, in order to obtain the performance promised by Shannon's theorems, it is necessary to use larger codes, and in general, a large code, even if it is linear, will be difficult to implement. For this reason, almost all block codes used in practice are in fact *cyclic codes*; cyclic codes form a very small and highly structured subset of the set of linear codes. In this chapter, we will give a general introduction to cyclic codes, discussing both the underlying mathematical theory (Section 8.1) and the basic hardware circuits used to implement cyclic codes (Section 8.2). In Section 8.3 we will show that Hamming codes can be implemented as cyclic codes, and in Sections 8.4 and 8.5 we will see how cyclic codes are used to combat *burst errors*. Our story will be continued in Chapter 9, where we will study the most important family of cyclic codes yet discovered: the BCH/ Reed–Solomon family.

We begin our studies with the innocuous-appearing definition of the class of cyclic codes.

Definition *An* (n, k) *linear code over a field F is said to be a cyclic code if, for every codeword* $\mathbf{C} = (C_0, C_1, \ldots, C_{n-1})$, *the right cyclic shift of* $\mathbf{C}$, *viz.* $\mathbf{C}^R = (C_{n-1}, C_0, \ldots, C_{n-2})$, *is also a codeword.*

As we shall see, there are many cyclic codes, but compared to linear codes,

167

they are quite scarce. For example, there are 11,811 linear $(7, 3)$ codes over $GF(2)$, but only two are cyclic!

Example 8.1 If F is any field, and n is an integer ≥ 3, there are always at least four cyclic codes of length n over F, usually called the four *trivial* cyclic codes:

- An $(n, 0)$ code, consisting of just the all-zero codeword, called the *no-information code*.
- An $(n, 1)$ code, consisting of all codewords of the form $(a, a, \ldots, a)$, for $a \in F$, called the *repetition code*.
- An $(n, n - 1)$ code, consisting of all vectors $(C_0, C_1, \ldots, C_{n-1})$ such that $\sum_i C_i = 0$, called the *single-parity-check code*.
- An (n, n) code, consisting of all vectors of length n, called the *no-parity code*.   □

For some values of n and F, the trivial cyclic codes described in Example 8.1 are the only cyclic codes of length n over F (e.g., $n = 19$ and $F = GF(2)$). However, there are often other, more interesting cyclic codes, as the next two examples illustrate.

Example 8.2 Consider the $(7, 3)$ linear code over $GF(2)$ defined by the generator matrix

$$G = \begin{bmatrix} 1 & 0 & 1 & 1 & 1 & 0 & 0 \\ 0 & 1 & 0 & 1 & 1 & 1 & 0 \\ 0 & 0 & 1 & 0 & 1 & 1 & 1 \end{bmatrix}.$$

This code has eight codewords. Denoting the rows of G by $\mathbf{C}_1, \mathbf{C}_2, \mathbf{C}_3$, the nonzero codewords are:

$$\mathbf{C}_1 = 1011100,$$

$$\mathbf{C}_2 = 0101110,$$

$$\mathbf{C}_3 = 0010111,$$

$$\mathbf{C}_1 + \mathbf{C}_2 = 1110010,$$

$$\mathbf{C}_1 + \mathbf{C}_3 = 1001011,$$

$$\mathbf{C}_2 + \mathbf{C}_3 = 0111001,$$

$$\mathbf{C}_1 + \mathbf{C}_2 + \mathbf{C}_3 = 1100101.$$

This code is in fact a cyclic code. To verify this, we need to check that the right cyclic shift of each codeword is also a codeword. For example, the right cyclic shift of $\mathbf{C}_1$ is $\mathbf{C}_2$. The complete list of right cyclic shifts follows:

$$\mathbf{C}_1 \rightarrow \mathbf{C}_2,$$

$$\mathbf{C}_2 \rightarrow \mathbf{C}_3,$$

$$\mathbf{C}_3 \rightarrow \mathbf{C}_1 + \mathbf{C}_3,$$

$$\mathbf{C}_1 + \mathbf{C}_2 \rightarrow \mathbf{C}_2 + \mathbf{C}_3,$$

$$\mathbf{C}_1 + \mathbf{C}_3 \rightarrow \mathbf{C}_1 + \mathbf{C}_2 + \mathbf{C}_3,$$

$$\mathbf{C}_2 + \mathbf{C}_3 \rightarrow \mathbf{C}_1,$$

$$\mathbf{C}_1 + \mathbf{C}_2 + \mathbf{C}_3 \rightarrow \mathbf{C}_1 + \mathbf{C}_2. \qquad \qquad \square$$

Example 8.3 Consider the $(4, 2)$ linear code over $GF(3)$ defined by the generator matrix

$$G = \begin{bmatrix} 1 & 0 & 2 & 0 \\ 1 & 1 & 2 & 2 \end{bmatrix}.$$

This code has nine codewords. Denoting the rows of G by $\mathbf{C}_1$ and $\mathbf{C}_2$ the nonzero codewords are:

$$\mathbf{C}_1 = 1020,$$

$$2\mathbf{C}_1 = 2010,$$

$$\mathbf{C}_2 = 1122,$$

$$\mathbf{C}_1 + \mathbf{C}_2 = 2112,$$

$$2\mathbf{C}_1 + \mathbf{C}_2 = 0102,$$

$$2\mathbf{C}_2 = 2211,$$

$$\mathbf{C}_1 + 2\mathbf{C}_2 = 0201,$$

$$2\mathbf{C}_1 + 2\mathbf{C}_2 = 1221.$$

This code is also a cyclic code. For example, the right cyclic shift of $\mathbf{C}_1$ is $2\mathbf{C}_1 + \mathbf{C}_2$. The complete list of right cyclic shifts follows:

$$\mathbf{C}_1 \rightarrow 2\mathbf{C}_1 + \mathbf{C}_2,$$

$$2\mathbf{C}_1 \rightarrow \mathbf{C}_1 + 2\mathbf{C}_2,$$

$$\mathbf{C}_2 \rightarrow \mathbf{C}_1 + \mathbf{C}_2,$$

$$\mathbf{C}_1 + \mathbf{C}_2 \rightarrow 2\mathbf{C}_2,$$

$$2\mathbf{C}_1 + \mathbf{C}_2 \rightarrow 2\mathbf{C}_1,$$

$$2\mathbf{C}_2 \rightarrow 2\mathbf{C}_1 + 2\mathbf{C}_2,$$

$$\mathbf{C}_1 + 2\mathbf{C}_2 \rightarrow \mathbf{C}_1,$$

$$2\mathbf{C}_1 + 2\mathbf{C}_2 \rightarrow \mathbf{C}_2. \qquad \square$$

The definition of a cyclic code may seem arbitrary, but in fact there are good reasons for it, which begin to appear if we introduce the notion of the generating function of a codeword. If $\mathbf{C} = (C_0, \ldots, C_{n-1})$ is a codeword, its *generating function* is defined to be the polynomial

$$C(x) = C_0 + C_1 x + \cdots + C_{n-1} x^{n-1},$$

where x is an indeterminate. The reason this is useful is that by using generating functions, we can give a simple algebraic characterization of the right cyclic shift of a codeword. In order to give this characterization, we need to define the important "mod" operator for integers and polynomials.[1]

Definition *If p and m are integers with $m > 0$, then "$p \bmod m$" denotes the remainder obtained when p is divided by m; thus $p \bmod m$ is the integer r such that $p - r$ is divisible by m, and $0 \leqslant r \leqslant m - 1$. Similarly, if $P(x)$ and $Q(x)$ are polynomials, $P(x) \bmod M(x)$ denotes the remainder when $P(x)$ is divided by $M(x)$; thus $P(x) \bmod M(x)$ is the unique polynomial $R(x)$ such that $P(x) - R(x)$ is divisible by $M(x)$ and $\deg R(x) < \deg M(x)$.* $\qquad \square$

Example 8.4 Here are some examples.

$$7 \bmod 5 = 2,$$

$$-6 \bmod 4 = 2,$$

$$4 \bmod 6 = 4,$$

$$21 \bmod 7 = 0,$$

$$x^3 \bmod x^2 = 0,$$

$$x^2 \bmod x^3 = x^2,$$

$$x^{1000} \bmod (x^2 + x + 1) = x,$$

$$(5x^2 + 1) \bmod (x^2 + 1) = -4 \qquad \text{(over the reals)},$$

$$(x + 1)^3 \bmod (x^2 + 1) = 0 \qquad \text{(over } GF(2)),$$

$$x^i \bmod (x^n - 1) = x^{i \bmod n}. \qquad \square$$

The following lemma lists the most important properties of the mod operator for polynomials.

Lemma 1
  (a) *If* $\deg P(x) < \deg M(x)$, *then* $P(x) \bmod M(x) = P(x)$.
  (b) *If* $M(x) | P(x)$, *then* $P(x) \bmod M(x) = 0$.
  (c) $(P(x) + Q(x)) \bmod M(x) = P(x) \bmod M(x) + Q(x) \bmod M(x)$.
  (d) $(P(x)Q(x)) \bmod M(x) = (P(x)(Q(x) \bmod M(x))) \bmod M(x)$.
  (e) *If* $M(x) | N(x)$, *then* $(P(x) \bmod N(x)) \bmod M(x) = P(x) \bmod M(x)$.

Proof Left as Problem 8.5. $\qquad \square$

Now we can give the promised algebraic characterization of the right cyclic shift operation.

Theorem 8.1 *If* $\mathbf{C} = (C_0, C_1, \ldots, C_n)$ *is a codeword with generating function* $C(x) = C_0 + C_1 x + \cdots + C_{n-1} x^{n-1}$, *then the generating function* $C^R(x)$ *for the right cyclic shift* $\mathbf{C}^R$ *is given by the formula*

$$C^R(x) = xC(x) \bmod (x^n - 1).$$

Proof Since $C(x) = C_0 + C_1 x + \cdots + C_{n-1} x^{n-1}$, we have

$$xC(x) = C_0 x + \cdots + C_{n-2} x^{n-1} + C_{n-1} x^n,$$

$$C^R(x) = C_{n-1} + C_0 x + \cdots + C_{n-2} x^{n-1}.$$

Hence $xC(x) - C^R(x) = C_{n-1}(x^n - 1)$. Since $\deg C^R(x) < \deg(x^n - 1)$, and $xC(x) - C^R(x)$ is a multiple of $x^n - 1$, the result follows from the definition of the mod operation.   □

The generating function for a codeword is so useful that we will often not bother to distinguish between a codeword and its generating function. Thus for example we might view an (n, k) linear code as a set of polynomials of (formal) degree $n - 1$, such that any linear combination of polynomials in the code is again in the code. From this viewpoint, by Theorem 8.1, a cyclic code is a linear code such that if $C(x)$ is a codeword, then $xC(x) \bmod (x^n - 1)$ is also. By repeatedly applying the right cyclic shift generation, we find that $x^i C(x) \bmod (x^n - 1)$ is a codeword, for all $i \geq 0$ (see Problem 8.6). The following theorem is a generalization of this observation. For convenience, we introduce the notation

$$[P(x)]_n$$

as a shorthand for

$$P(x) \bmod (x^n - 1).$$

Theorem 8.2 *If C is an (n, k) cyclic code, and if $C(x)$ is a codeword in C, then for any polynomial $P(x)$, $[P(x)C(x)]_n$ is also a codeword in C*

Proof Suppose $P(x) = \sum_{i=0}^{m} P_i x^i$. Then

$$[P(x)C(x)]_n = \left[\left(\sum_{i=0}^{m} P_i x^i \right) C(x) \right]_n$$

$$= \sum_{i=0}^{m} P_i [x^i C(x)]_n$$

by Lemma 1(c). But by the remarks immediately preceding the statement of this theorem, $[x^i C(x)]_n$ is a codeword for each i, and so, since the code is linear, the linear combination $\sum P_i [x^i C(x)]_n$ is also a codeword.   □

Example 8.5 Consider the $(7, 3)$ cyclic code of Example 8.2. The codeword

$\mathbf{C}_1 + \mathbf{C}_3$, viewed as a polynomial, is $1 + x^3 + x^5 + x^6$. According to Theorem 8.2, if we multiply this polynomial by any other polynomial, and reduce the result mod $(x^7 - 1)$, the resulting polynomial will also be in the code. For example:

$$[(1 + x)(1 + x^3 + x^5 + x^6)]_7 = x + x^3 + x^4 + x^5$$

$$= \mathbf{C}_2,$$

$$[(1 + x^{53} + x^{100})(1 + x^3 + x^5 + x^6)]_7 = 1 + x + x^4 + x^6$$

$$= \mathbf{C}_1 + \mathbf{C}_2 + \mathbf{C}_3,$$

$$[(1 + x^2 + x^3)(1 + x^3 + x^5 + x^6)]_7 = 0. \qquad \square$$

The key to the design and analysis of cyclic codes is the *generator polynomial*.

Definition *If C is a cyclic code, a nonzero polynomial of lowest degree in C is called a generator polynomial for C. The symbol $g(x)$ is usually reserved to denote a generator polynomial.*

Example 8.6 In the code of Example 8.2, the codeword $\mathbf{C}_1$, viewed as a polynomial, has lowest degree among all nonzero codewords, and so $g(x) = 1 + x^2 + x^3 + x^4$ is the generator polynomial for this code. In the code of Example 8.3, there are two lowest-degree polynomials, viz. $\mathbf{C}_1 = 2x^2 + 1$ and $2\mathbf{C}_1 = x^2 + 2$. The first part of the following lemma shows that the generator polynomial for a cyclic code is always unique up to multiplication by scalars, and so we are justified in referring to *the* generator polynomial of a cyclic code, and assuming it is monic. Thus for the code of Example 8.3, normally one would say that $g(x) = x^2 + 2$ is the generator polynomial. $\qquad \square$

Lemma 2 *Suppose that C is a cyclic code with generator polynomial $g(x)$.*

(a) *If $g'(x)$ is another generator polynomial, then $g'(x) = \lambda g(x)$, for some nonzero element $\lambda \in F$.*

(b) *If $P(x)$ is a polynomial such that $[P(x)]_n$ is a codeword, then $g(x)$ divides $P(x)$.*

Proof To prove (a), suppose that $g(x) = g_r x^r + \cdots + g_0$ and $g'(x) = g'_r x^r + \cdots + g'_0$, with $g_r \neq 0$ and $g'_r \neq 0$. Then if $\lambda = g'_r / g_r$ the polynomial

$g''(x) = g'(x) - \lambda g(x)$ has degree less than r and is in C. But r is the lowest possible degree among *nonzero* codewords in C, so that $g''(x) = 0$, i.e., $g'(x) = \lambda g(x)$.

To prove (b), let $Q(x)$ and $R(x)$ be the quotient and remainder obtained when $P(x)$ is divided by $g(x)$, i.e.,

$$P(x) = Q(x)g(x) + R(x), \tag{8.1}$$

with deg $R < \deg g$. Reducing each of these polynomials mod $x^n - 1$, and using the fact that deg $R < \deg g \leq n - 1$ (the latter inequality since $g(x)$ is a codeword), we obtain

$$R(x) = [P(x)]_n - [Q(x)g(x)]_n.$$

But $[P(x)]_n$ is a codeword by assumption, and $[Q(x)g(x)]_n$ is a codeword by Theorem 8.2. Thus, since C is linear, $R(x)$ is a codeword also. But deg $R < \deg g$, and $g(x)$ is a nonzero codeword of least degree, so $R(x) = 0$, and (8.1) becomes

$$P(x) = Q(x)g(x)$$

which shows that $g(x)$ divides $P(x)$.   □

We can now state and prove the main theorem about cyclic codes. It establishes a one-to-one correspondence between cyclic codes of length n and monic divisors of $x^n - 1$.

Theorem 8.3
   (a) *If C is an (n, k) cyclic code over F, then its generator polynomial is a divisor of $x^n - 1$. Futhermore, the vector $\mathbf{C} = (C_0, C_1 \ldots, C_{n-1})$ is in the code if and only if the corresponding generating function $C(x) = C_0 + C_1 x + \cdots + C_{n-1}x^{n-1}$ is divisible by $g(x)$. If k denotes the dimension of C, then $k = n - \deg g(x)$.*
   (b) *Conversely, if $g(x)$ is a divisor of $x^n - 1$, then there is an (n, k) cyclic code with $g(x)$ as its generator polynomial and $k = n - \deg g(x)$, namely, the set of all vectors $(C_0, C_1, \ldots, C_{n-1})$ whose generating functions are divisible by $g(x)$.*

Proof of (a) First, let $P(x) = x^n - 1$ in Lemma 2(b); $[P(x)]_n = 0$, which is of course a codeword, and so $g(x)$ divides $x^n - 1$. Next, by Theorem 8.2, any vector of length n whose generating function is a multiple of $g(x)$ is a codeword. Conversely, if $C(x) = C_0 + C_1 x + \cdots + C_n x^{n-1}$ is a codeword, then $[C(x)]_n = C(x)$ and so Lemma 2 implies that $g(x)$ divides $C(x)$. Finally, the assertion about the degree of $g(x)$ follows since $C(x) = C_0 + C_1 x + \cdots + C_{n-1} x^{n-1}$ is a multiple of $g(x)$ if and only if $C(x) = g(x)I(x)$, where $\deg I \leqslant n - 1 - \deg g$.

Proof of (b) Suppose that $g(x)$ is a divisor of $x^n - 1$. Then $C(x) = (C_0, C_1, \ldots, C_{n-1})$ is a multiple of $g(x)$ if and only if $C(x) = g(x)I(x)$, where $\deg g + \deg I \leqslant n - 1$. Thus the set of all such words is an (n, k) linear code, where $k = n - \deg g$. To show that this code is cyclic, we must show that the right cyclic shift of any codeword is also a codeword. Thus let $I(x)g(x)$ be any codeword; by Theorem 8.1, its right cyclic shift is $[xI(x)g(x)]_n$. But since $g(x)$ divides $x^n - 1$, we have

$$[xI(x)g(x)]_n \bmod g(x) = [xI(x)g(x)] \bmod g(x) \qquad \text{(by Lemma 1(e))}$$

$$= 0 \qquad \text{(by Lemma 1(b)),}$$

which proves that $[xI(x)g(x)]_n$ is a multiple of $g(x)$, so the code is indeed a cyclic code. $\qquad \Box$

Theorem 8.3 shows the importance of the generator polynomial of a cyclic code. A closely related, and equally important, polynomial is the *parity-check* polynomial for a cyclic code, which is denoted by $h(x)$, and defined by

$$h(x) = \frac{x^n - 1}{g(x)}.$$

The following corollaries to Theorem 8.3 give explicit descriptions of generator and parity-check matrices for a cyclic code, in terms of $g(x)$ and $h(x)$.

Corollary 1 *If C is an (n, k) cyclic code with generator polynomial $g(x) = g_0 + g_1 x + \cdots + g_r x^r$ (with $r = n - k$), and parity-check polynomial $h(x) = h_0 + h_1 x + \cdots + h_k x^k$, then the following matrices are generator and parity-check matrices for C:*

$$G_1 = \begin{bmatrix} g_0 & g_1 & \cdots & \cdots & g_r & 0 & \cdots & \cdots & 0 \\ 0 & g_0 & g_1 & \cdots & \cdots & g_r & 0 & \cdots & 0 \\ \vdots & & & & & & & & \vdots \\ 0 & \cdots & \cdots & 0 & g_0 & g_1 & \cdots & \cdots & g_r \end{bmatrix} = \begin{bmatrix} g(x) \\ xg(x) \\ \vdots \\ x^{k-1}g(x) \end{bmatrix},$$

$$H_1 = \begin{bmatrix} h_k & h_{k-1} & \cdots & \cdots & h_0 & 0 & \cdots & \cdots & 0 \\ 0 & h_k & h_{k-1} & \cdots & \cdots & h_0 & 0 & \cdots & 0 \\ \vdots & & & & & & & & \vdots \\ 0 & \cdots & \cdots & 0 & h_k & h_{k-1} & \cdots & \cdots & h_0 \end{bmatrix} = \begin{bmatrix} \tilde{h}(x) \\ x\tilde{h}(x) \\ \vdots \\ x^{r-1}\tilde{h}(x) \end{bmatrix},$$

where $\tilde{h}(x) = h_k + h_{k-1} + \cdots + h_0 x^k$ is $h(x)$'s "reciprocal" polynomial. Furthermore, if the vector $\mathbf{I} = (I_0, I_1, \ldots, I_{k-1})$ is encoded as $\mathbf{C} = \mathbf{I}G_1$ (cf. Eq. (7.1)), then the generating functions $I(x) = I_0 + I_1 x + \cdots + I_{k-1}x^{k-1}$ and $C(x) = C_0 + C_1 x + \cdots + C_{n-1}x^{n-1}$ are related by

$$C(x) = I(x)g(x).$$

Proof The ith row of G_1, by definition, is $x^i g(x)$, for $i = 0, 1, \ldots, k-1$. Each of these k vectors is in C by Theorem 8.2, and they are linearly independent since each row has a different degree. Since C is k-dimensional, it follows that G_1 is a generator matrix for C. Furthermore, the vector $\mathbf{I}G_1$, where $\mathbf{I} = (I_0, I_1, \ldots, I_{k-1})$, has generating function $I_0 g(x) + I_1 xg(x) + \cdots + I_{n-1}x^{n-1}g(x) = I(x)g(x)$.

To prove that H_1 is a parity-check matrix for C, note that the inner product of the ith row of G_1 and the jth row of H_1 is the coefficient of x^{k-i+j} in the product $g(x)h(x)$. But $g(x)h(x) = x^n - 1$, and since the index $k - i + j$ ranges from 1 (when $i = k - 1$ and $j = 0$) to $n - 1$ (when $i = 0$ and $j = r - 1$), each of these inner products is zero. Thus each row of H_1 is in the nullspace of C; but the nullspace of C is r-dimensional, and H_1 has r linearly independent rows, so H_1 is in fact a parity-check matrix for C. $\square$

The matrices in Corollary 1 are sometimes useful, but more often, the "systematic" matrices in the following Corollary 2 are used.

Corollary 2 *Let C be an (n, k) cyclic code with generator polynomial $g(x)$. For $i = 0, 1, \ldots, k - 1$, let $G_{2,i}$ be the length n vector whose generating function is $G_{2,i}(x) = x^{r+i} - x^{r+i} \bmod g(x)$. Then the $k \times n$ matrix*

$$G_2 = \begin{bmatrix} G_{2,0} \\ G_{2,1} \\ \vdots \\ G_{2,k-1} \end{bmatrix}$$

is a generator matrix for C. Similarly, if $H_{2,j}$ is the length r vector whose generating function is $H_{2,j}(x) = x^j \bmod g(x)$, then the $r \times n$ matrix

$$H_2 = [H_{2,0}^T, H_{2,1}^T, \ldots, H_{2,n-1}^T]$$

is a parity-check matrix for $\mathcal{C}$. Furthermore, if the vector $\mathbf{I} = (I_0, I_1, \ldots, I_{k-1})$ is encoded as $\mathbf{C} = \mathbf{I}G_2$, then the generating functions $I(x)$ and $C(x)$ are related by

$$C(x) = x^r I(x) - [x^r I(x)] \bmod g(x).$$

Also, if the syndrome of the vector $\mathbf{R} = (R_0, R_1, \ldots, R_{n-1})$ is calculated as $\mathbf{S}^T = H_2 \mathbf{R}^T$, then the generating functions $R(x)$ and $S(x)$ are related by

$$S(x) = R(x) \bmod g(x).$$

Proof The ith row of G_2 is a multiple of $g(x)$ and so a codeword of $\mathcal{C}$, because

$$[x^{r+i} - x^{r+i} \bmod g(x)] \bmod g(x) = x^{r+i} \bmod g(x) - x^{r+i} \bmod g(x)$$

$$= 0$$

by Lemma 1 parts (c) and (a). Since G_2 has k linearly independent rows (the last k columns of G_2 form a $k \times k$ identity matrix), each of which is a codeword, G_2 is a generator matrix of C. Furthermore, the vector $\mathbf{I}G_2$, where $\mathbf{I} = (I_0, I_1, \ldots, I_{k-1})$, has generating function $I_0(x^r - x^r \bmod g(x)) + I_1(x^{r+1} - x^{r+1} \bmod g(x)) + \cdots + I_{k-1}(x^{r+k-1} - x^{r+k-1} \bmod g(x)) = x^r I(x) - [x^r I(x)] \bmod g(x)$.

To prove the assertion about H_2, note first that H_2 has rank r (its first r columns form an $r \times r$ identity matrix). Furthermore, if $\mathbf{R} = (R_0, R_1, \ldots, R_{n-1})$ is a received word, then $H_2 \mathbf{R}^T$ is an $r \times 1$ column vector with generating function $\sum_{j=0}^{n-1} R_j(x^j \bmod g(x)) = (\sum_{j=0}^{n-1} R_j x^j) \bmod g(x) = R(x) \bmod g(x)$. If $R(x)$ is a codeword, then $R(x) \bmod g(x) = 0$ by Theorem 8.3(a), and so $H_2 \mathbf{R}^T = \mathbf{0}$ for any codeword. Thus shows that H_2 is a parity-check matrix for C. On the other hand, if $\mathbf{R}$ is an arbitrary received vector, its syndrome with respect to H_2 is $\mathbf{S} = H_2 \mathbf{R}^T$, which, as we have just seen, has generating function $S(x) = R(x) \bmod g(x)$, as asserted. $\square$

Note *The form of the syndrome relative to the parity-check matrix H_2 as described in Corollary 2 is so natural that it is common to refer to it simply as the remainder syndrome of the vector $\mathbf{R}$ with respect to the cyclic code with generator polynomial $g(x)$.* $\square$

Example 8.7 In Example 8.6 we saw that the generator polynomial for the $(7, 3)$ cyclic code of Example 8.2 is $g(x) = x^4 + x^3 + x^2 + 1$. The corresponding parity-check polynomial is then $h(x) = (x^7 + 1)/(x^4 + x^3 + x^2 + 1) = x^3 + x^2 + 1$. The eight codewords, as multiples of $g(x)$, are:

$$\mathbf{C}_0 = 0 \cdot g(x),$$

$$\mathbf{C}_1 = 1 \cdot g(x),$$

$$\mathbf{C}_2 = x \cdot g(x),$$

$$\mathbf{C}_3 = x^2 \cdot g(x),$$

$$\mathbf{C}_4 = (1 + x) \cdot g(x),$$

$$\mathbf{C}_5 = (1 + x^2) \cdot g(x),$$

$$\mathbf{C}_6 = (x + x^2) \cdot g(x),$$

$$\mathbf{C}_7 = (1 + x + x^2) \cdot g(x).$$

The generator and parity-check matrices described in Corollary 1 are

$$G_1 = \begin{bmatrix} 1 & 0 & 1 & 1 & 1 & 0 & 0 \\ 0 & 1 & 0 & 1 & 1 & 1 & 0 \\ 0 & 0 & 1 & 0 & 1 & 1 & 1 \end{bmatrix} = \begin{bmatrix} g(x) \\ xg(x) \\ x^2 g(x) \end{bmatrix},$$

$$H_1 = \begin{bmatrix} 1 & 1 & 0 & 1 & 0 & 0 & 0 \\ 0 & 1 & 1 & 0 & 1 & 0 & 0 \\ 0 & 0 & 1 & 1 & 0 & 1 & 0 \\ 0 & 0 & 0 & 1 & 1 & 0 & 1 \end{bmatrix} = \begin{bmatrix} \tilde{h}(x) \\ x\tilde{h}(x) \\ x^2 \tilde{h}(x) \\ x^3 \tilde{h}(x) \end{bmatrix}.$$

The generator and parity-check matrices described in Corollary 2 are

$$G_2 = \begin{bmatrix} 1 & 0 & 1 & 1 & 1 & 0 & 0 \\ 1 & 1 & 1 & 0 & 0 & 1 & 0 \\ 0 & 1 & 1 & 1 & 0 & 0 & 1 \end{bmatrix} = \begin{bmatrix} x^4 - x^4 \bmod g(x) \\ x^5 - x^5 \bmod g(x) \\ x^6 - x^6 \bmod g(x) \end{bmatrix},$$

$$H_2 = \begin{bmatrix} 1 & 0 & 0 & 0 & 1 & 1 & 0 \\ 0 & 1 & 0 & 0 & 0 & 1 & 1 \\ 0 & 0 & 1 & 0 & 1 & 1 & 1 \\ 0 & 0 & 0 & 1 & 1 & 0 & 1 \end{bmatrix}$$

$$= [1, x, x^2, x^3, x^4 \bmod g, x^5 \bmod g, x^6 \bmod g].$$

Note the appearance of a 3×3 identity matrix on the right side of G_2, and the 4×4 identity matrix on the left side of H_2. This is what makes these matrices "systematic." If we wish to obtain generator and parity-check matrices of the form $G = [I_k A]$ and $H = [-A^T I_{n-k}]$, as promised in Theorem 7.1, we can use the cyclic property of the code, cyclically shifting the rows of G_2 three places to the right, and those of H_2 three places to the right, thus obtaining

$$G_3 = \begin{bmatrix} 1 & 0 & 0 & 1 & 0 & 1 & 1 \\ 0 & 1 & 0 & 1 & 1 & 1 & 0 \\ 0 & 0 & 1 & 0 & 1 & 1 & 1 \end{bmatrix},$$

$$H_3 = \begin{bmatrix} 1 & 1 & 0 & 1 & 0 & 0 & 0 \\ 0 & 1 & 1 & 0 & 1 & 0 & 0 \\ 1 & 1 & 1 & 0 & 0 & 1 & 0 \\ 1 & 0 & 1 & 0 & 0 & 0 & 1 \end{bmatrix}.$$

(For another way to describe G_3 and H_3, see Problem 8.9.) If we encode the vector $\mathbf{I} = [101]$ using G_1, the resulting codeword is $(1 + x^2)(1 + x^2 + x^3 + x^4) = 1 + x^3 + x^5 + x^6 = [1001011]$. On the other hand, if we use G_2 instead, the codeword is $x^4(1 + x^2) - [x^4(1 + x^2)] \bmod (x^4 + x^3 + x^2 + 1) = x^6 + x^4 + x + 1 = [11001010]$. The syndrome of the vector $\mathbf{R} = [1010011]$ with respect to H_1 is $[1101]$. (For an efficient way to make this computation, see Problem 8.10.) On the other hand, if we use H_2 instead, we find that the "remainder" syndrome is $R(x) \bmod g(x) = (1 + x^2 + x^5 + x^6) \bmod (1 + x^2 + x^3 + x^4) = x^3 + x^2$, i.e., $\mathbf{S} = [0011]$.   □

Example 8.8 In Example 8.6 we saw that the generator polynomial for the $(4, 2)$ $GF(3)$-cyclic code of Example 8.3 is $g(x) = x^2 + 2 = x^2 - 1$. Then the parity-check polynomial is $(x^4 - 1)/(x^2 - 1) = x^2 + 1$. The generator and parity-check matrices of Corollary 1 to Theorem 8.3 are then

$$G_1 = \begin{bmatrix} 2 & 0 & 1 & 0 \\ 0 & 2 & 0 & 1 \end{bmatrix},$$

$$H_1 = \begin{bmatrix} 1 & 0 & 1 & 0 \\ 0 & 1 & 0 & 1 \end{bmatrix}$$

and

$$G_2 = \begin{bmatrix} 2 & 0 & 1 & 0 \\ 0 & 2 & 0 & 1 \end{bmatrix},$$

$$H_2 = \begin{bmatrix} 1 & 0 & 1 & 0 \\ 0 & 1 & 0 & 1 \end{bmatrix}.$$

It is only a coincidence that $G_1 = G_2$ and $H_1 = H_2$ in this case. But see Problem 8.21. ☐

According to Theorem 8.3, the cyclic codes of length n over a given field F are in one-to-one correspondence with the monic divisors of $x^n - 1$ over F. Plainly then, in order to study cyclic codes over F it is useful to know how to factor $x^n - 1$ over F. Although we shall not make a systematic study of the factorization of $x^n - 1$ here, in Table 8.1 we give the factorization of $x^n - 1$ over $GF(2)$, for $1 \leqslant n \leqslant 31$. This table contains all the information necessary for a study of cyclic codes of length $\leqslant 31$ over $GF(2)$.

Example 8.9 Let's use Table 8.1 to list all possible binary codes of length 7. According to Table 8.1, $x^7 - 1$ factors into three distinct irreducible factors: $x^7 - 1 = (x + 1)(x^3 + x + 1)(x^3 + x^2 + 1)$, and so $x^7 - 1$ has $2^3 = 8$ distinct divisors $g(x)$. This yields the following list:

(n, k)	$g(x)$	comment
$(7, 7)$	1	no parity code
$(7, 6)$	$x + 1$	overall parity-check code
$(7, 4)$	$x^3 + x + 1$	Hamming code
$(7, 4)$	$x^3 + x^2 + 1$	Hamming code
$(7, 3)$	$(x + 1)(x^3 + x + 1)$	Example 8.2 code
$(7, 3)$	$(x + 1)(x^3 + x^2 + 1)$	Example 8.2 code "reversed"
$(7, 1)$	$(x^3 + x + 1)(x^3 + x^2 + 1)$	repetition code
$(7, 0)$	$x^7 + 1$	no information code

For example, we assert that the $(7, 4)$ cyclic code with $g(x) = x^3 + x + 1$ is a Hamming code. To see why this is so, we note that the parity-check polynomial is $(x^7 + 1)/(x^3 + x + 1) = x^4 + x^2 + x + 1$. Thus by Corollary 1 to Theorem 8.3, a parity-check matrix for this code is

$$H = \begin{bmatrix} 1 & 0 & 1 & 1 & 1 & 0 & 0 \\ 0 & 1 & 0 & 1 & 1 & 1 & 0 \\ 0 & 0 & 1 & 0 & 1 & 1 & 1 \end{bmatrix}.$$

The columns of H are nonzero and distinct, and so by the definition in Section 7.4, this is indeed a Hamming code. We leave the investigation of the other "comments" as Problem 8.23. ☐

In Example 8.9, we saw that $g(x) = x^3 + x + 1$ generates a $(7, 4)$ Hamming code. But according to Table 8.1, $g(x)$ divides not only $x^7 - 1$, but also $x^{14} - 1$; in fact, $x^3 + x + 1 | x^n - 1$ for any n which is a multiple of 7 (see Problem 8.24). Thus by Theorem 8.3(b), $g(x)$ generates a whole family of cyclic codes, with parameters $(7, 4)$, $(14, 11)$, $(21, 18)$, ... However, all of these codes except the first contain a vector whose generating function is $x^7 - 1$, and so have minimum weight equal to 2. Hence these codes are not interesting for error correction; we shall call them *improper* cyclic codes. (See Problem 8.26.) In the rest of this chapter, when we refer to an (n, k) cyclic code with generator polynomial $g(x)$, we shall assume it is a *proper* cyclic code, i.e., one for which n is the *smallest* positive integer such that $g(x) | x^n - 1$. This integer is sometimes called the *period* of $g(x)$, since it is the period of the sequence $[x^i \bmod g(x)]_{i \geq 0}$. See Problem 8.27 for a discussion of cyclic codes for which the *parity-check* polynomial has period less than n.) With this convention established, we are ready to continue our study of cyclic codes.

8.2 Shift-register encoders for cyclic codes

Now that we know by Theorem 8.3 that every cyclic code is characterized by its generator polynomial, we can begin to see why cyclic codes are much easier to implement than arbitrary linear codes. In this section we shall show, in fact, that every cyclic code can be encoded with a simple finite-state machine called a *shift-register encoder*.

Recall that an encoding algorithm for an (n, k) linear code, cyclic or not, is a rule for mapping the set of length-k information sequences $(I_0, I_1, \ldots, I_{k-1})$ onto the set of length-n codewords $(C_0, C_1, \ldots, C_{n-1})$, or equivalently for mapping the information polynomials $I(x) = (I_0, I_1 x + \cdots + I_{k-1} x^{k-1}$ onto the set of code polynomials $C(x) = C_0 + C_1 x + \cdots + C_{n-1} x^{n-1}$. If the code is cyclic, Theorem 8.3 tells us that $C(x)$ is a codeword if and only if $C(x)$ is a multiple of $g(x)$, and so, as we saw in Corollary 1 to Theorem 8.3, one way to change an information polynomial $I(x)$ into a code polynomial is simply to multiply $I(x)$ by $g(x)$:

$$I(x) \rightarrow I(x)g(x). \tag{8.2}$$

It turns out that polynomial multiplication is easy to implement using something called *shift-register logic*, and we shall now make a brief study of this subject.[2]

Figure 8.1 is an abstract representation of a machine capable of multiplying

Table 8.1 Factorization of $x^n - 1 (= x^n + 1)$ into powers of irreducible polynomials over $GF(2)$, for $1 \leqslant n \leqslant 31$.

n	$x^n + 1 =$
1	$(x + 1)$
2	$(x + 1)^2$
3	$(x + 1)(x^2 + x + 1)$
4	$(x + 1)^4$
5	$(x + 1)(x^4 + x^3 + x^2 + x + 1)$
6	$(x + 1)^2(x^2 + x + 1)^2$
7	$(x + 1)(x^3 + x + 1)(x^3 + x^2 + 1)$
8	$(x + 1)^8$
9	$(x + 1)(x^2 + x + 1)(x^6 + x^3 + 1)$
10	$(x + 1)^2(x^4 + x^3 + x^2 + x + 1)^2$
11	$(x + 1)(x^{10} + x^9 + x^8 + x^7 + x^6 + x^5 + x^4 + x^3 + x^2 + x + 1)$
12	$(x + 1)^4(x^2 + x + 1)^4$
13	$(x + 1)(x^{12} + x^{11} + x^{10} + x^9 + x^8 + x^7 + x^6 + x^5 + x^4 + x^3 + x^2 + x + 1)$
14	$(x + 1)^2(x^3 + x + 1)^2(x^3 + x^2 + 1)^2$

15    $(x+1)(x^2+x+1)(x^4+x+1)(x^4+x^3+1)(x^4+x^3+x^2+x+1)$

16    $(x+1)^{16}$

17    $(x+1)(x^8+x^5+x^4+x^3+1)(x^8+x^7+x^6+x^4+x^2+x+1)$

18    $(x+1)^2(x^2+x+1)^2(x^6+x^3+1)^2$

19    $(x+1)(x^{18}+x^{17}+x^{16}+\cdots+x+1)$

20    $(x+1)^4(x^4+x^3+x^2+x+1)^4$

21    $(x+1)(x^2+x+1)(x^3+x+1)(x^6+x^4+x^2+x+1)(x^6+x^5+x^4+x^2+1)$

22    $(x+1)^2(x^{10}+x^9+x^8+x^7+x^6+x^5+x^4+x^3+x^2+x+1)^2$

23    $(x+1)(x^{11}+x^9+x^7+x^6+x^5+x+1)(x^{11}+x^{10}+x^6+x^5+x^4+x^2+1)$

24    $(x+1)^8(x^2+x+1)^8$

25    $(x+1)(x^4+x^3+x^2+x+1)(x^{20}+x^{15}+x^{10}+x^5+1)$

26    $(x+1)^2(x^{12}+x^{11}+x^{10}+x^9+x^8+x^7+x^6+x^5+x^4+x^3+x^2+x+1)^2$

27    $(x+1)(x^2+x+1)(x^6+x^3+1)(x^{18}+x^9+1)$

28    $(x+1)^4(x^3+x+1)^4(x^3+x^2+1)^4$

29    $(x+1)(x^{28}+x^{27}+\cdots+x+1)$

30    $(x+1)^2(x^2+x+1)^2(x^4+x^3+1)^2(x^4+x^3+x^2+x+1)^2$

31    $(x+1)(x^5+x^2+1)(x^5+x^3+1)(x^5+x^3+x^2+x+1)(x^5+x^4+x^2+x+1)(x^5+x^4+x^3+x+1)(x^5+x^4+x^3+x^2+1)$

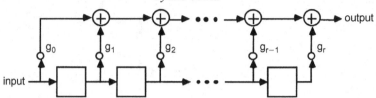

Figure 8.1 A shift-register circuit for multiplying by $g(x) = g_0 + g_1x + \cdots + g_rx^r$. Alternatively, this is an encoder for an (n, k) cyclic code with generator polynomial $g(x)$, with $r = n - k$.

an *arbitrary* polynomial $I(x)$ by a *fixed* polynomial $g(x)$. Before proving that the circuit works, we had better explain its various components.

The circuit in Figure 8.1 is build by connecting together three types of components: *flip-flops, adders*, and *constant multipliers*. The most important of these is the flip-flop, sometimes called a *delay* element:

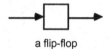

a flip-flop

A flip-flop is a device which can store one element from the field F. Not pictured in our simplified circuit diagrams, but an important part of them, is an external clock which generates a timing signal ("tick") every t_0 seconds.[3] When the clock ticks, the contents of the flip-flop are shifted out of the flip-flop in the direction of the arrow, through the circuit, until the next flip-flop is reached. Here the signal stops until the next tick. The contents of the shift register can be modified by the circuit between successive flip-flops, which is where the other two logic elements come in. The two-input *adder*, which looks like this

an adder

– is a device that computes the sum of its two input signals. (The circuit is always arranged in such a way that after every tick each input to the adder receives exactly one signal, so that the adder's output is always unambiguous.) Finally we come to the simplest circuit element, the *constant multiplier:*

constant multiplier

This device simply multiplies its input by the constant a, with no delay.

Now let's see if we can understand why the circuit of Figure 8.1 can be used for polynomial multiplication. We first write out in detail the formulas for the coefficients of the product $C(x) = C_0 + C_1 x + \cdots + C_{n-1} x^{n-1}$, where $C(x) = I(x)g(x)$:

$$C_0 = I_0 g_0,$$

$$C_1 = I_0 g_1 + I_1 g_0,$$

$$C_2 = I_0 g_2 + I_1 g_1 + I_2 g_0,$$

$$\vdots$$

$$C_j = I_0 g_j + I_1 g_{-1} + \cdots + I_j g_0,$$

$$\vdots$$

$$C_{n-1} = I_{k-1} g_r.$$

(8.3)

The flip-flops of Fig. 8.1 are initially filled with 0's. We then feed in the sequence $I_0, \ldots, I_{k-1}$, followed by $r = n - k$ 0's, one bit every tick, to the shift register via the input arrow. Let us now study the behavior of the circuit at each tick of the clock:

Tick 0:   Input: I_0
              Shift-register contents: $[0, 0, 0, \ldots, 0]$
              Output: $I_0 g_0$

Tick 1:   Input: I_1
              Shift-register contents: $[I_0, 0, 0, \ldots, 0]$
              Output: $I_0 g_1 + I_1 g_0$

                      $\vdots$

Tick j: Input: I_j
              Shift-register contents: $[I_{j-1}, \ldots, I_1, I_0, \ldots, 0]$
              Output: $I_0 g_j + I_1 g_{j-1} + \cdots + I_j g_0$

                      $\vdots$

Tick n − 1:   Input: 0
              Shift-register contents: $[0, \ldots, 0 I_{k-1}]$
              Output: $I_{k-1} g_r$

Hence if the circuit in Figure 8.1 is intialized by placing r 0's in the flip-flops, and given the n-symbol input sequence $(I_0, I_1, \ldots, I_{k-1}, 0, \ldots, 0)$ the output sequence will indeed be $(C_0, C_1, \ldots, C_{n-1})$, where the C_j's are defined by Eq. (8.3), and so this circuit can be used as an encoder for the (n, k) cyclic code with generator polynomial $g(x)$.

Comment *The preceeding description is for a "generic" shift-register circuit, capable of performing arithmetic in an arbitrary field F. The required devices (flip-flop, adders, multipliers) are not "off-the-shelf" items, except when the field is GF(2). In this case, however, these devices are very simple indeed:*

$$
\begin{aligned}
\text{flip-flop} \quad &= \quad D \text{ flip-flop,} \\
\text{adder} \quad &= \quad \text{XOR gate,} \\
\text{0-multiplier} \quad &= \quad \text{no connection,} \\
\text{1-multiplier} \quad &= \quad \text{direct wire.}
\end{aligned}
$$

Example 8.10 The generator polynomial for the (7, 3) binary cyclic code of Example 8.2 is $g(x) = x^4 + x^3 + x^2 + 1$, as we saw in Example 8.6. The corresponding shift-register encoder is shown in Figure 8.2. If (I_0, I_1, I_2) are the three information bits to be encoded, the four flip-flops should be initialized to 0, and the seven bits $(I_0, I_1, I_2, 0, 0, 0, 0)$ should be used as inputs to the encoder. These seven input bits will then produce seven output bits $(C_0, C_1, C_2, C_3, C_4, C_5, C_6)$, which is the corresponding codeword. For example, if the information bits are $(1, 0, 0)$, the corresponding codeword will be the "impulse response" $(1, 0, 1, 1, 1, 0, 0)$, which is in fact the generator polynomial $1 + x^2 + x^3 + x^4$ (and also the codeword $\mathbf{C}_1$ from Example 8.2). $\qquad\qquad\square$

The encoders of Figures 8.1 and 8.2 could hardly be simpler, but they are unfortunately not *systematic*, i.e., the information bits $(I_0, I_1, \ldots, I_{k-1})$ do not appear unchanged in the corresponding codeword $(C_0, C_1, \ldots, C_{n-1})$. It is possible, however, to design a systematic shift-register encoder for any cyclic code which is only slightly more complex than the nonsystematic one.

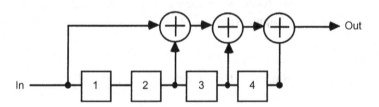

Figure 8.2 Nonsystematic encoder for the (7, 3) cyclic code with generator polynomial $g(x) = 1 + x^2 + x^3 + x^4$.

The idea is to use the result of Theorem 8.3, Corollary 2, which says that if $I(x)$ is an information polynomial, then

$$I(x) \to x^r I(x) - [x^r I(x)] \bmod g(x) \tag{8.4}$$

is a systematic encoding rule for a cyclic code with generator polynomial $g(x)$. The central component in such an encoder will be a "mod $g(x)$" circuit, where $g(x)$ is th e code's generator polynomial. Figure 8.3(a) shows such a circuit, if $g(x) = x^r + g_{r-1}x^{r-1} + \cdots + g_0$.

In order to see why the circuit of Figure 8.3(a) is able to perform a "mod $g(x)$" calculation, we focus on the shift-register contents $[s_0, s_1, \ldots, s_{r-1}]$, which we call the *state vector* of the machine, and on the corresponding generating function $S(x) = s_0 + s_1 + \cdots + s_{r-1}x^{r-1}$, which we call the *state polynomial*. We begin with a lemma which explains how the state changes in reponse to an input.

Lemma 3 *If the circuit in Figure 8.3(a) has state polynomial $S(x)$, and the input is s, then the next state polynomial will be*

$$S'(x) = (s + xS(x)) \bmod g(x).$$

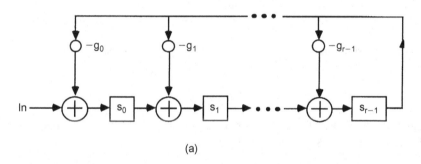

(a)

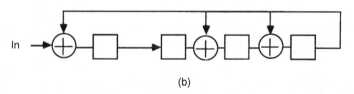

(b)

Figure 8.3 (a). A general "mod $g(x)$" shift-register circuit, where $g(x) = x^r + g_{r-1}x^{r-1} + \cdots + g_0$. (b) A specific "mod $x^4 + x^3 + x^2 + 1$" shift register, where all arithmetic is mod 2.

Proof If the present state vector is $\mathbf{S} = [s_0, \ldots, s_{r-1}]$ and the input is s, then by our rules describing the building blocks of the circuit, the next state vector will be, by definition,

$$\mathbf{S}' = [s - g_0 s_{r-1}, s_0 - g_1 s_{r-1}, \ldots, s_{r-2} - g_{r-1} s_{r-1}].$$

Hence the next state polynomial is

$$S'(x) = s + s_0 x + \cdots + s_{r-2} x^{r-1} - s_{r-1}(g_0 + g_1 x + \cdots + g_{r-1} x^{r-1})$$

$$= s + xS(x) - s_{r-1} g(x).$$

Thus $S'(x)$ is a polynomial of degree $< \deg g(x)$ such that $(s + xS(x)) - S'(x)$ is a multiple of $g(x)$, and so $S'(x) = (s + xS(x)) \bmod g(x)$, by Lemma 1(c). $\square$

Theorem 8.4 *If the circuit of Figure 8.3(a) is initialized by setting $s_0 = s_1 = \cdots = s_{r-1} = 0$, and then given the input sequence $a_0, a_1, a_2, \ldots$ (a_0 is input at the 0th tick, a_1 at the first tick, etc.), then after the tth tick, the state polynomial will be*

$$S_t(x) = \sum_{j=0}^{t} a_j x^{t-j} \bmod g(x).$$

Proof We use induction on t. For $t = 0, 1, \ldots, r - 1$, the state vector is

$$\mathbf{S}_t = \left[a_t, a_{t-1}, \ldots, a_0, \overbrace{0, \ldots, 0}^{r-t-1} \right],$$

and the statement of the theorem is a tautology. Assuming then that the theorem is true for $S_t(x)$, we consider $S_{t+1}(x)$:

$$S_{t+1}(x) = a_{t+1} + xS_t(x) \bmod g(x) \qquad \text{(Lemma 3)}$$

$$= a_{t+1} + x \left(\sum_{j=0}^{t} a_j x^{t-j} \bmod g(x) \right) \bmod g(x) \qquad \text{(induction hypothesis)}$$

$$= a_{t+1} + \left(\sum_{j=0}^{t} a_j x^{t+1-j} \right) \bmod g(x) \qquad \text{(Lemma 1(d))}$$

$$= \sum_{j=0}^{t+1} a_j x^{t+1-j} \bmod g(x) \qquad \text{(Lemma 1(a) and (c)),}$$

as asserted. $\square$

Theorem 8.4 explains why the circuit of Figure 8.3(a) is called a "mod $g(x)$ circuit." Our next goal is to use it, together with the encoding rule (8.4), to build a systematic shift-register encoder for a cyclic code with generator polynomial $g(x)$. The encoding rule (8.4) requires us to compute

$$[x^r I(x)] \bmod g(x) = \sum_{j=r}^{n-1} I_{j-r} x^j \bmod g(x).$$

According to Theorem 8.4, to compute this, we can give the mod $g(x)$ circuit the n-symbol input sequence

$$\left(I_{k-1}, I_{k-2}, \ldots, I_0, \overbrace{0, 0, \ldots, 0}^{r} \right).$$

However, since a systematic encoder could output the k information symbols unchanged on ticks 0 through $k-1$, it would like to start outputting the parity-checks, i.e., the coefficients of $[x^r I(x)] \bmod g(x)$, on the kth tick, but will have to wait for r more ticks before $[x^r I(x)] \bmod g(x)$ is ready. Fortunately, it is possible to avoid this r-tick "down time" by using the circuit shown in Figure 8.4(a).

In the circuit of Figure 8.4(a), the input bits are fed into the *right* side of the shift register, rather than the left, as in Figure 8.3. The result is that if the input stream is $a_0, a_1, \ldots$, after the tth tick, the shift register's state polynomial will be $\sum_{j=0}^{t} a_j x^{r+t-j} \bmod g(x)$, rather than $\sum_{j=0}^{t} a_j x^{t-j} \bmod g(x)$. (To verify this assertion, do Problem 8.31.) Thus if the encoder inputs the k information symbols $I_{k-1}, I_{k-2}, \ldots, I_0$ to the shift register of Figure 8.4, the coefficients of $[x^r I(x)] \bmod g(x)$ will be ready, starting with the kth tick, as desired. A complete systematic encoder based on the shift register of Figure 8.4(a) is shown in Figure 8.5(a). Note that as compared to the nonsystematic encoder of Figure 8.1, in the encoder of Figure 8.5(a), the information symbols are sent to the encoder in the *reverse* order, viz., $I_{k-1}, \ldots, I_0$, and the codeword components are also sent to the channel in the reverse order, viz., $C_{n-1}, C_{n-2}, \ldots, C_1, C_0$.

Example 8.11 If we apply the general construction described in Figure 8.5(a) to the special case of $g(x) = x^4 + x^3 + x^2 + 1$, over the field $GF(2)$, we get the encoder depicted in Figure 8.5(b). If, for example, the information sequence $(I_2, I_1, I_0) = (1, 1, 0)$, followed by $(0, 0, 0, 0)$ is used as the input sequence to the encoder circuit of Figure 8.5(b), the output will be $(1, 1, 0, 1, 0, 0, 1)$, as detailed in the following table.

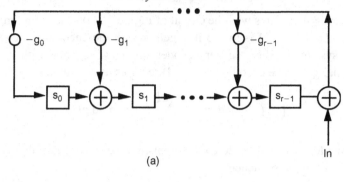

(a)

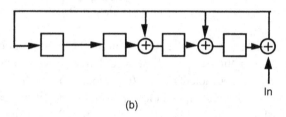

(b)

Figure 8.4 (a) A general shift register needed to build an n-tick systematic encoder for a cyclic code with generator polynomial $g(x) = g_0 + g_1x + \cdots + g_rx^r$. (b) Special case $g(x) = x^4 + x^3 + x^2 + 1$.

Tick	Input	SR contents	Output
0	1	1011	1
1	1	0101	1
2	0	1001	0
3	0	0100	1
4	0	0010	0
5	0	0001	0
6	0	0000	1

Recalling that the components appear in reverse order, we see that this codeword is the same as codeword $\mathbf{C}_1 + \mathbf{C}_3$ in Example 8.2. $\quad\square$

We will conclude this section with a brief discussion of a third type of shift-register encoder for a cyclic code, which is based on the code's *parity-check polynomial*. Recall from Section 8.1 that if C is an (n, k) cyclic code with

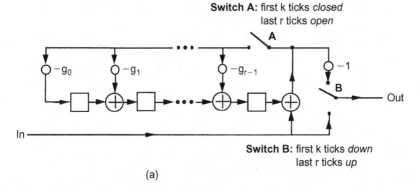

(a)

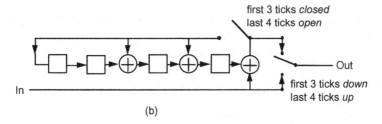

(b)

Figure 8.5 (a) An n-tick systematic encoder for a cyclic code with generator polynomial $g(x) = g_0 + g_1 x + \cdots + g_{r-1} x^{r-1} + x^r$. (b) Special case $g(x) = x^4 + x^3 + x^2 + 1$.

generator polynomial $g(x)$, its parity-check polynomial $h(x) = h_0 + h_1 x + \cdots + h_k x^k$ is defined as

$$h(x) = \frac{x^n - 1}{g(x)}. \tag{8.5}$$

Since every codeword $C(x)$ is by Theorem 8.3 a multiple of $g(x)$, it follows from (8.5) that $C(x)h(x)$ is a multiple of $g(x)h(x) = x^n - 1$, i.e.,

$$[C(x)h(x)]_n = 0. \tag{8.6}$$

Equation (8.6) places strong conditions on the coefficients C_i of the codeword $C(x)$, as the following theorem shows.

Theorem 8.5 *For $i = 0, 1, \ldots, n - 1$, we have*

$$\sum_{j=0}^{k} h_j C_{(i-j) \bmod n} = 0.$$

Proof We have

$$h(x)C(x) = \left(\sum_{j=0}^{k} h_j x^j \right) \left(\sum_{m=0}^{n-1} C_m x^m \right)$$

$$= \sum_{j=0}^{k} h_j \left(\sum_{m=0}^{n-1} C_m x^{j+m} \right).$$

If we reduce this expression $\mod(x^n - 1)$, and use the fact that $x^i \mod (x^n - 1) = x^{i \bmod n}$ (see Example 8.4), we find

$$[h(x)C(x)]_n = \sum_{j=0}^{k} h_j \left(\sum_{m=0}^{n-1} C_m x^{(j+m) \bmod n} \right). \tag{8.7}$$

Since by (8.6), $[h(x)C(x)]_n = 0$, it follows that each coefficient of $[h(x)C(x)]_n$ is zero as well. But from (8.7) if follows that for $i = 0, 1 \ldots, n-1$, the coefficient of x^i in $[h(x)C(x)]_n$ is

$$\sum_{j=0}^{k} h_j \cdot \text{coefficient of } x^i \text{ in } \left\{ \sum_{m=0}^{n-1} C_m x^{(j+m) \bmod n} \right\}.$$

If $m \in \{0, 1, \ldots, n-1\}$, the exponent $(j + m) \bmod n$ will equal i if and only if $m = (i - j) \bmod n$ (see Problem 8.4), and so the coefficient of x^i in $[h(x)C(x)]_n$ is

$$\sum_{j=0}^{k} h_j C_{(i-j) \bmod n},$$

which must therefore be zero.   □

Corollary *If $h_0 = 1$, then for $i = k, k+1, \ldots, n-1$,*

$$C_i = - \sum_{j=0}^{k} h_j C_{i-j}.$$

Proof If $k \leqslant i \leqslant n-1$, then for $j = 0, 1 \ldots, k$ we have $(i - j) \bmod n = i - j$, and Theorem 8.5 becomes

$$\sum_{j=0}^{k} h_j C_{i-j} = 0.$$

If $h_0 = 1$, this last equation can be rearranged to give the result stated.   □

The corollary to Theorem 8.5 says that once the first k components $C_0, C_1, \ldots, C_{k-1}$ of a codeword are known, the remaining r components $C_k, C_{k+1}, \ldots, C_{n-1}$ can be computed by *linear recursion*, i.e., each new component is a fixed linear combination of the previous k components. This fact leads immediately to a shift-register encoder, because it is easy to design a shift-register circuit that can implement a given linear recursion; see Figure 8.6.

Figure 8.6(a) shows a k-stage shift-register circuit capable of generating any sequence $(S_0, S_1, S_2 \ldots)$ that satisfies the linear recursion $S_t = -\sum_{j=1}^{k} h_j S_{t-j}$, where $h_1, h_2, \ldots, h_k$ are fixed constants. The circuit must be initalized by loading the k flip-flops with the k initial values $S_0, S_1, \ldots, S_{k-1}$ (with S_0 occupying the rightmost flip-flop). Then, for $t \geqslant k$, if the shift-register contents are $(S_{t-1}, \ldots, S_{t-k})$, after the next tick, the shift-register contents will be $(S_t, S_{t-1}, \ldots, S_{t-k+1})$. Figure 8.6(b) shows the special case $h(x) = x^3 + x^2 + 1$, corresponding to the recursion $S_t = S_{t-2} + S_{t-3}$, for $t \geqslant 3$, over the binary field $GF(2)$.

It should be clear how to use the shift-register circuit of Figure 8.6(a) to build a systematic encoder for a cyclic code with parity-check polynomial

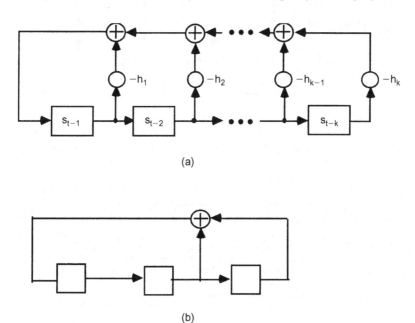

(a)

(b)

Figure 8.6 (a) A shift-register circuit that implements the kth-order linear recursion $S_t = -\sum_{j=1}^{k} h_j S_{t-j}$. (b) The special case $S_t = S_{t-2} + S_{t-3}$, corresponding to $h(x) = 1 + x^2 + x^3$.

$h(x) = 1 + h_1 x + \cdots + h_k x^k$ (see Figure 8.7(a)). During the first k ticks the switch is in the "down" position and the k information symbols are simultaneously sent out to the channel, and loaded into the k-stage register. Then the switch is put in the "up" position, and during the remaining r ticks, the shift-register circuit calculates the remaining r codeword components via the recursion $C_i = -\sum_{j=1}^{k} h_j C_{i-j}$. Notice that the systematic "$h(x)$ encoder" of Figure 8.7 differs from the systematic "$g(x)$ encoder" of Figure 8.5 in two important ways: in the $g(x)$ encoder, the information symbols $(I_0, I_1, \ldots, I_{k-1})$ occupy codeword components $C_r, C_{r+1}, \ldots, C_{n-1}$, and the codeword symbols are sent to the channel in the reverse order $C_{n-1}, C_{n-2}, \ldots, C_0$, whereas in the $h(x)$ encoder the information symbols occupy codeword components $C_0, C_1, \ldots, C_{k-1}$ and the codeword symbols are sent to the channel in the natural order $C_0, C_1, \ldots, C_{n-1}$.

Example 8.12 If we apply the general construction of Figure 8.7(a) to the particular $(7, 3)$ binary cyclic code with $g(x) = x^4 + x^3 + x^2 + 1$, and

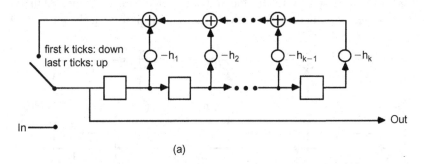

(a)

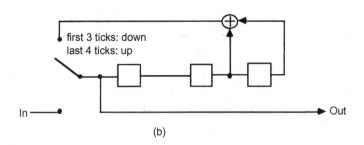

(b)

Figure 8.7 (a) A systematic shift-register encoder for an (n, k) cyclic code with parity-check polynomial $h(x) = 1 + h_1(x) + \cdots + h_k x^k$. (b) Special case $h(x) = 1 + x^2 + x^3$.

$h(x) = x^3 + x^2 + 1$, we get the encoder shown in Figure 8.7(b). Thus for example, if the information symbols are $(I_0, I_1, I_2) = (1, 1, 0)$, the encoder of Figure 8.7(b) implements the recursion $C_i = C_{i-2} + C_{i-3}$ and produces the codeword $(1, 1, 0, 0, 1, 0, 1)$, which is codeword $\mathbf{C}_1 + \mathbf{C}_2 + \mathbf{C}_3$ in Example 8.2. Notice also that the $g(x)$ encoder for this code, illustrated in Figure 8.5(b), needs four flip-flops and 3 mod-2 adders whereas the $h(x)$ encoder in Figure 8.7(b) only needs three flip-flops and one mod-2 adder. As a general rule the $h(x)$ encoder will be simpler when $k < r$, i.e. $k < n/2$. ☐

8.3 Cyclic Hamming codes[4]

In Section 7.4 we defined a binary Hamming code of length $2^m - 1$ to be any linear code whose parity-check matrix has as columns the $2^m - 1$ nonzero binary vectors of length m, *arranged in any order*. For example, the following parity-check matrix defines a $(7, 4)$ Hamming code (cf. the matrix H_3 in Section 7.4):

$$H = \begin{bmatrix} 0 & 1 & 1 & 1 & 1 & 0 & 0 \\ 1 & 0 & 1 & 1 & 0 & 1 & 0 \\ 1 & 1 & 0 & 1 & 0 & 0 & 1 \end{bmatrix}.$$

There are of course $(2^m - 1)!$ ways to order these columns, and although any one of these orderings produces a perfect single-error-correcting code, some orderings are better than others from an implementational standpoint. In fact, we shall see in this section that it is possible to choose an ordering that produces a *cyclic* version of the Hamming code, which leads to a simple shift-register implementation of both the encoder and the decoder. To describe the appropriate ordering, we must assume the reader to be familiar with the finite field $GF(2^m)$, in which each element is represented by a binary vector of length m. (In Appendix C we give a summary of the needed facts.)

To obtain a cyclic Hamming code of length $2^m - 1$, begin with a primitive root $\alpha \in GF(2^m)$ and use it to define a linear code $\mathcal{C}$ as follows. A vector $\mathbf{C} = (C_0, C_1, \ldots, C_{n-1})$[5] with components in $GF(2)$ is in $\mathcal{C}$ if and only if

$$C_0 + C_1\alpha + \cdots + C_{n-1}\alpha^{n-1} = 0. \tag{8.8}$$

Alternatively, $\mathcal{C}$ is the binary linear code defined by the $1 \times n$ parity-check matrix

$$H = [1 \quad \alpha \quad \alpha^2 \quad \ldots \quad \alpha^{n-1}]. \tag{8.9}$$

We shall soon prove that the code defined by (8.8) or (8.9) is a cyclic Hamming code. For now, the important thing to notice about this definition is

Table 8.2 The powers of α in $GF(8)$, where $\alpha^3 = \alpha + 1$.

i	α^i
0	001
1	010
2	100
3	011
4	110
5	111
6	101

that the codeword components are in the binary field $GF(2)$ but the entries in the parity-check matrix are in the extension field $GF(2^m)$. There are, as we shall see, many advantages to this "two-field" approach. Still, it is possible to define the same code via a more conventional (and more complicated) binary parity-check matrix, by replacing each of the powers of α in H by its representation as an m-bit column vector. For example, with $m = 3$, if α is a primitive root in $GF(2^3)$ satisfying $\alpha^3 = \alpha + 1$, then Table 8.2 shows the powers of α represented as three-dimensional vectors. Thus the 1×7 $GF(8)$-matrix H in (8.9) defines the same binary code as the following 3×7 binary matrix H':

$$H' = \begin{bmatrix} 0 & 0 & 1 & 0 & 1 & 1 & 1 \\ 0 & 1 & 0 & 1 & 1 & 1 & 0 \\ 1 & 0 & 0 & 1 & 0 & 1 & 1 \end{bmatrix}.$$

The following theorem is the main result of this section.

Theorem 8.6 *The code defined above (i.e., by (8.8) or (8.9)) is an $(n, n - m)$ binary cyclic code with generator polynomial $g(x)$, the minimal polynomial of α. Furthermore, its minimum distance is 3, so that it is a Hamming code.*

Proof It is clear that the code is a binary linear code of length n. We need to show that it is cyclic, and that $g(x)$ is the generator polynomial. To show that it is cyclic, note that if we multiply equation (8.8) by α, and use the fact that $\alpha^n = 1$, we get

$$C_{n-1} + C_0\alpha + \cdots + C_{n-2}\alpha^{n-2} = 0.$$

Thus if $\mathbf{C}$ satisfies (8.8), so does $\mathbf{C}^R$, and so the code is cyclic. To show that $g(x)$ is the generator polynomial, we note that (8.8) is equivalent to $C(\alpha) = 0$, where $C(x) = C_0 + C_1x + \cdots + C_{n-1}x^{n-1}$ is the code polynomial, and observe that this is true if and only if $C(x)$ is a multiple of α's minimal polynomial $g(x)$. Thus the code consists of all polynomials of degree $n - 1$ or less which are multiples of $g(x)$, and hence by Theorem 8.3, $g(x)$ is the code's generator polynomial.

To complete the proof, we need to show that the code's minimum distance $d_{\min}$ is 3. Since the code is linear, $d_{\min} = w_{\min}$. We use the definition (8.8); if there were a codeword of weight 1, then $\alpha^i = 0$ for some i, which is impossible. Similarly, a word of weight 2 could exist if and only if $\alpha^i + \alpha^j = 0$ for some i and j with $0 \leqslant i < j \leqslant n - 1$. Dividing by α^i, this becomes $1 + \alpha^{j-i} = 0$, which is impossible, because the smallest positive power of α equal to 1 is α^n. Finally we note that there are many words of weight 3; for example, if $1 + \alpha = \alpha^j$, then there is a codeword of weight 3 with nonzero components C_0, C_1, and C_j. $\qquad\square$

Now that we know that there are cyclic Hamming codes, it follows from the results in Section 8.2 that it is possible to build simple shift-register encoders for Hamming codes, based on the generator polynomial. Since the generator polynomial for a Hamming code must be a primitive polynomial, in order to implement a cyclic Hamming code of length $2^m - 1$, it is necessary to have a primitive polynomial of degree m. In Table 8.3 we list one primitive polynomial of degree m, for $1 \leqslant m \leqslant 12$. For a given value of m, there will in general be many primitive polynomials of degree m, but the ones in Table 8.3 are chosen to have the fewest possible nonzero coefficients, which will lead to shift-register encoders with the fewest possible mod-2 adders.

More important than the encoding simplicity of Hamming codes, however, is the fact it is also possible to build simple *decoders* for them. We illustrate such a decoder for the (7, 4) Hamming code with generator polynomial $g(x) = x^3 + x + 1$ in Figure 8.8. The decoder consists of three main parts, two shift register circuits, and an AND gate. The upper shift register is a "mod $g(x)$" circuit, and the lower shift register is a "mod $x^n + 1$" circuit. The AND gate outputs a 1 if the upper shift register contains the pattern $10 \ldots 0$, and otherwise it ouputs 0. We will assume that the transmitted codeword is $C(x) = (C_0, C_1, \ldots, C_{n-1})$ and the received word is $R(x) = (R_0, R_1, \ldots, R_{n-1})$, with $R(x) = C(x) + E(x)$, where $E(x) = (E_0, E_1, \ldots, E_{n-1})$ is the error pattern.

In the decoding circuit of Figure 8.8, the noisy codeword $(R_0, R_1, \ldots, R_{n-1})$ is clocked in from the left, in the reverse order (i.e., R_{n-1} goes first), so that (by Theorem 8.4) after n ticks the upper shift register contains $R(x) \bmod g(x)$, and the lower shift register contains $R(x) \bmod x^n + 1$, i.e.,

Table 8.3 Some primitive polynomials—possible generator polynomials for Hamming codes.

m	Primitive polynomial of degree m
1	$x + 1$
2	$x^2 + x + 1$
3	$x^3 + x + 1$
4	$x^4 + x + 1$
5	$x^5 + x^2 + 1$
6	$x^6 + x + 1$
7	$x^7 + x + 1$
8	$x^8 + x^7 + x^2 + x + 1$
9	$x^9 + x^4 + 1$
10	$x^{10} + x^3 + 1$
11	$x^{11} + x^2 + 1$
12	$x^{12} + x^6 + x^4 + x + 1$

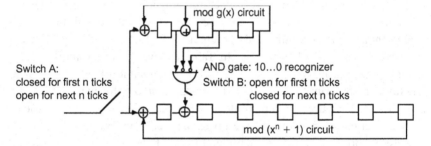

Figure 8.8 A decoding circuit for the (7, 4) Hamming code with generator polynomial $g(x) = x^3 + x + 1$.

$(R_0, R_1, \ldots, R_{n-1})$. Since $R(x) = C(x) + E(x)$, and $C(x) \bmod g(x) = 0$, since $C(x)$ is a codeword, in fact the upper shift register will contain $E(x) \bmod g(x)$. If there are no errors, then $E(x) = 0$ and the upper shift register will contain all zeros. If there is one error in the eth position then $E(x) = x^e$, where $0 \leqslant e \leqslant n - 1$.

At this point, switch A is opened, switch B is closed, and the two shift registers run autonomously for n more ticks. We shall call these n ticks the *decoding cycle*. At the end of the decoding cycle, if there is at most one error, the original codeword $C(x)$ will appear in the lower shift register. Let us see why this is so.

After the tth tick of the decoding cycle, the contents of the upper shift register will be $x^t E(x) \bmod g(x)$ and the lower shift register will contain $x^t R(x) \bmod x^n + 1$, i.e., the tth right cyclic shift $(R_{n-t}, \ldots, R_{n-t-1})$ of the received word. The two shift registers are connected via switch B and the AND gate, which outputs a 1 if and only if the upper shift register contains $10 \ldots 0$, i.e., iff $x^t E(x) \bmod g(x) = 1$. If there are no errors, $E(x) = 0$ and so the AND gate will never be triggered, so that after the decoding cycle, the lower shift register will contain the received word, with no changes. However, if there is one error, in the eth position, so that $E(x) = x^e$, then after $(n - e) \bmod n$ ticks of the decoding cycle, the upper shift register will contain $x^{n-e} x^e \bmod g(x) = x^n \bmod g(x) = 1$, since $g(x)$ is a divisor of $x^n - 1$. At this point the lower shift register will contain $(R_e, R_{e+1}, \ldots, R_{e-1})$ and on the next tick the AND gate will output a 1 and complement the erroneous component R_e of the received word, i.e., the error will be corected. Then after $(e - 1) \bmod n$ further ticks, the received word will have completed its circular journey around the lower shift register, and will appear, with its error corrected, in the lower shift register.

In Section 8.5, we will see how to generalize the circuit of Figure 8.8 in order to build a decoder for a cyclic *burst*-error-correcting code. The underlying theory for burst-error correction will be presented in Section 8.4. (Incidentally, the *dual* codes of cyclic Hamming codes are also interesting and important. They are covered briefly in Problem 8.41.)

8.4 Burst-error correction

On many channels of practical importance, errors, when they occur, tend to occur in *bursts*. Physically, a burst of errors occurs when for some reason the channel noise severity increases for a brief time, and then returns to normal. In this section we will see how cyclic codes can be used to detect and correct error bursts.

We begin with a purely mathematical definition of an error burst. If a codeword $\mathbf{C}$ is transmitted, and is received as $\mathbf{R} = \mathbf{C} + \mathbf{E}$, then the error vector $\mathbf{E}$ is called a *burst of length b* if the nonzero components of $\mathbf{E}$ are confined to b consecutive components. For example $\mathbf{E} = (010000110)$ is a burst of length 7:

$$\begin{array}{ccccccccc} & * & * & * & * & * & * & * & \\ 0 & 1 & 0 & 0 & 0 & 0 & 1 & 1 & 0 \end{array}$$

(In the display, the $*$'s mark the error burst.) Since we will be correcting bursts of errors with cyclic codes, for technical reasons we also need to define a *cyclic burst.* An error vector $\mathbf{E}$ is called a *cyclic burst of length b* if its nonzero components are confined to b cyclically consecutive components. For example, the error vector $\mathbf{E} = (010000110)$ cited above as a burst of length 7 is also a cyclic burst of length 5:

$$
\begin{array}{ccccccccc}
* & * & & & & & * & * & * \\
0 & 1 & 0 & 0 & 0 & 0 & 1 & 1 & 0
\end{array}
$$

In the rest of this section, "burst" will always mean "cyclic burst".

It is useful to have a compact description of a burst-error vector, and so we introduce the notion of the *pattern* and *location* of a burst. If $\mathbf{E}$ is a nonzero burst-error vector, its *burst pattern* is the string of symbols beginning with the first nonzero symbol in $\mathbf{E}$ and ending with the last nonzero symbol. The burst's *location* is the index of the first nonzero symbol in the burst. For example, the vector $\mathbf{E} = (010000110)$, viewed as a burst of length 7, has burst pattern 1000011 and burst location 1 (assuming the components are numbered $0, 1, \ldots, 8$). Unfortunately, the "burst pattern–burst location" description of most vectors $\mathbf{E}$ isn't unique, since any nonzero symbol in $\mathbf{E}$ can be taken as the first symbol in a cyclic burst. Thus an error vector of weight w will have w burst descriptions. For example, $\mathbf{E} = (010000110)$ has three burst descriptions:

Pattern	Location
1000011	1
11001	6
100100001	7

This ambiguity is annoying but usually not serious, since the correctable bursts that occur in practice tend to be quite short, and a short burst can have only one short description, as the following theorem shows.

Theorem 8.7 *Suppose* $\mathbf{E}$ *is an error vector of length* n *with two burst descriptions* (pattern$_1$, location$_1$) *and* (pattern$_2$, location$_2$). *If* length (pattern$_1$) $+$ length(pattern$_2$) $\leqslant n + 1$, *then the two descriptions are identical, i.e.,* pattern$_1$ $=$ pattern$_2$ *and* location$_1$ $=$ location$_2$.

Proof As noted above, if $\mathbf{E}$ has weight w, then $\mathbf{E}$ has exactly w different burst descriptions. If $w = 0$ or 1, therefore, there is nothing to prove, so we assume $w \geqslant 2$.

For a given burst description of **E**, the pattern will contain all of **E**'s nonzero components, and so the components of **E** *not* included in the pattern will form a cyclic run of 0's, beginning just after the last nonzero component in the pattern, and continuing until just before the first nonzero component in the pattern. The set of indices corresponding to this run of 0's we call the *zero run* associated with the given burst description. For example, for **E** = (01000010), as we saw above, there are three burst descriptions. In the first of these descriptions, i.e., 1000011, the pattern begins at position 1 and ends at position 7. Thus the zero run associated with this burst description is (0, 8). Altogether there are three zero runs, one for each of the burst descriptions of **E**:

Pattern	Location	Zero run
1000011	1	$(8, 0)$
11001	6	$(2, 3, 4, 5)$
100100001	7	none

(In the last burst description, the zero run is empty, which we indicate by the word "none.") Plainly the zero runs associated with different burst descriptions are disjoint, and the total length of all of the zero runs is $n - w$, where w is the weight of **E**. To prove the theorem, we note that if the two burst descriptions

$$(\text{pattern}_1, \text{location}_1) \text{ and } (\text{pattern}_2, \text{location}_2)$$

of **E** are different, then since their zero runs are disjoint they account for $(n - \text{length}(\text{pattern}_1)) + (n - \text{length}(\text{pattern}_2))$ zeros in **E**. But since $\text{length}(\text{pattern}_1) + \text{length}(\text{pattern}_2) \leq n + 1$, this number is $\geq n - 1$, which contradicts the fact that **E** has weight ≥ 2. Thus the burst descriptions must be identical. □

Corollary *An error vector* **E** *can have at most one description as a burst of length* $\leq (n + 1)/2$.

Proof Two distinct descriptions of length $\leq (n + 1)/2$ would contadict Theorem 8.7. □

With the help of Theorem 8.7, we can now count the number of burst patterns of a given length.

Theorem 8.8 *Over a two-letter alphabet, there are exactly $n2^{b-1} + 1$ vectors of length n which are bursts of length $\leq b$, provided $1 \leq b \leq (n+1)/2$.*

Proof By the corollary to Theorem 8.7, if $b \leq (n+1)/2$, a nonzero burst of length b has a unique description (as a burst of length $\leq b$). There are n possibilities for the location. The pattern must start with a 1 and have length $\leq b$, which means that the possible patterns are in one-to-one correspondence with the 2^{b-1} binary strings of length b which begin with a 1. Thus there are 2^{b-1} possible patterns, and so a total of $n \cdot 2^{b-1}$ nonzero bursts of length $\leq b$. Adding 1 to this number to account for the all-zero burst, we get $n2^{b-1} + 1$, as asserted.   □

The next two theorems give useful bounds on the size of codes which correct burst errors. For simplicity, we shall call a code capable of correcting all burst-error patterns of length $\leq b$ a *burst-b-error-correcting code*.

Theorem 8.9 (*The Hamming bound for burst-error correction.*) *If $1 \leq b \leq (n+1)/2$, a binary burst-b-error-correcting code has at most $2^n/(n2^{b-1} + 1)$ codewords.*

Proof By Theorem 8.8 there are $n2^{b-1} + 1$ burst-error patterns of length $\leq b$. If there are M codewords, there are then $M(n2^{b-1} + 1)$ words which differ from a codeword by a burst of length $\leq b$. These words must all be distinct, and so $M(n2^{b-1} + 1) \leq 2^n$.   □

Corollary (the Abramson bounds). *If $1 \leq b \leq (n+1)/2$, a binary linear (n, k) linear burst-b-error-correcting code must satisfy*

$$n \leq 2^{r-b+1} - 1 \text{ (strong Abramson bound)},$$

where $r = n - k$ is the code's redundancy. An alternative formulation is

$$r \geq \lceil \log_2(n+1) \rceil + (b - 1) \text{ (weak Abramson bound)}.$$

Proof For a linear (n, k) code there are $M = 2^k$ codewords, and so by Theorem 8.9, $2^k \leq 2^n/(n2^{b-1} + 1)$. Rearranging this, we get $n \leq 2^{r-b+1} - 2^{-b+1}$. Since n must be an integer, this bound can be improved to $n \leq 2^{r-b+1} - 1$, the strong Abramson bound. Rearranging *this* to obtain a bound on r, we get the weak Abramson bound.   □

Theorem 8.10 *If* $b \leqslant n/2$, *a binary burst-b-error-correcting code has at most* 2^{n-2b} *codewords.*

Proof If $M > 2^{n-2b}$, then by the pigeon-hole principle there must be two distinct codewords which agree in their first $n - 2b$ coordinates. These two codewords can then be represented schematically as follows:

$$X = \overbrace{* \; * \; * \; * \; * \; * \; * \; *}^{n-2b} \; \overbrace{A \; A \; A \; A \; A \; A}^{2b}$$

$$Y = * \; * \; * \; * \; * \; * \; * \; * \; B \; B \; B \; B \; B \; B$$

where "$*$" denotes agreement, and the A's and B's are arbitrary. But then the word

$$Z = * \; * \; * \; * \; * \; * \; * \; * \; A \; A \; A \; B \; B \; B$$

differs from both X and Y by a burst of length $\leqslant b$, a contradiction. □

Corollary (the Reiger bound). *If* $0 \leqslant b \leqslant n/2$, *a binary* (n, k) *linear burst-b-error-correcting code must satisfy*

$$r \geqslant 2b$$

where $r = n - k$ *is the code's redundancy.*

Proof The number of codewords in an (n, k) binary linear code is 2^k, which by Theorem 8.10 must be $\leqslant 2^{n-2b}$. This is equivalent to the statement of the corollary. □

We are now in a position to discuss a series of examples of burst-error-correcting codes. In each case, the code will be cyclic and meet either the strong Abramson or Reiger bound (which apply to all linear codes, not just cyclic codes). In this discussion, when we say that a particular bound (either the Abramson bound or the Reiger bound) is *tight*, we mean that there exists a code whose redundancy is equal to the value of the bound. If no such code exists, we will say that the bound is *loose*.

Example 8.13 The $(n, 1)$ binary repetition code with $g(x) = x^{n-1} + x^{n-2} + \cdots + x + 1$, where n is odd, can correct all error patterns of weight $\leqslant (n - 1)/2$, and so is a burst-$((n - 1)/2)$-error-correcting code. Since $r = n - 1$, the Reiger bound is tight. □

Example 8.14 The (n, n) code, consisting of all possible codewords of length n, is a cyclic code with $g(x) = 1$. It is (trivially) a $b = 0$ burst-error-correcting code, and since $r = 0$ too, the Reiger bound is again tight. □

Example 8.15 Any binary Hamming code, with $n = 2^m - 1$, $r = m$, and $g(x) = $ a primitive polynomial of degree m, is a $b = 1$ burst-error-correcting code. (Any error vector of weight 1 is ipso facto a burst of length 1.) The strong Abramson bound is tight for all these codes. □

Example 8.16† Any cyclic Hamming code from which the codewords of odd weight have been removed is a $b = 2$ burst-error-correcting code called an *Abramson code*. These codes are cyclic codes with generator polynomials of the form $g(x) = (x + 1)p(x)$, where $p(x)$ is a primitive polynomial. (See Problem 8.55.) The smallest Abramson code is the $(7, 3)$ cyclic code with $g(x) = (x + 1)(x^3 + x + 1)$ and parity-check matrix

$$H = \begin{pmatrix} 1 & \alpha & \alpha^2 & \alpha^3 & \alpha^4 & \alpha^5 & \alpha^6 \\ 1 & 1 & 1 & 1 & 1 & 1 & 1 \end{pmatrix},$$

where α is a primitive root satisfying the equation $\alpha^3 + \alpha + 1 = 0$ in GF(8), To verify that this code is indeed a $b = 2$ burst-error-connecting code, we need to check that the syndromes of all bursts of length $b \leqslant 2$ are distinct. For $b = 0$ (the all-zero error pattern), the syndrome is $\binom{0}{0}$. For $b = 1$, a burst with description $(1, i)$ has syndrome $\binom{\alpha^i}{1}$. For $b = 2$, a burst description $(11, i)$ has syndrome $\binom{\alpha^i(\alpha+1)}{0}$. These $1 + 2n$ syndromes are all distinct, and so the code is indeed a $b = 2$ burst-error-correcting code. Note finally that if $g(x) = (x + 1)p(x)$, where $p(x)$ is a primitive polynomial of degree m, then $n = 2^m - 1$, $r = m + 1$, and $b = 2$, so that the strong Abramson bound is tight. (For $m = 3$, i.e. the $(7, 3)$, $b = 2$ code, the Reiger bound is tight, too, but for all larger values of m the Reiger bound is loose.) □

Example 8.17[6] The $(15, 9)$ binary cyclic code with generator polynomial $g(x) = (x^4 + x + 1)(x^2 + x + 1) = x^6 + x^5 + x^4 + x^3 + 1$ turns out to be a $b = 3$ burst-error-correcting code, for which, therefore, both the strong Abramson and Reiger bounds are tight. To prove that this is so, we use the parity-check matrix

$$H = \begin{pmatrix} 1 & \alpha & \alpha^2 & \cdots & \alpha^{14} \\ 1 & \omega & \omega^2 & \cdots & \omega^{14} \end{pmatrix},$$

where α is a primitive root in GF(16) satisfying $\alpha^4 + \alpha + 1 = 0$, and ω is an element of order three in $GF(16)$ satisfying $\omega^2 + \omega + 1 = 0$, and check that the syndromes of all error bursts of length $b \leqslant 3$ are distinct.

The all-zero pattern has syndrome $\binom{0}{0}$. For $b = 1$, a burst with description $(1, i)$ has syndrome $\binom{\alpha^i}{\omega^i}$. For $b = 2$, a burst with description $(11, i)$ has syndrome $\binom{\alpha^i(1+\alpha)}{\omega^i(1+\omega)} = \binom{\alpha^{i+4}}{\omega^{i+2}}$, since $1 + \alpha = \alpha^4$ and $1 + \omega = \omega^2$. For $b = 3$, there are two possible patterns, viz. 101 and 111. A burst with description $(101, i)$ has syndrome $\binom{\alpha^i(1+\alpha^2)}{\omega^i(1+\omega^2)} = \binom{\alpha^{i+8}}{\omega^{i+1}}$, since $1 + \alpha^2 = \alpha^8$ and $1 + \omega^2 = \omega$, and a burst with description $(111, i)$ has syndrome $\binom{\alpha^i(1+\alpha+\alpha^2)}{\omega^i(1+\omega+\omega^2)} = \binom{\alpha^{i+10}}{0}$, since $1 + \alpha + \alpha^2 = \alpha^{10}$ and $1 + \omega + \omega^2 = 0$. Plainly the all-zero pattern and the pattern 111 cannot be confused with anything else, because of the "0" in the second component of the syndrome. To distinguish between the patterns 1, 11, and 101, we need to look a bit deeper. For any one of these three patterns, the syndrome will be of the form $\binom{\alpha^s}{\omega^t}$; to distinguish between them, we look at $(s - t) \bmod 3$. If the pattern is 1, then $s = i$, $t = i$, and $(s - t) \bmod 3 = 0$; if the pattern is 11, then $s = i + 4$, $t = i + 2$, and $(s - t) \bmod 3 = 2$; and if the pattern is 101, then $s = i + 8$, $t = i + 1$, and

Syndrome	Burst description	
$\binom{0}{0}$	$\varnothing$	
$\binom{\alpha^s}{0}$	$(111, s - 10 \bmod 15)$	
$\binom{\alpha^s}{\omega^t}$	$(1, s)$	if $(s - t) \bmod 3 = 0$
	$(101, s - 8 \bmod 15)$	if $(s - t) \bmod 3 = 1$
	$(11, s - 4 \bmod 15)$	if $(s - t) \bmod 3 = 2$

$(s - t) \bmod 3 = 1$. Thus the 61 syndromes of the bursts of length $\leqslant 3$ are distinct, and the following table summarizes the relationship between syndromes and burst descriptions.

For example, the sydrome $\binom{\alpha^{11}}{1}$ has $s = 11$, $t = 0$, and so corresponds to the burst description $(11, 7)$, i.e., the error pattern is (0000000110000000). $\square$

The mathematical details of Example 8.17 are somewhat intricate, and do not easily generalize. (But see Problem 8.63.) Nevertheless, researchers over the years have succeeded in finding a large number of cyclic burst-error-correcting codes which meet the strong Abramson bound; these are usually called *optimum* burst-error-correcting codes. In Table 8.3 we list a number of such codes.

Table 8.4 Some cyclic burst-error-correcting codes that meet the strong Abramson bound, i.e., satisfy $n = 2^{r-b+1} - 1$.

Generator polynomial	(n, k)	b
$(x^3 + x + 1)(x + 1)$	$(7, 3)$	2
$(x^4 + x + 1)(x + 1)$	$(15, 10)$	2
$(x^4 + x + 1)(x^2 + x + 1)$	$(15, 9)$	3
$(x^5 + x^2 + 1)(x + 1)$	$(31, 25)$	2
$(x^6 + x + 1)(x + 1)$	$(63, 56)$	2
$(x^6 + x + 1)(x^2 + x + 1)$	$(63, 55)$	3
$(x^7 + x + 1)(x + 1)$	$(127, 119)$	2
$(x^8 + x^7 + x^2 + x + 1)(x + 1)$	$(255, 246)$	2
$(x^8 + x^7 + x^2 + x + 1)(x^2 + x + 1)$	$(255, 245)$	3
$(x^9 + x^7 + x^6 + x^3 + x^2 + x + 1)(x + 1)$	$(511, 501)$	2
$(x^9 + x^7 + x^6 + x^3 + x^2 + x + 1)(x^3 + x + 1)$	$(511, 499)$	4
$(x^{10} + x^5 + x^3 + x^2 + 1)(x + 1)$	$(1023, 1012)$	2
$(x^{10} + x^5 + x^3 + x^2 + 1)(x^2 + x + 1)$	$(1023, 1011)$	3
$(x^{10} + x^5 + x^3 + x^2 + 1)(x^2 + x + 1)(x + 1)$	$(1023, 1010)$	4
$(x^{11} + x + 1)(x + 1)$	$(2047, 2035)$	2
$(x^{12} + x^{11} + x^9 + x^8 + x^7 + x^5 + x^2 + x + 1)(x + 1)$	$(4095, 4082)$	2
$(x^{12} + x^{11} + x^9 + x^8 + x^7 + x^5 + x^2 + x + 1)(x^2 + x + 1)$	$(4095, 4081)$	3
$(x^{12} + x^{11} + x^9 + x^8 + x^7 + x^5 + x^2 + x + 1)(x^2 + x + 1)(x + 1)$	$(4095, 4080)$	4
$(x^{15} + x^{13} + x^{10} + x^9 + x^5 + x^3 + x^2 + x + 1)(x^3 + x + 1)(x + 1)$	$(32767, 32748)$	5

The next example illustrates the important *interleaving* technique, which is a simple way of boosting the burst-error-connecting ability of a code. (See also Problem 7.32.)

Example 8.18 Consider again the $(7, 3)$ $b = 2$ Abramson code with $g(x) = x^4 + x^3 + x^2 + 1$ (see Example 8.16). Let **A**, **B**, and **C** be any three words from this code, which we can display as a 3×7 array:

$$
\begin{array}{ccccccc}
A_0 & A_1 & A_2 & A_3 & A_4 & A_5 & A_6 \\
B_0 & B_1 & B_2 & B_3 & B_4 & B_5 & B_6 \\
C_0 & C_1 & C_2 & C_3 & C_4 & C_5 & C_6
\end{array}.
$$

The following length-21 vector, which is built by reading the above array by columns, is called the *interleaving* of **A**, **B**, and **C**:

$$A_0 \; B_0 \; C_0 \; A_1 \; B_1 \; C_1 \; A_2 \; B_2 \; C_2 \; A_3 \; B_3 \; C_3 \; A_4 \; B_4 \; C_4 \; A_5 \; B_5 \; C_5 \; A_6 \; B_6 \; C_6.$$

Let us suppose that this long codeword is transmitted over a bursty channel and suffers a burst of length 6, indicated by $*$'s:

$$A_0 \; B_0 \; C_0 \; A_1 \; B_1 \; C_1 \; A_2 \; B_2 \; C_2 \; A_3 \; * \; * \; * \; * \; * \; * \; B_5 \; C_5 \; A_6 \; B_6 \; C_6.$$

It is possible to correct this burst of errors, simply by "de-interleaving" this long codeword into its component codewords, because after de-interleaving, no one of the three codewords will have suffered a burst larger than two:

$$
\begin{array}{ccccccc}
A_0 & A_1 & A_2 & A_3 & * & * & A_6 \\
B_0 & B_1 & B_2 & * & * & B_5 & B_6 \\
C_0 & C_1 & C_2 & * & * & C_5 & C_6
\end{array}.
$$

The code consisting of all possible interleavings of three codewords from the $(7, 3)$ Abramson code is called the *depth-3 interleaving* of the original code. It is a $(21, 9)$ linear code; and the argument above shows that it is in fact a $b = 6$ burst-error-correcting code. More generally, for any positive integer j, the depth-j interleaving of the $(7, 3)$ Abramson code is a $(7j, 3j)$ $b = 2j$ burst-error-correcting code. Note that each one of these codes satisfies the Reiger bound with equality (since $r = 4j$ and $b = 2j$), so that in a certain sense, no loss of efficiency results when a code is interleaved. $\square$

A straightforward generalization of the argument given in Example 8.18 leads to the following important theorem.

Theorem 8.11 *If C is an (n, k) linear burst-b-error correcting code, then the depth-j interleaving of C is a (nj, kj) burst-bj-error-correcting code.*

It is not obvious, but it is nevertheless true, that if we interleave a *cyclic* code to depth j, the resulting code is also cyclic. The next theorem spells this out.

Theorem 8.12 *If C is an (n, k) cyclic code with generator polynomial $g(x)$, then the depth-j interleaving of C is an (nj, kj) cyclic code with generator polynomial $g(x^j)$.*

Proof To prove the theorem, we introduce the symbol " $\wr$ " to denote the interleaving operation. Thus the vector obtained by interleaving the j code-words $\mathbf{C}_0, \mathbf{C}_1, \ldots, \mathbf{C}_{j-1}$ is denoted by $\mathbf{C}_0 \wr \mathbf{C}_1 \wr \cdots \wr \mathbf{C}_{j-1}$. An easy lemma, whose proof we leave as Problem 8.65, shows how to compute the right cyclic shift of an interleaved codeword:

$$[\mathbf{C}_0 \wr \mathbf{C}_1 \wr \ldots \wr \mathbf{C}_{j-1}]^R = [\mathbf{C}_{j-1}^R \wr \mathbf{C}_0 \wr \ldots \wr \mathbf{C}_{j-2}]. \tag{8.10}$$

Equation (8.10) shows that the depth-j interleaving of a cyclic code is cyclic, since if $\mathbf{C}_0, \mathbf{C}_1, \ldots, \mathbf{C}_{j-1}$ are words from a fixed cyclic code, then so are $\mathbf{C}_{j-1}^R, \mathbf{C}_0, \ldots, \mathbf{C}_{j-2}$. Since the redundancy of the interleaved code is rj, its generator polynomial will be its unique monic polynomial of degree rj. But if $g(x)$ is the generator polynomial for the original code, then the interleaved codeword $[g(x) \wr 0 \wr \ldots \wr 0]$, which is the polynomial $g(x^j)$, has degree rj, and so $g(x^j)$ must be the generator polynomial for the interleaved code. $\square$

Example 8.19 Starting with the $(7, 3)$ $b = 2$ Abramson code with generator polynomial $g(x) = x^4 + x^3 + x^2 + 1$, and using the interleaving technique, we can produce an infinite family of cyclic burst-error-correcting codes, viz. the $(7j, 3j)$ $b = 2j$ codes, with generator polynomials $g_j(x) = x^{4j} + x^{3j} + x^{2j} + 1$. Similarly, starting with the cyclic $(15, 19)$ $b = 3$ code of Example 8.17, we obtain another infinite family, viz. the $(15j, 9j)$ $b = 3j$ cyclic codes with generator polynomials $g(x) = x^{6j} + x^{5j} + x^{4j} + x^{3j} + 1$. Note that every code in each of the families meets the Reiger bound (Corollary to Theorem 8.10). $\square$

Example 8.20 In the following table, we consider binary cyclic $n = 15$ burst-error-correcting codes, for $b = 1, 2, \ldots, 7$.

For each such value of b, we list the lower bounds on the needed redundancy using the (weak) Abramson bound (r_A) and the Reiger bound (r_R). By chance, it turns out that in every case there is a cyclic code whose redundancy is $\max(r_A, r_R)$, and we have listed the generator polynomial $g(x)$ for such a code in each case. The cases $b = 1, 2, 3$ we have already consid-

b	r_A	r_R	$g(x)$	Comment
1	4	2	$x^4 + x + 1$	Hamming code
2	5	4	$(x^4 + x + 1)(x + 1)$	Abramson code (Example 8.16)
3	6	6	$(x^4 + x + 1)(x^2 + x + 1)$	Example 8.17
4	7	8	$(x^4 + x + 1)(x^4 + x^3 + x^2 + x + 1)$	Problem 8.58
5	8	10	$(x^4 + x + 1)(x^2 + x + 1)(x^4 + x^3 + 1)$ $= x^{10} + x^5 + 1$	(3, 1) interleaved $\times$ 5
6	9	12	$(x^4 + x + 1)(x^4 + x^3 + x^2 + x + 1)$ $(x^4 + x^3 + 1) = x^{12} + x^9 + x^6$ $+ x^3 + 1$	Problem 8.60
7	10	14	$x^{14} + x^{13} + \cdots + x + 1$	repetition code

ered; the case $b = 4$ we leave as Problem 8.58. To obtain the $b = 5$ code, we interleave the (3, 1) repetition code with $g(x) = x^2 + x + 1$ to depth 3, using Theorem 8.12. The case $b = 6$ is left as Problem 8.60. Finally, the case $b = 7$ is simply the $n = 15$ repetition code. □

If we combine the codes found in Table 8.4 (or in more extensive tables which have been found by elaborate algebraic or ad hoc computer methods) with the interleaving technique, we can produce many good burst-error-correcting cyclic codes, and indeed some of the burst-error-correcting codes used in practice have been constructed this way. However, there is another, quite different, general approach to designing burst-error-correcting codes which has had great success as well, called the *Fire code* method, which we shall now study.

Fire codes differ in a number of ways from the other cyclic burst-error-correcting codes we have studied. The most important difference is their guaranteed ability to *detect* many error bursts that are too long to *correct*. Indeed, codes are *strong* burst-error-correcting codes in the sense of the following definition.

Definition *An (n, k) cyclic code with generator polynomial $g(x)$ is said to be a strong burst-b-error-correcting code, if, whenever $\mathbf{Z}_1$ and $\mathbf{Z}_2$ are burst-error vectors with the same syndrome, and with burst descriptions (pattern$_1$, location$_1$) and (pattern$_2$, location$_2$) such that length(pattern$_1$) + length(pattern$_2$) $\leqslant 2b$, then $\mathbf{Z}_1 = \mathbf{Z}_2$.*

The following theorem shows that strong burst-error-correcting codes are capable of simultaneous burst-error correction and detection.

Theorem 8.13 *If C is a strong burst-b-error-correcting code, then for any pair of nonnegative integers b_1 and b_2 such that $b_1 \leq b_2$ and $b_1 + b_2 = 2b$, it is possible to design a decoder for C that will correct all bursts of length $\leq b_1$, while at the same time detecting all burst of length $\leq b_2$.*

Proof This follows from Theorem 7.4 and the definition of strong burst correction. To see this, let $\mathcal{E}$ denote the set of error bursts of length $\leq b_1$, and let $\mathcal{F}$ denote the set of error bursts of length $\leq b_2$. Then if $\mathbf{Z}_1 \in \mathcal{E}$ and $\mathbf{Z}_2 \in \mathcal{F}$, we know that $\mathbf{Z}_1$ and $\mathbf{Z}_2$ have burst descriptions such that length(pattern$_1$) $\leq b_1$ and length(pattern$_2$) $\leq b_2$. But $b_1 + b_2 = 2b$, and so by the definition of a strong burst-b-error-correcting code, $\mathbf{Z}_1$ and $\mathbf{Z}_2$ have different syndromes. Thus by Theorem 7.4, C has the advertised capability. $\square$

Example 8.21† Most burst-b-error-correcting codes, including all of the codes in Table 8.3, are *not* strong. For example, consider the $(7, 3)$ $b = 2$ Abramson code. Its parity-check matrix can be taken to be

$$H = \begin{pmatrix} 1 & \alpha & \alpha^2 & \alpha^3 & \alpha^4 & \alpha^5 & \alpha^6 \\ 1 & 1 & 1 & 1 & 1 & 1 & 1 \end{pmatrix},$$

where $\alpha^3 + \alpha + 1 = 0$ in $GF(8)$, as we say in Example 8.16. This code corrects all bursts of length ≤ 2; but it cannot correct all bursts of length ≤ 1 while detecting all bursts of length ≤ 3, since e.g., the error patterns (1110000) and (0000010) have the same syndrome, viz. $\binom{\alpha^5}{1}$. (See Problem 8.68.) $\square$

Example 8.22 As a degenerate example of a strong burst-b-error-correcting code, consider the $(n, 0)$ "no information" cyclic code with generator polynomial $x^n - 1$. This code has only one codeword (the all-zero codeword) and so it is not useful for transmitting information; nevertheless, all error patterns have distinct syndromes, and so it is a strong burst-n-error-correcting code. (This code will be used as a building block for the codes to be defined in the corollary to Theorem 8.14, which follows.) Similarly, the $(n, 1)$ binary repetition code with $g(x) = (x^n + 1)/(x + 1)$ is a strong burst-b-error-correcting code with $b = (n - 1)/2$, if n is odd. (See Problem 8.66.) $\square$

The following theorem gives a general, construction for cyclic burst-b-error-correcting codes, both strong and weak.

Theorem 8.14 *(the Fire construction). Suppose $g_1(x)$ is the generator polynomial for an (n_1, k_1) cyclic code which is a (strong) burst-b_1-error-correcting code, $g_2(x)$ is the generator polynomial for an (n_2, k_2) cyclic code, and all of $g_2(x)$'s irreducible factors have degree $\geq m$, and finally that $g_1(x)$ and $g_2(x)$ are relatively prime. Then $g(x) = g_1(x)g_2(x)$ is the generator polynomial for an (n, k) cyclic code which is a (strong) burst-b-error-correcting code, where:*

$$n = \text{lcm}(n_1, n_2),$$

$$k = n - \deg(g_1) - \deg(g_2),$$

$$b = \min(b_1, m, (n_1 + 1)/2).$$

Proof By Theorem 8.3(a), $g_1(x)|x^{n_1} - 1$ and $g_2(x)|x^{n_2} - 1$. Thus since both $x^{n_1} - 1$ and $x^{n_2} - 1$ divide $x^n - 1$, where $n = \text{lcm}(n_1, n_2)$, and since $g_1(x)$ and $g_2(x)$ are relatively prime, it follows that $g(x) = g_1(x)g_2(x)|x^n - 1$, and so by Theorem 8.3(b), $g(x)$ generates an (n, k) cyclic code, where $k = n - \deg(g) = n - \deg(g_1) - \deg(g_2)$. This much is easy. The heart of the proof deals with the assertion about the burst-error-correcting capability of the code. We will assume that the code generated by $g_1(x)$ is strong, and show that the resulting longer code is also strong. The proof for a "weak" $g_1(x)$ is left as Problem 8.69.

Thus let $\mathbf{Z}_1$ and $\mathbf{Z}_2$ be two error vectors of length n with the same syndrome relative to the cyclic code with generator polynomial $g(x)$, and with burst descriptions (P_1, i) and (P_2, j) such that $\text{length}(P_1) + \text{length}(P_2) \leq 2b$. We need to show that $\mathbf{Z}_1 = \mathbf{Z}_2$.

Let's denote the generating functions for the two burst *patterns* by $P_1(x)$ and $P_2(x)$. It follows that the generating functions for the error vectors $\mathbf{Z}_1$ and $\mathbf{Z}_2$ are $[x^i P_1(x)]_n$ and $[x^j P_2(x)]_n$, respectively, and so the remainder syndromes are $[x^i P_1(x)]_n \bmod g(x)$ and $[x^j P_2(x)]_n \bmod g(x)$. But since $g(x)|x^n - 1$, Lemma 1(e) implies that the two syndromes can in fact be written as

$$S_1(x) = x^i P_1(x) \bmod g(x),$$

$$S_2(x) = x^j P_2(x) \bmod g(x).$$

Since $\mathbf{Z}_1$ and $\mathbf{Z}_2$ are assumed to have the same syndrome, we therefore have

$$x^i P_1(x) \equiv x^j P_2(x) \quad (\text{mod } g(x)). \tag{8.11}$$

Since $g(x) = g_1(x)g_2(x)$, then also

$$x^i P_1(x) \equiv x^j P_2(x) \quad (\text{mod } g_1(x)).$$

But length(P_1) + length(P_2) $\leqslant 2b$ by assumption, and $2b \leqslant 2b_1$ by the definition of b. Thus since $g_1(x)$ generates a strong burst-b_1-error-correcting code, it follows that from the viewpoint of the cyclic code generated by $g_1(x)$, the error vectors $\mathbf{Z}_1$ and $\mathbf{Z}_2$ are identical, i.e.

$$x^i P_1(x) = x^j P_2(x) \quad (\text{mod } x^{n_1} - 1).$$

But length(P_1) + length(P_2) $\leqslant 2b \leqslant n_1 + 1$, and so by Theorem 8.7,

$$P_1 = P_2, \tag{8.12}$$

$$i = j \quad (\text{mod } n_1). \tag{8.13}$$

In view of this, (8.11) implies

$$(x^i - x^j) P_1(x) \equiv 0 \quad (\text{mod } g_2(x)). \tag{8.14}$$

But $2 \cdot$ length(P_1) $\leqslant 2b$, and so $P_1(x)$ has degree $\leqslant b - 1$, which is less than m, the degree of the smallest-degree divisor of $g_2(x)$. Thus $P_1(x)$ and $g_2(x)$ are relatively prime, and so we can cancel $P_1(x)$ from (8.14), obtaining

$$x^i - x^j \equiv 0 \quad (\text{mod } g_2(x)). \tag{8.15}$$

But since the sequence $x^t \bmod g(x)$, has period n_2, (8.15) implies $i \equiv j$ (mod n_2). Combining this fact with (8.13), and using the fact that $n = \text{lcm}(n_1, n_2)$, we have $i \equiv j$ (mod n); but since i and j both lie in the range $0, 1, \ldots, n - 1$, it follows that $i = j$. Since we already showed that $P_1 = P_2$, it follows that the error vectors $\mathbf{Z}_1$ and $\mathbf{Z}_2$ are identical, and this completes the proof. $\qquad\square$

Corollary (the classical Fire codes) *Let* $g(x) = (x^{2b-1} - 1) f(x)$, *where* $f(x)$ *is an irreducible polynomial which is not a divisor of* $x^{2b-1} - 1$, *of degree* $m \geqslant b$ *and period* n_0. *Then* $g(x)$ *is the generator polynomial for an* $(n, n - 2b + 1 - m)$ *cyclic code which is a strong burst-b corrector, where* $n = \text{lcm}(2b - 1, n_0)$. *This code is called a Fire code, in honor of Philip Fire, its discoverer.*

Proof This follows immediately from Theorem 8.14, by taking $g_1(x) = (x^{2b-1} - 1)$ and $g_2(x) = f(x)$ (In Example 8.22 we saw that the $g_1(x)$ code is a strong burst-b corrector.) $\qquad\square$

Example 8.23 According to the corollary to Theorem 8.14, the binary cyclic code with $g(x) = (x^3 + 1)(x^3 + x + 1) = x^6 + x^4 + x + 1$ generates a $(21, 15)$ strong $b = 2$ burst-error-correcting Fire code. Thus it is possible to devise a decoder for this code that corrects all bursts of length $\leqslant 2$; or to correct all bursts of length $\leqslant 1$ and detect all bursts of length $\leqslant 3$; or to detect all bursts of length $\leqslant 4$. Note that this code meets neither the strong Abramson bound nor the Reiger bound; on the other hand, codes that do meet these bounds are apparently never strong burst correctors. Also note that if we take $g_1(x) = x^4 + x^3 + x^2 + 1$, which generates the (weak) (7, 3) $b = 2$ Abramson code, and $g_2(x) = x^2 + x + 1$, Theorem 8.14 implies that $g(x) = g_1(x)g_2(x)$ generates a (weak) (21, 15) $b = 2$ burst-error-correcting code. But $g_1(x)g_2(x) = x^6 + x^4 + x + 1$, and we have already seen that this polynomial is strong! $\qquad\square$

Example 8.24 The polynomial $P_{35}(x) = x^{35} + x^{23} + x^8 + x^2 + 1$ is a primitive binary polynomial of degree 35, and so by the corollary to Theorem 8.14, $g(x) = (x^{13} + 1)P_{35}(x) = x^{48} + x^{36} + x^{35} + x^{23} + x^{21} + x^{15} + x^{13} + x^8 + x^2 + 1$ generates a strong cyclic $(13(2^{35} - 1), \quad 13(2^{35} - 1) - 48) = (446,676,598,771, 446,676,598,723)$ $b = 7$ Fire code. This particular Fire code is quite famous, because IBM has used a "shortened" version of it in many of its disk drives. What is a shortened cyclic code? In general, if $g(x)$ has degree r and generates an $(n, n - r)$ cyclic code, then for any $n_0 \leqslant n$, $g(x)$ also generates an $(n_0, n_0 - r)$ shortened cyclic code. This code consists of all vectors $\mathbf{C}$ of length n_0 whose generating function $C(x)$ is a multiple of $g(x)$. In the case of the IBM Fire code, $n_0 = 152,552$, so the code used is actually a (152,552, 152,504) shortened cyclic $b = 7$ strong bursterror-correcting code. The IBM decoder is very conservative and only attempts to correct bursts of length $\leqslant 4$, and so, since the code is strong, it thereby gains the ability to detect all bursts of length $\leqslant 10$. In fact, however, because the code has been so drastically shortened, a computer-aided calculation shows that the IBM decoder will actually detect all burst-error patterns of length $\leqslant 26$, and all but an infinitesmal fraction of longer bursts. $\qquad\square$

We conclude this section with a short table, Table 8.5, listing some useful binary Fire codes. For each of these codes, $g(x) = (x^{2b-1} + 1)P_b(x)$, where $P_b(x)$ is a primitive polynomial of degree b. The redundancy is thus $r = 3b - 1$, although the redundancy required by the weak Abramson and Reiger bounds is typically considerably smaller than this. This "extra" redundancy is apparently the price that must be paid in order that the codes be strong burst correctors.

Table 8.5 Some Fire codes.

(n, k)	b	Generator polynomial
$(35, 27)$	3	$(x^5 + 1)(x^3 + x + 1) = x^8 + x^6 + x^5 + x^3 + x + 1$
$(105, 94)$	4	$(x^7 + 1)(x^4 + x + 1) = x^{11} + x^8 + x^7 + x^4 + x + 1$
$(279, 265)$	5	$(x^9 + 1)(x^5 + x^2 + 1) = x^{14} + x^{11} + x^9 + x^5 + x^2 + 1$
$(693, 676)$	6	$(x^{11} + 1)(x^6 + x + 1) = x^{17} + x^{12} + x^{11} + x^6 + x + 1$
$(1651, 1631)$	7	$(x^{13} + 1)(x^7 + x + 1) = x^{20} + x^{14} + x^{13} + x^7 + x + 1$
$(255, 232)$	8	$(x^{15} + 1)(x^8 + x^4 + x^3 + x^2 + 1) = x^{23} + x^{19} + x^{18} + x^{17} + x^{15} + x^8 + x^4 + x^3 + x^2 + 1$
$(8687, 8661)$	9	$(x^{17} + 1)(x^9 + x^4 + 1) = x^{26} + x^{21} + x^{17} + x^9 + x^4 + 1$
$(19437, 19408)$	10	$(x^{19} + 1)(x^{10} + x^3 + 1) = x^{29} + x^{22} + x^{19} + x^{10} + x^3 + 1$

8.5 Decoding burst-error-correcting cyclic codes

In section 8.2 we saw how to design a shift-register encoder for an arbitrary cyclic code. As a rule it is much harder to design a corresponding *decoder*, but in the important special case of burst-error-correcting cyclic codes, there is a simple decoding algorithm called the *burst-trapping algorithm* which naturally lends itself to shift-register implementation. In this section we will see how burst trapping works.

Here is the underlying idea. Suppose that $g(x)$ generates and (n, k) cyclic code C, and that a transmitted codeword $C(x)$ is received as $R(x)$, where

$$R(x) = C(x) + E(x),$$

$E(x)$ being the error pattern. If the decoder computes the remainder syndrome $S(x)$ defined by

$$S(x) = R(x) \bmod g(x),$$

then the vector obtained by subtracting $S(x)$ from $R(x)$, viz.

$$\hat{C}(x) = R(x) - S(x), \tag{8.16}$$

is guaranteed to be a codeword. This is because $\hat{C}(x) \bmod g(x) = R(x) \bmod g(x) - S(x) \bmod g(x) = 0$. Thus if C can correct all errors in the set $\mathcal{E}$, and $S(x)$ lies in $\mathcal{E}$, then the decoder can safely assert that $\hat{C}(x)$ is the actual transmitted codeword, since no other codeword can differ from $R(x)$ by an error pattern in $\mathcal{E}$. This much is true for any cyclic code. Let's now consider the special case of a cyclic *burst*-error-correcting code.

Suppose then that C is a burst-b-error-correcting code, and that the syndrome $S(x)$ satisfies the two conditions

$$S(0) \neq 0,$$

$$\deg(S(x)) \leq b - 1. \tag{8.17}$$

What this means is that the *syndrome itself* is a (left-justified) burst of length $\leq b$, and so by the above discussion the decoder can safely assert that $\hat{C}(x)$, as defined in (8.16), is the actual transmitted codeword. Decoding is therefore easy if (8.17) is satisfied. Unfortunately, however, this only happens if the error vector is a burst located at position 0.

Surprisingly, however, this simple idea can be made to work even if the burst is located in a position *other* than 0, by taking advantage of the "cyclicness" of the code. Here's how: if the error vector $E(x)$ is a nonzero burst of length $\leq b$, then $E(x)$ has a unique burst description of the form

$(P(x), i_0)$ with $P(0) \neq 0$ and $\deg(P) \leqslant b - 1$. Then $E(x) = [x^{i_0} P(x)]_n$ and the remainder syndrome is

$$S(x) = [x^{i_0} P(x)]_n \bmod g(x),$$

where $g(x)$ is the code's generator polynomial. This situation is depicted in Figure 8.9(a). As we already noted, if the burst is located in position 0, i.e., if $i_0 = 0$, then $S(x) = P(x)$ and the error burst can be corrected immediately. If $i_0 \neq 0$, however, it is possible to shift $E(x)$ cyclically to the right until the burst pattern $P(x)$ is located, or "trapped", in position 0 as shown in Figure 8.9(b). The number of cyclic shifts required to do this is the unique integer j_0 in the range $0 \leqslant j_0 \leqslant n - 1$ such that $i_0 + j_0 \equiv 0 \pmod{n}$, which is

$$j_0 = (-i_0) \bmod n.$$

If now $R_{j_0}(x)$ denotes the j_0th cyclic shift of $R(x)$, we have

$$R_{j_0}(x) = C_{j_0}(x) + E_{j_0}(x),$$

where $C_{j_0}(x)$ and $E_{j_0}(x)$ are the j_0th cyclic shifts of $C(x)$ and $E(x)$, respectively. Now $C_{j_0}(x)$ is a codeword since the code is cyclic, and $E_{j_0}(x) = P(x)$ by the definition of j_0, and so if $S_{j_0}(x)$ denotes the remainder syndrome of $R_{j_0}(x)$, we have

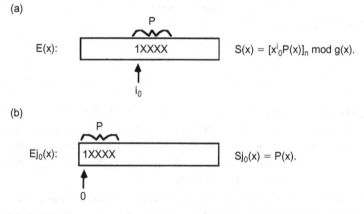

Figure 8.9 (a). The burst-error vector $E(x) = [x^{i_0} P(x)]_n$. (b) After shifting $\mathbf{E}(x) j_0$ units to the right, where $j_0 = (n - i_0) \bmod n$, the error pattern is "trapped" at location 0, where the corresponding syndrome is $P(x)$, a left-justified burst of length $\leqslant b - 1$.

$$S_{j_0}(x) = R_{j_0}(x) \bmod g(x)$$

$$= (C_{j_0}(x) + E_{j_0}(x)) \bmod g(x)$$

$$= P(x).$$

This means that $S_{j_0}(x)$ satisfies the conditions (8.17), and so the decoder can safely assert that $\hat{C}_{j_0}(x)$, defined by

$$\hat{C}_{j_0}(x) = R_{j_0}(x) - S_{j_0}(x),$$

is the j_0th cyclic shift of the actual transmitted codeword.

It follows that if the decoder successively computes $S_0(x)$, $S_1(x)$, ..., and tests each of these polynomials for the conditions (8.17), eventually the burst error will be "trapped," and it can be corrected. Indeed, the burst-error *pattern* will be given by $S_{j_0}(x)$, and the burst-error *location* will be given by the formula $i_0 = (-j_0) \bmod n$, where j_0 is the number of shifts required to trap the error. At this point, we could design a fairly simple decoder using these ideas. However, before doing so, we wish to note that the calculation of the successive $S_j(x)$'s can be considerably simplified by using the following result.

Theorem 8.15 (*Meggitt's lemma*) For $j \geqslant 0$ define

$$S_j(x) = [x^j R(x)]_n \bmod g(x),$$

i.e., $S_j(x)$ is the remainder syndrome of the jth cyclic shift of $R(x)$. Then for $j \geqslant 0$,

$$S_{j+1}(x) = [xS_j(x)] \bmod g(x).$$

Proof First note that by Lemma 1(e),

$$S_j(x) = [x^j R(x)] \bmod g(x),$$

since $g(x)|x^n - 1$. Then

$$[xS_j(x)] \bmod g(x) = [x([x^j R(x)] \bmod g(x))] \bmod g(x)$$

$$= [x^{j+1} R(x)] \bmod g(x) \qquad \text{by Lemma 1(d)}$$

$$= S_{j+1}(x). \qquad \qquad \square$$

Example 8.25 We illustrate these ideas using the $(7, 3)$ $b = 2$ Abramson code with $g(x) = x^4 + x^3 + x^2 + 1$. Suppose the received vector is $\mathbf{R} =$

[1010011], i.e., $R(x) = x^6 + x^5 + x^2 + 1$. Then $S_0(x) = R(x) \bmod g(x) = (x^6 + x^5 + x^2 + 1) \bmod (x^4 + x^3 + x^2 + 1) = x^3 + x^2$. Using Meggitt's lemma, we successively complete $S_1(x)$, $S_2(x)$, etc.:

$$S_1(x) = [xS_0(x)] \bmod g(x)$$

$$= (x^4 + x^3) \bmod (x^4 + x^3 + x^2 + 1)$$

$$= x^2 + 1.$$

Similarly,

$$S_2(x) = x^3 + x,$$

$$S_3(x) = (x^4 + x^2) \bmod g(x) = x^3 + 1,$$

$$S_4(x) = (x^4 + x) \bmod g(x) = x^3 + x^2 + x + 1,$$

$$S_5(x) = (x^4 + x^3 + x^2 + x) \bmod g(x) = x + 1.$$

Now we stop, since $S_5(x)$ satisfies the conditions (8.17), and conclude that the burst-error *pattern* is 11, and the burst-error *location* is $(-5) \bmod 7 = 2$. Thus the error vector is $E = [0011000]$, and the corrected codeword is $\mathbf{R} + \mathbf{E} = [1001011]$. $\qquad\square$

In Figure 8.10 we display a complete decoding algorithm for a burst-b-error-correcting cyclic code, including the simplification afforded by Meggitt's lemma. In this algorithm, the syndrome $S_0(x)$ is first computed at line 3. The for loop at lines 4–8 successively tests the shifted syndrome $S_0(x)$, $S_1(x)$, ..., $S_{n-1}(x)$ for the condition (8.17). If this condition is satisfied, the burst of errors is corrected at line 6. Then the next cyclic shift $R_{j+1}(x)$ of the received word is completed at line 7, and the next syndrome $S_{j+1}(x)$ is computed (using Meggitt's lemma) at line 8. After n cyclic shifts, the original received sector, minus the burst-error pattern, is output at line 9. (The algorithm also works if there are *no* errors; see Problem 8.75.)

The algorithm described in Figure 8.10 lends itself to implementation by shift-register logic. We illustrate this in Figures 8.11 and 8.12, for two different burst-error-correcting cyclic codes, the $(7, 3)$ $b = 2$ Abramson code of Example 8.16 and the $(15, 9)$ $b = 3$ code of Example 8.17.

In the decoding circuits in Figures 8.11 and 8.12, there are three main components: a mod-$g(x)$ shift register, a mod-$x^n - 1$ lower shift register, and a "100 ... 0 recognizer" circuit correcting the two shift registers. Initially, each flip-flop contains 0, switch A is in the *closed* position, and switch B is in

```
/** Burst-Error Trapping Decoding Algorithm **/
{
1.      input R(x);
2.      R₀(x) ← R(x);
3.      S₀(x) ← R₀(x) mod g(x);
4.      for (j = 0 to n − 1) {
5.          if (Sⱼ(0) ≠ 0 and deg Sⱼ(x) ≤ b − 1)
6.              Rⱼ(x) ← Rⱼ(x) − Sⱼ(x);
7.          Rⱼ₊₁(x) ← [xRⱼ(x)] mod xⁿ − 1;
8.          Sⱼ₊₁(x) ← [xSⱼ(x)] mod g(x);
        }
9.      output Rₙ(x);
}
```

Figure 8.10 "Full-cycle-clock-around" burst-trapping decoding algorithm for cyclic burst-error correction.

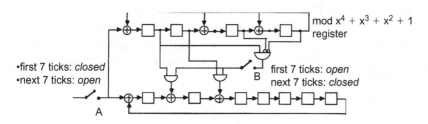

Figure 8.11 Complete decoding of circuit for the $(7, 3)$ $b = 2$ Abramson code with $g(x) = x^4 + x^3 + x^2 + 1$ (see Example 8.16). After 14 ticks, the decoding is complete, and the decoded word appears in the lower register.

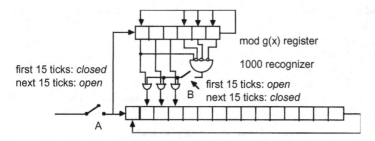

Figure 8.12 Simplified diagram for a complete decoding circuit for the $(15, 9)$ $b = 3$ cyclic code with $g(x) = x^6 + x^5 + x^4 + x^3 + 1$ (see Example 8.17). After 30 ticks, the decoding is complete, and the decoded word appears in the lower register.

the *open* position. The symbols of the received vector are then clocked in from the left in the order $R_{n-1}, R_{n-2}, \ldots, R_0$. After n ticks, the upper shift register will contain $S_0(x) = R(x) \bmod g(x)$ (by Theorem 8.4) and the lower shift will contain $R(x)$. Next, switch A is *opened*, and switch B is *closed*, thus allowing the upper shift register to communicate to the lower shift register via the $100 \ldots 0$ recognizer, and the decoder operates for n more ticks. We call these n ticks the *decoding cycle*. After the jth tick of the decoding cycle, the upper shift register will contain $[x^j S_0(x)] \bmod g(x) = S_j(x)$, and the lower register will contain $[x^j R(x)]_n = R_j(x)$. If conditions (8.17) are satisfied, then the leftmost flip-flop of the upper register will contain a 1, and the rightmost $r - b$ flip-flop, will all contain 0's. This will cause the $100 \ldots 0$ recognizer to output a 1, and the trapped error pattern $S_j(x)$ will be added into $R_j(x)$ at the next tick. After the n ticks of the decoding cycle have been completed, the symbols in $R(x)$ will have made a complete circuit around the $\bmod(x^n - 1)$ register, and be returned to their original position, with the error burst corrected.

In general, circuits like these in Figures 8.11 and 8.12 are called "full-period-clock-around" decoders. They require exactly $2n$ clock ticks to correct any burst of length $\leqslant b$, with hardware complexity that is $O(n)$. They are in widespread use in practice. However, for drastically shortened cyclic burst-error-correcting codes (e.g. Example 8.24), a modification of this circuit may be required—see Problem 8.77.

Problems

8.1 Show that the number of k-dimensional subspaces of an n-dimensional vector space over the field $GF(q)$ is exactly

$$\frac{(q^n - 1)(q^{n-1} - 1) \ldots (q^{n-k+1} - 1)}{(q^k - 1)(q^{k-1} - 1) \ldots (q - 1)},$$

and use this formula to verify that there are exactly 11,811 binary (7, 3) linear codes.

8.2 Calculate the following:
 (a) $10^{10} \bmod 12$;
 (b) $x^{10^{10}} \bmod (x^{12} - 1)$;
 (c) $(x^{15} - 1) \bmod x^{10} - 1$.

8.3 Is the "mod" operation associative and/or commutative? That is, are the identities

$$(P \bmod Q) \bmod R = P \bmod (Q \bmod R),$$

$$P \bmod Q = Q \bmod P$$

true in general?

8.4 Show that if $m \in \{0, 1, \ldots, n-1\}$, then $(j+m) \bmod n = i$ if and only if $m = (i-j) \bmod n$.

8.5 (a) Prove Lemma 1, part (a).
(b) Prove Lemma 1, part (b).
(c) Prove Lemma 1, part (c).
(d) Prove Lemma 1, part (d).
(e) Prove Lemma 1, part (e).

8.6 Give a formal proof that if $\mathcal{C}$ is a cyclic code, and if $C(x) \in \mathcal{C}$, then for all $i \geqslant 1$, $x^i C(x) \bmod x^n - 1 \in \mathcal{C}$. *Hint*: Use Lemma 1(d).]

8.7 Find the G_1 and H_1 matrices for the (8, 4) cyclic code over $GF(3)$ with generator polynomial $g(x) = (x^2 + 1)(x^2 + x + 1)$. (See Corollary 1 to Theorem 8.3.)

8.8 Find the G_2 and H_2 matrices for the (8, 4) cyclic code over $GF(3)$ with generator polynomial $g(x) = (x^2 - 1)(x^2 + x + 1)$. (See Corollary 2 to Theorem 8.3.)

8.9 (a) Show that the following matrices are generator parity-check matrices for an (n, k) cyclic code with parity-check polynomial $h(x)$ (and "reversed" parity-check polynomial $\tilde{h}(x)$):

$$G_3 = [x^j \tilde{h}(x)]_{j=0}^{n-1} \quad \text{(columns)},$$

$$H_3 = [x^{i+k} - x^{i+k} \tilde{h}(x)]_{i=0}^{r-1} \quad \text{(rows)}.$$

(b) Use this result to find generator and parity-check matrices for the (7, 3) binary cyclic code with $g(x) = x^4 + x^3 + x^2 + 1$.

8.10 This problem concerns the matrix H_1 of Corollary 1 to Theorem 8.3.
(a) If the syndrome $\mathbf{S}$ of the vector $\mathbf{R} = (R_0, R_1, \ldots, R_{n-1})$ is calculated as $\mathbf{S}^T = H_1 \mathbf{R}^T$, show that the generating functions $R(x) = R_0 + R_1 x + \cdots + R_{n-1} x^{n-1}$ and $S(x) = S_0 + S_1 x + \cdots + S_{r-1} x^{r-1}$ are related by

$$S(x) = \frac{[R(x)h(x)] \bmod x^n - [R(x)h(x)] \bmod x^k}{x^k}$$

(which is a just clumsy way of saying that $(S_0, S_1, \ldots, S_{r-1})$ are the coefficients of $x^k, x^{k+1}, \ldots, x^{n-1}$ in the product $R(x)h(x)$).
(b) Using the result of part (a) find the syndrome of the vector $\mathbf{R} = [1001011]$ with respect to the "H_1" matrix for the cyclic code with $g(x) = x^4 + x^3 + x^2 + 1$. (Cf. Example 8.7.)

8.11 Show that the dual code of a cyclic code is also a cyclic code. If an (n, k) cyclic code has generator polynomial $g(x)$, and parity-check polynomial $h(x)$, what are the generator and parity-check polynomials of the dual code? [*Hint*: Refer to Corollary 1 of Theorem 8.3.] Illustrate your results by finding the generator and parity-check polynomials for the dual code for the cyclic code of Examples 8.2 and 8.4.

8.12 Write each of the 9 codewords in the (4, 2) cyclic code of Example 8.3 as a multiple of the generator polynomial $g(x) = x^2 + 2$.

8.13 This problem asks you to count the number of binary block codes codes of length 7 with 16 codewords, with various restrictions.
(a) How many such codes, total, are there?

(b) How many such *linear* codes?

(c) How many such *linear* codes, with $d_{min} = 3$?

(d) How many such *cyclic* codes?

(e) How many such *cyclic* codes, with $d_{min} = 3$?

8.14 Consider the (7, 4) cyclic Hamming code with $g(x) = x^3 + x + 1$, as used to correct *erasures*. Since its minimum distance is 3, we know that it will correct all patterns of 2 or fewer erasures. However, it will also correct some patterns of *three* erasures. For example, the codeword $g(x) = [1101000]$ with erasures in positions 1, 3 and 6 is $\mathbf{R} = [1 \ * \ 0 \ * \ 00 \ *]$. This pattern of three erasures is correctable, because, of the 16 codewords in the code, only $[1101000]$ agrees with $\mathbf{R}$ in the unerased positions. However, not all patterns of three erasures are correctable: e.g., $[* \ * \ 0 \ * \ 000]$ could be either $[1101000]$ or $[0000000]$. Here is the problem: of the $\binom{7}{3} = 35$ possible patterns of three erasures, how many are correctable, and how many are not?

8.15 How many binary cyclic codes of length n are there, for $1 \le n \le 20$? How many of these are improper?

8.16 How many binary cyclic codes of length 63 are there? How many are improper?

8.17 For which values of n in the range $3 \le n \le 20$ are there exactly 4 cyclic codes of length n over F_2?

8.18 What fraction of linear codes of length 4 over F_2 are cyclic?

8.19 Find a formula for the number of cyclic codes of length 2^m over F_2.

8.20 Over the field $GF(3)$, the polynomial $x^8 - 1$ factors as $x^8 - 1 = (x + 1)(x + 2)(x^2 + 1)(x^2 + x + 1)(x^2 + 2x + 1)$. For each k in the range $0 \le k \le 8$, determine the number of (8, k) cyclic codes over $GF(3)$.

8.21 In Example 8.8, we saw that for the (4, 2) cyclic code over $GF(3)$ with $g(x) = x^2 - 1$, the generator and parity-check matrices of Corollaries 1 and 2 of Theorem 8.3 are equal, i.e., $G_1 = G_2$ and $H_1 = H_2$. Can you find a general class of cyclic codes for which this property holds?

8.22 Suppose that over a given field F, the polynomial $x^n - 1$ factors as

$$x^n - 1 = P_1(x)^{e_1} P_2(x)^{e_2} \cdots P_M(x)^{e_M},$$

where the $P_i(x)$'s are distinct irreducible polynomials. In terms of $e_1, e_2, \ldots, e_M$, how many cyclic codes of length n over F are there?

8.23 Explain each of the "comments" in the table in Example 8.9.

8.24 Prove that over the field $GF(2)$, $x^3 + x + 1 | x^{7m} - 1$, for all $m \ge 1$.

8.25 There are eight fourth-degree binary polynomials of the form $g(x) = x^4 + g_3 x^3 + g_2 x^2 + g_1 x + 1$. For each of these polynomials answer the following:

(a) What is the period of $g(x)$?

(b) If n is the period you found in part (a), what are the k and d for the corresponding cyclic code?

8.26 (K. Sivarajan). Show that a binary cyclic code is capable of correcting a single error, i.e., $d_{min} \ge 3$, if and only if it is a proper cyclic code.

8.27 At the end of Section 8.1, we discussed briefly *improper* cyclic codes, i.e., those for which the period of the generator polynomial $g(x)$ is less than n. In this problem, we will investigate cyclic codes for which the *parity-check* polynomial has period less than n. Thus let C be an (n, k) cyclic code, and let $x^n - 1 = g(x)h(x)$, where $g(x)$ is the generator polynomial and $h(x)$ is the parity-check polynomial, for C. Suppose further that the period of $h(x)$ is $n_0 < n$, and that C_0 is the (n_0, k) cyclic code with generator polynomial $g(x) = (x^{n_0} - 1)/h(x)$.

(a) What is the parity-check polynomial for C_0?

(b) If d_0 is the minimum distance of C_0, what is the minimum distance of C?

(c) Show that $C = (n/n_0)C_0$, where "jC_0" denotes the code of length n obtained from C_0 by repeating each codeword j times.

8.28 For each j in the range $1 \leqslant j \leqslant 16$, find the length, dimension, and minimum distance of the (proper) binary cyclic code with generator polynomial $(x + 1)^j$. [*Hint*: The result of the previous problem will be helpful.]

8.29 If $g(x)$ has period n and degree r, then for any $m \geqslant 1$, $g(x)|x^{mn} - 1$ and so by Theorem 8.3(b), $g(x)$ generates an $(nm, nm - r)$ cyclic code. (For $m \geqslant 2$, this code is improper.)

(a) Show that a vector $C = [C_0, C_1, \ldots, C_{nm-1}]$ is in this code if and only if $\sum_{i=0}^{nm-1} C_i x^{i \bmod n} \equiv 0 \pmod{g(x)}$.

(b) Show that the condition in part (a) is equivalent to saying that

$$[C_0, C_1, \ldots, C_{n-1}] + [C_n, C_{n+1}, \ldots, C_{2n-1}]$$

$$+ \cdots + [C_{(m-1)n}, C_{(m-1)n+1}, \ldots, C_{mn-1}]$$

is in the $(n, n - r)$ proper cyclic code generated by $g(x)$.

(c) How many words of weight 2 are in this code?

8.30 In Figure 8.3 we see a "mod $g(x)$" circuit for a *monic* polynomial $g(x)$. Explain how to modify this circuit for a non-monic polynomial $g(x) = g_0 + g_1 x + \cdots + g_r x^r$, with $g_r \neq 0, 1$.

8.31 (a) Show that for the circuit of Figure 8.4(a), the relationship between the present state polynomial $S(x)$ and the next state polynomial $S'(x)$ is

$$S'(x) = (xS(x) + sx^r) \bmod g(x),$$

where s is the input signal.

(b) Use the result of part (a) to show that if the circuit of Figure 8.4(a) is intialized with $s_0 = s_1 = \cdots = s_{r-1} = 0$, and then given the input sequence $a_0, a_1, \ldots$, then after the tth tick the state polynomial will be

$$S_t(x) = \sum_{j=0}^{t} a_j x^{r+t-j} \bmod g(x).$$

8.32 Consider the polynomial $g(x) = x^3 + 2x^2 + 2$ over the three-element field $GF(3)$.

(a) What is the period of this polynomial?

(b) If n denotes the period you found in part (a), find a parity-check matrix for the cyclic code of length n generated by $g(x)$.

(c) Does this code have any words of weight 2? If not, explain why not. If so, explicity list all such words.

(d) Draw a systematic shift-register encoder for this code.

8.33 Design three different shift-register encoders for the (15, 7) binary cyclic code with $g(x) = x^8 + x^7 + x^6 + x^4 + 1$.

8.34 This problem concerns the (31, 26) cyclic Hamming code with $g(x) = x^5 + x^2 + 1$.
  (a) Find a systematic (i.e., of the form $[A|I_5]$) parity-check matrix for this code.
  (b) Design a decoding circuit for this code.
  (c) Use your decoder from part (b) to decode the received word

$$\mathbf{R} = [1111111111110000000000000000000000].$$

8.35 Find a formula for the number of words of weight 3 in the $(2^m - 1, 2^m - m - 1)$ cyclic Hamming code defined by the parity-check matrix (8.9).

8.36 Let $g(x)$ be a polynomial of degree r over a finite field with q elements, such that $g(0) \neq 0$.
  (a) Show that the period of $g(x)$ is $\leq q^r - 1$.
  (b) Show that the period of $g(x)$ is $q^r - 1$, then $g(x)$ is necessarily irreducible. [*Note*: An irreducible polynomial of degree r and period $q^r - 1$ is called a *primitive* polynomial.]

8.37 Consider the (15, 11) binary cyclic Hamming code with generator polynomial $g(x) = x^4 + x + 1$.
  (a) Write out a 4×15 binary parity-check matrix.
  (b) Write out a 1×15 $GF(16)$ parity-check matrix.

8.38 Prove that if α is a primitive root in the field $GF(2^m)$ with minimal polynomial $g(x)$, and if the expansion of α^j is

$$\alpha^j = \sum_{i=0}^{m-1} h_{i,j}\alpha^i \qquad \text{for } j = 0, 1, \ldots, n-1,$$

then the $m \times 2^m - 1$ matrix $H = (h_{i,j})$ for $i = 0, 1, \ldots, m-1, j = 0, 1, \ldots, 2^m - 2$ is the parity-check matrix for a cyclic Hamming code with polynomial $g(x)$.

8.39 Let $g(x)$ be a binary primitive polynomial of degree m, and let H be a $m \times 2^m - 1$ binary matrix which is a parity-check matrix for a $(2^m - 1, 2^m - m - 1)$ Hamming code. How many of the $(2^m - 1)!$ possible permutations of the columns of H yield a parity-check matrix for the *cyclic* Hamming code with $g(x)$ as the generator polynomial?

8.40 In Table 8.3, except for the $m = 1$ entry, there are no polynomials of even weight. Explain why this is so, i.e., why no polynomial of even weight can ever be primitive.

8.41 (Dual codes to cyclic Hamming codes—also called *cyclic simplex codes*, or *maximal-length shift-register code*.) Let C_k be an (n, k) (proper) cyclic code whose parity-check polynomial $h(x)$ is a primitive polynomial of degree k, and let G_1 be the generator matrix for C_k described in Corollary 1 to Theorem 8.3.
  (a) Show that $n = 2^k - 1$.
  (b) Describe an efficient shift-register encoder for C_k.

(c) Show that the set of nonzero codewords of C is exactly equal to the set of cyclic shifts of the first row of G_1. Illustrate this result for C_4 using the primitive polynomial $x^4 + x + 1$ from Table 8.3.

(d) Find the weight enumerator for C_k.

8.42 Find a formula for the number of *ordinary* (i.e., not cyclic) bursts of length b, over a two-letter alphabet (cf. Theorem 8.8), for $1 \leqslant b \leqslant n$.

8.43 Generalize Theorem 8.8 to a q-letter alphabet.

8.44 Extend Theorem 8.8 to cover the case $b > (n + 1)/2$.

8.45 Generalize Theorem 8.9 and its corollary to a q-letter alphabet.

8.46 Generalize Theorem 8.10 and its corollary to cover the case of a q-letter alphabet.

8.47 Why is the case $b = 0$ not covered in Theorem 8.9?

8.48 Why do you suppose the two versions of the Abramson bound are called "strong" and "weak"?

8.49 Are the Abramson bounds (corollary to Theorem 8.9) ever tight for the binary repetition codes?

8.50 Is the Reiger bound ever tight for the Abramson codes?

8.51 In some burst-error-correcting applications, it is sufficient to correct bursts that occur only in certain special locations. The most important example of this is what is called *phased burst-error correction*. In phased burst-error correction, the block length n is a multiple of b, and bursts of length b may occur only at locations which are multiples of b. A code capable of correcting all phased bursts of length b is called a *phased* burst-b-error-correcting code. For example, a phased burst-3-error-correcting code of length 12 must be able to correct any burst of length 3 which occurs at locations 0, 3, 6, or 9.

(a) Show that Theorem 8.10, and hence the Reiger bound, applies to phased burst-error-correcting codes.

(b) Show that if a code is capable of correcting t phased bursts of length b, then $r \geqslant 2tb$.

8.52 Consider a (20, 12) binary linear code in which the parity-checks are described by the following array:

$$\begin{bmatrix} C_0 & C_4 & C_8 & C_{12} & C_{16} \\ C_{17} & C_1 & C_5 & C_9 & C_{13} \\ C_{14} & C_{18} & C_2 & C_6 & C_{10} \\ C_{11} & C_{15} & C_{19} & C_3 & C_7 \end{bmatrix}.$$

In a given codeword $[C_0, C_1, \ldots, C_{19}]$, the components occupying the upper left 3×4 portion of the array carry the information, and the components forming the right column and bottom row of the array are parity positions. The parity check rules are that the sum of the components in each row and column of the array is zero. Thus for example, $C_0 + C_4 + C_8 + C_{12} + C_{16} = 0$ (first row sum), and $C_4 + C_1 + C_{18} + C_{15} = 0$ (second column sum).

(a) Show that this code can correct all phased bursts of length 4.

(b) Show that the corresponding 3×4 array code is *not* capable of correcting all phased burst of length 3.

(c) Generalize, and show that a $b \times (b + 1)$ array code can correct all phased bursts of length b, if and only if $b + 1$ is a prime number.

8.53 (Burton codes).

(a) Let $g(x) = (x^b - 1)p_b(x)$, where $p_b(x)$ is an irreducible polynomial of degree b and period n_1. Show that $g(x)$ generates an $(n, n - 2b)$ cyclic code with $n = \text{lcm}(n_1, b)$ which is capable of correcting all b-*phased bursts* of length b. (See Problem 8.51.)

(b) Consider the sequence of $(nj, (n - 2b)j)$ cyclic codes obtained by interleaving the codes in part (a), and let b_j denote the corresponding (ordinary, i.e., unphased) burst-error-correcting capability. Show that $\lim_{j \to \infty} b_j / 2b_j = \frac{1}{2}$, so that these code-asymptotically meet the Reiger bound.

8.54 Consider a binary cyclic $b = 3$ burst-error-correcting code of length 31.

(a) Use the Abramson and the Reiger bounds to estimate the minimum needed redundancy.

(b) Do you think there is such a code with the redundancy predicted in part (a)? Explain fully.

8.55 Show that if $g(x)$ is the generator polynomial for an (n, k) cyclic code, and if $(x + 1)$ is not a divisor of $g(x)$, then $g'(x) = (x + 1)g(x)$ is the generator polynomial for an $(n, k - 1)$ code, which equals the original code minus its words of odd weight.

8.56 Decode the following noisy codewords from the $(7, 3)$ Abramson code of Example 8.16:

(a) [0101010];
(b) [1111111];
(c) [1110010];
(d) [1101000];
(e) [1011111].

8.57 Decode the following noisy codewords from the $(15, 9)$ $b = 3$ burst-error-correcting code of Example 8.17:

(a) [000001001111000];
(b) [101110001100000];
(c) [111001000001111];
(d) [110100000101010];
(e) [111100000000000].

8.58 Verify the assertion, made in Example 8.20, that the polynomial $g(x) = (x^4 + x + 1)(x^4 + x^3 + x^2 + x + 1)$ generates a $(15, 7)$ $b = 4$ burst-error-correcting cyclic code.

8.59 Does either of the following two polynomials generate a $(15, 7)$ $b = 4$ burst-error-correcting cyclic code?

(a) $g(x) = (x^4 + x^3 + 1)(x^4 + x^3 + x^2 + x + 1)$;
(b) $g(x) = (x^4 + x + 1)(x^4 + x^3 + 1)$.

8.60 Verify the assertion, made in Example 8.20, that the polynomial $g(x) = x^{12} + x^9 + x^6 + x^3 + 1$ generates a $(15, 3)$ $b = 6$ burst-error-correcting cyclic code.

8.61 In the $(15, 9)$ code of Example 8.17, 61 of the 64 cosets are accounted for by the bursts of length ≤ 3. Describe the syndromes of the missing three cosets. What is the length of the shortest burst pattern in each of these cosets?

8.62 In Example, 8.17, we saw that $g(x) = (x^4 + x + 1)(x^2 + x + 1)$ generates a $(15, 9)$ $b = 3$ burst-error-correcting cyclic code. Does $g'(x) = (x^4 + x + 1)(x^2 + x^3 + 1)$ work as well? How about $g''(x) = (x^4 + x^3 + x^2 + x + 1)(x^2 + x + 1)$?

8.63 Generalize Example 8.17 by showing that the $(2^m - 1, 2^m - m - 3)$ binary cyclic code with generator polynomial $g(x) = p(x)(x^2 + x + 1)$, where $p(x)$ is a primitive polynomial of degree m such that $1 + x \equiv x^a \pmod{p(x)}$, where a mod $3 \neq 2$, is a $b = 3$ burst-error-correcting code that meets the Abramson bound.

8.64 If the code $\mathcal{C}$ meets the Abramson bound, will the depth-j interleaving of $\mathcal{C}$ also meet the Abramson bound? If so, give a proof. If not, give a counterexample.

8.65 Prove the result (8.10) about the right cyclic shift of an interleaved vector.

8.66 Prove that the $(n, 1)$ binary repetition code with $g(x) = (x^n + 1)/(x + 1)$ is a strong burst-b-error-correcting code with $b = (n - 1)/2$, if n is odd.

8.67 Give a formal proof of Theorem 8.11.

8.68 (See Examples 8.11 and 8.21). For the $(7, 3)$ Abramson code, show that every burst with pattern 111 has the same syndrome as some burst with pattern 1.

8.69 Prove Theorem 8.14 in the case that the code generated by $g_1(x)$ is a weak (i.e. not strong) burst-b-error-correcting code.

8.70 As mentioned in Example 8.24, if $g(x)$ is a polynomial of degree r and period n, and if $n_0 < n$, the $(n_0, n_0 - r)$ *shortened* cyclic code with generator polynomial $g(x)$ consists of all vectors $[C_0, C_1, \ldots, C_{n_0-1}]$ whose generating function $C_0 + C_1 x + \cdots + C_{n_0-1} x^{n_0-1}$ is a multiple of $g(x)$.
(a) Show that this is a linear code.
(b) If $g(x) = x^6 + x^4 + x + 1$ (see Example 8.23), find generator and parity-check matrices for the $(16, 10)$ shortened cyclic code generated by $g(x)$.

8.71 Why is there no entry for $b = 2$ in Table 8.5?

8.72 For each of the n's and b's in Table 8.5, estimate the redundancy needed, using the Abramson and Reiger bounds, and compare this value with the actual redundancy of the given Fire code.

8.73 Consider the $(35, 27)$ $b = 3$ Fire code from Table 8.5. Decode the following received words, using a "$b_1 = 3$, $b_2 = 3$" decoder (see Theorem 8.13):

$$\mathbf{R}_1 = [11010110111010110111010110100000111],$$

$$\mathbf{R}_2 = [01001110101101000001101011011010110].$$

Now assume a "$b_1 = 2$, $b_2 = 4$" decoder, and decode the same two received words.

8.74 Explain in detail how you would modify a burst-trapping decoding circuit for a burst-b-error-correcting Fire code, so that it is capable of *correcting* all bursts of length $\leq b_1$, and also *detecting* all bursts of length between $b_1 + 1$ and $2b - b_1$ (inclusive). (Here b_1 is an integer satisfying $1 \leq b_1 < b$.) Give a sketch for your modified decoder.

8.75 Show that if *no* errors occur, the algorithm in Figure 8.10 will still work.

8.76 Explain how to modify the algorithm of Figure 8.10 so that a burst of length $\geq b + 1$ which does not have the same syndrome as a burst of length $\leq b$ will be detected.

8.77 The decoding algorithm described in Figure 8.10 is for an $(n, n - r)$ burst-error-correcting cyclic code with generator polynomial $g(x)$. It requires just n executions of the for loop at lines 4–8 to complete the decoding of one length-n codeword. It will also work for a *shortened* $(n_0, n_0 - r)$ cyclic code with the same generator polynomial, but it still requires n, rather that n_0, executions of the for loop. If the code has been drastically shortened, this is unacceptable. For example, the IBM code described in Example 8.24 has $n_0 = 152{,}552$ but $n = 446{,}676{,}598{,}771$, and a decoder that required $446{,}676{,}598{,}771$ steps to decode one word of length $152{,}552$ would be useless. Fortunately, there is a simple modification of the decoding algorithm that allows the decoding of the shortened code in just n_0 steps. It is described in the following pseudocode listing. (Note the changes from the Figure 8.10 algorithm at lines 7 and 8.)

```
/* Burst-Error-Trapping-Decoding Algorithm for
   Shortened Cyclic Codes */
   {
```

1.　　input $R(x)$:
2.　　$R_0(x) \leftarrow R(x)$;
3.　　$S_0(x) \leftarrow R_0(x) \bmod g(x)$;
4.　　for $(j = 0$ to $n_0 - 1)$ {
5.　　　　if $(S_j(0) \neq 0$ and deg $S_j(x) \leqslant b - 1)$
6.　　　　　$R_j(x) \leftarrow R_j(x) - S_j(x)$;
7.　　　　$R_{j+1}(x) \leftarrow [x^{-1} R_j(x)] \bmod x^{n_0} - 1$;
8.　　　　$S_{j+1}(x) \leftarrow [x^{-1} S_j(x)] \bmod g(x)$;
　　　　}
9.　　output $R_{n_0}(x)$;
　　}

(a) Explain why this algorithm works.

(b) Consider the $(30, 16)$ *shortened* Fire code with $g(x) = (x^5 + x^2 + 1)(x^9 + 1)$, and use the given algorithm to decode the following received word:

$$\mathbf{R} = [111101001110100011011011110010].$$

(c) Describe a shift-register implememtation of this algorithm. [*Hint*: You will require both an "$S(x) \rightarrow xS(x) \bmod g(x)$" shift register and an "$S(x) \rightarrow x^{-1} S(x) \bmod g(x)$" shift register.]

Notes

1 (p. 170). This use of "mod" as a *binary operator* is closely related to, but not identical with, the more familiar use of "mod" as an *equivalence relation*. Thus $Q(x) \equiv P(x) \pmod{M(x)}$ (the equivalence relation use of "mod") means only that $Q(x) - P(x)$ is divisible by $M(x)$, whereas $Q(x) = P(x) \bmod M(x)$ (the binary operator) means that $Q(x) - P(x)$ is divisible by $M(x)$ *and* deg $Q(x) <$ deg $M(x)$.

2 (p. 181). The reader who becomes interested in the fascinating topic of logical circuits is encouraged to read Berlekamp [14], Chapter 2.

3 (p. 184). With today's available logic, t_0 can be as small as several picoseconds (pico $= 10^{-12}$).

4 (p. 195). This section assumes the reader to be familiar with the theory of finite fields, and may be omitted on first reading.

5 (p. 195). Here and hereafter in this section, we adopt the convention that $n = 2^m - 1$.

6 (p. 204). These examples assume a familiarity with the theory of finite fields, and may be omitted on first reading.

9

BCH, Reed–Solomon, and related codes

9.1 Introduction[1]

In Chapter 7 we gave one useful generalization of the (7, 4) Hamming code of the Introduction: the family of $(2^m - 1, 2^m - m - 1)$ single-error-correcting Hamming codes. In Chapter 8 we gave a further generalization, to a class of codes capable of correcting a single *burst* of errors. In this chapter, however, we will give a far more important and extensive generalization, the multiple-error-correcting BCH[2] and *Reed–Solomon* codes.

To motivate the general definition, recall that the parity-check matrix of a Hamming code of length $n = 2^m - 1$ is given by (see Section 7.4)

$$H = [\mathbf{v}_0 \quad \mathbf{v}_1 \quad \cdots \quad \mathbf{v}_{n-1}], \tag{9.1}$$

where $(\mathbf{v}_0, \mathbf{v}_1, \ldots, \mathbf{v}_{n-1})$ is some ordering of the $2^m - 1$ nonzero (column) vectors from $V_m = GF(2)^m$. The matrix H has dimensions $m \times n$, which means that it takes m parity-check bits to correct one error. If we wish to correct *two* errors, it stands to reason that m more parity checks will be required. Thus we might guess that a matrix of the general form

$$H_2 = \begin{bmatrix} \mathbf{v}_0 & \mathbf{v}_1 & \cdots & \mathbf{v}_{n-1} \\ \mathbf{w}_0 & \mathbf{w}_1 & \cdots & \mathbf{w}_{n-1} \end{bmatrix},$$

where $\mathbf{w}_0, \mathbf{w}_1, \ldots, \mathbf{w}_{n-1} \in V_m$, will serve as the parity-check matrix for a two-error-correcting code of length n. Since however, the $\mathbf{v}_i$'s are distinct, we may view the correspondence $\mathbf{v}_i \to \mathbf{w}_i$ as a *function* from V_m into itself, and write H_2 as

$$H_2 = \begin{bmatrix} \mathbf{v}_0 & \mathbf{v}_1 & \cdots & \mathbf{v}_{n-1} \\ \mathbf{f}(\mathbf{v}_0) & \mathbf{f}(\mathbf{v}_1) & \cdots & \mathbf{f}(\mathbf{v}_{n-1}) \end{bmatrix}. \tag{9.2}$$

230

But how should the function $\mathbf{f}$ be chosen? According to the results of Section 7.3, H_2 will define a two-error-correcting code iff the syndromes of the $1 + n + \binom{n}{2}$ error pattern of weights 0, 1 and 2 are all distinct. Now any such syndrome is a sum of a (possibly empty) subset of columns of H_2, and so is a vector in V_{2m}. But to be consistent with our present viewpoint let us break the syndrome $\mathbf{s} = (s_1, \ldots, s_{2m})$ in two halves: $\mathbf{s} = (\mathbf{s}_1, \mathbf{s}_2)$, where $\mathbf{s}_1 = (s_1, \ldots, s_m)$ and $\mathbf{s}_2 = (s_{m+1}, \ldots, s_{2m})$ are both in V_m. With this convention, the syndrome of the all-zero pattern is $(\mathbf{0}, \mathbf{0})$; a single error in position i has $\mathbf{s} = (\mathbf{v}_i, \mathbf{f}(\mathbf{v}_i))$; a pair of errors at positions i and j gives $\mathbf{s} = (\mathbf{v}_i + \mathbf{v}_j, \mathbf{f}(\mathbf{v}_i) + \mathbf{f}(\mathbf{v}_j))$. We can unify these three cases by defining $\mathbf{f}(\mathbf{0}) = \mathbf{0}$ (notice that since $\mathbf{0}$ is not a column of H, $\mathbf{f}$ has not yet been defined at $\mathbf{0}$); then the condition that these syndromes are all distinct is that the system of equations

$$\mathbf{u} + \mathbf{v} = \mathbf{s}_1,$$

$$\mathbf{f}(\mathbf{u}) + \mathbf{f}(\mathbf{v}) = \mathbf{s}_2 \tag{9.3}$$

has at most one solution $(\mathbf{u}, \mathbf{v})$ for each pair of vectors from V_m. (Naturally we do not regard the solution $(\mathbf{u}, \mathbf{v})$ as distinct from $(\mathbf{v}, \mathbf{u})$.)

Now we must try to find a function $\mathbf{f} : V_m \to V_m$, $\mathbf{f}(\mathbf{0}) = \mathbf{0}$, with the above property. We could try a linear mapping $\mathbf{f}(\mathbf{v}) = T\mathbf{v}$ for some linear transformation T, but this doesn't work (see Prob 9.1); so $\mathbf{f}$ must be nonlinear. To describe nonlinear functions of vectors $\mathbf{v} \in V_m$, we need to know that it is possible to define a multiplication on the vectors of V_m, which when combined with the vector addition makes V_m into a field. (The field is the Galois field $GF(2^m)$; the properties of finite fields that we shall need are stated in Appendix C.) Using this fact, it is easy to see (see Prob 9.2) that every function $\mathbf{f} : V_m \to V_m$ can be represented by a polynomial. Polynomials of degree $\leqslant 2$ don't work (see Prob 9.1); but $\mathbf{f}(\mathbf{v}) = \mathbf{v}^3$ does, as we shall shortly see. Hence (we change notation to emphasize that from now on we regard the elements of V_m not as m-dimensional vectors over $GF(2)$, but as scalars from $GF(2^m)$) if $(\alpha_0, \alpha_1, \ldots, \alpha_{n-1})$ is an arbitrary ordering of the nonzero elements of $GF(2^m)$, then the matrix

$$H_2 = \begin{bmatrix} \alpha_0 & \alpha_1 & \cdots & \alpha_{n-1} \\ \alpha_0^3 & \alpha_1^3 & \cdots & \alpha_{n-1}^3 \end{bmatrix} \tag{9.4}$$

is the parity-check matrix of a two-error-correcting binary code of length $n = 2^m - 1$. Equivalently, $\mathbf{C} = (C_0, C_1, \ldots, C_{n-1}) \in V_n$ is a codeword in the code with parity-check matrix H_2 iff $\sum_{i=0}^{n} C_i \alpha_i = \sum_{i=0}^{n} C_i \alpha_i^3 = 0$. Since as a matrix over $GF(2)$, H_2 has $2m$ rows (which are linearly independent for

$m \geqslant 3$; see Prob. 9.5), the dimension of the code is $\geqslant n - 2m$ $= 2^m - 1 - 2m$.

The *proof* that the matrix H_2 in (9.4) does indeed define a two-error-correcting code, as well as the generalization to t-error-correcting codes, is given in the following celebrated theorem.

Theorem 9.1 *Let* $(\alpha_0, \alpha_1, \ldots, \alpha_{n-1})$ *be a list of* n *distinct nonzero elements of* $GF(2^m)$, *and let* t *be a positive integer* $\leqslant (n - 1)/2$. *Then the* $t \times n$ *matrix*

$$
H = \begin{bmatrix}
\alpha_0 & \alpha_1 & \cdots & \alpha_{n-1} \\
\alpha_0^3 & \alpha_1^3 & \cdots & \alpha_{n-1}^3 \\
\alpha_0^5 & \alpha_1^5 & \cdots & \alpha_{n-1}^5 \\
\vdots & & & \\
\alpha_0^{2t-1} & \alpha_1^{2t-1} & \cdots & \alpha_{n-1}^{2t-1}
\end{bmatrix}
$$

is the parity-check matrix of a binary (n, k) *code capable of correcting all error patterns of weight* $\leqslant t$, *with dimension* $k \geqslant n - mt$.

Proof A vector $\mathbf{C} = (C_0, \ldots, C_{n-1}) \in V_n$ will be a codeword iff $H\mathbf{C}^T = \mathbf{0}$, which is equivalent to the following system of t linear equations in the C_i's:

$$
\sum_{i=0}^{n-1} C_i \alpha_i^j = 0, \qquad j = 1, 3, \ldots, 2t - 1. \tag{9.5}
$$

Squaring the jth equation in (9.5), we get $0 = (\sum C_i \alpha_i^j)^2 = \sum C_i^2 \alpha_i^{2j} = \sum C_i \alpha_i^{2j}$ (since $(x + y)^2 = x^2 + y^2$ in characteristic 2 and $x^2 = x$ in $GF(2)$). Hence an equivalent definition of a codeword is the following system of $2t$ equations:

$$
\sum_{i=0}^{n-1} C_i \alpha_i^j = 0, \qquad j = 1, 2, \ldots, 2t. \tag{9.6}
$$

It follows that we could equally well use the $2t \times n$ parity-check matrix

$$
H' = \begin{bmatrix}
\alpha_0 & \alpha_1 & \cdots & \alpha_{n-1} \\
\alpha_0^2 & \alpha_1^2 & \cdots & \alpha_{n-1}^2 \\
\vdots & & & \\
\alpha_0^{2t} & \alpha_1^{2t} & \cdots & \alpha_{n-1}^{2t}
\end{bmatrix}
$$

to describe the code. According to Theorem 7.3, H, will be the parity-check matrix of a t-error-correcting code iff every subset of $2t$ or fewer columns of H' is linearly independent. Now a subset of r columns from H', where $r \leqslant 2t$, will have the form

$$B = \begin{bmatrix} \beta_1 & \cdots & \beta_r \\ \beta_1^2 & \cdots & \beta_r^2 \\ \vdots & & \\ \beta_1^{2t} & \cdots & \beta_r^{2t} \end{bmatrix},$$

where $\beta_1, \beta_2, \ldots, \beta_r$ are distinct nonzero elements of $GF(2)$. Now consider the matrix $\mathbf{B}'$ formed from the first r rows of β:

$$B' = \begin{bmatrix} \beta_1 & \cdots & \beta_r \\ \vdots & & \\ \beta_1^r & \cdots & \beta_r^r \end{bmatrix}.$$

The matrix B' is nonsingular, since its determinant is

$$\det(\mathbf{B}') = \beta_1 \ldots \beta_r \det \begin{bmatrix} 1 & \cdots & 1 \\ \beta_1 & \cdots & \beta_r \\ \vdots & & \\ \beta_1^{r-1} & \cdots & \beta_r^{r-1} \end{bmatrix}$$

$$= \beta_1 \ldots \beta_r \prod_{i<j} (\beta_j - \beta_i) \neq 0$$

by the Vandermonde determinant theorem (see Prob. 9.3). Hence the columns of $\mathbf{B}'$, let alone those of $\mathbf{B}$, cannot be linearly dependent, and so the code does correct all error patterns of weight $\leqslant t$. To verify the bound $k \geqslant n - mt$ on the dimension, observe that the original parity-check matrix H, viewed as a matrix with entries from $GF(2)$ rather than $GF(2^m)$, has dimensions $mt \times n$. And by the results of Section 7.1, this means that the dual code has dimension $\leqslant mt$, and so the code itself has dimension $\geqslant n - mt$. $\square$

The codes described in Theorem 9.1 are called *BCH codes*, in honor of their inventors Bose, Ray-Chaudhuri, and Hocquenghem. These codes are important, not so much because of Theorem 9.1 itself (other codes can have higher rates and larger minimum distances), but rather because there are efficient encoding and, especially, decoding algorithms for them. In the next section, we will see that if we choose exactly the right ordering $(\alpha_0, \alpha_1, \ldots, \alpha_{n-1})$, BCH codes magically become cyclic codes, and so by the results of Chapter 8, the encoding automatically becomes simple. Additionally, this "cyclic" view of BCH codes will allow us to refine our estimates of the codes' dimensions. Then in Sections 9.3–9.5, we will fully describe one version of Berlekamp's famous decoding algorithm for BCH codes.

9.2 BCH codes as cyclic codes

Recall the definition of a t-error-corercting BCH code of length $n = 2^m - 1$: $\mathbf{C} = (C_0, \ldots, C_{n-1})$ is a codeword iff $\sum_{i=0}^{n-1} C_i \alpha_i^j = 0$ for $j = 1, 3, \ldots, 2t-1$ (equivalently, for $j = 1, 2, 3, \ldots, 2t$), where $(\alpha_0, \alpha_1, \ldots, \alpha_{n-1})$ is a list of n distinct nonzero elements of $GF(2^m)$. If the list is chosen properly, the code becomes a cyclic code, and thereby inherits all the implementational machinery available for cyclic codes. These "cyclic" lists are those of the form

$$(1, \alpha, \ldots, \alpha^{n-1}),$$

where n is a divisor of $2^m - 1$ and α is an element of $GF(2^m)$ of order n. With respect to such a list, the definition becomes: $\mathbf{C} = (C_0, C_1, \ldots, C_{n-1})$ is a codeword iff

$$\sum_{i=0}^{n-1} C_i \alpha^{ij} = 0, \qquad \text{for } j = 1, 3, \ldots, 2t-1 \text{ (or } j = 1, 2, 3, \ldots, 2t). \quad (9.7)$$

In this realization, the BCH code becomes a *cyclic code*, in the sense of Chapter 8. To see that this is so, let $C(x) = C_0 + C_1 x + \cdots + C_{n-1} x^{n-1}$ be the generating function for the codeword $\mathbf{C}$; then (9.7) becomes

$$C(\alpha^j) = 0, \qquad j = 1, 2, \ldots, 2t. \quad (9.8)$$

Now let $\mathbf{C}^R$ be the right cyclic shift of the codeword $\mathbf{C}$; its generating function is, by Theorem 8.1, $C^R(x) = xC(x) \bmod (x^n - 1)$, which means that $C^R(x) = xC(x) + M(x)(x^n - 1)$ for some polynomial $M(x)$. Thus for $j = 1, 2, \ldots, 2t$,

$$C^R(\alpha^j) = \alpha^j C(\alpha^j) + M(\alpha^j)(\alpha^{jn} - 1).$$

But $C(\alpha^j) = 0$ by (9.8), and $\alpha^{jn} - 1 = 0$ since $\alpha^n = 1$. It follows that $C^R(\alpha^j) = 0$ for $j = 1, 2, \ldots, 2t$, so that $\mathbf{C}^R$ is also in the BCH code defined by (9.7), which means that the code is cyclic.

It now follows from Theorem 8.3 that every BCH code is characterized by its generator polynomial $g(x)$. But how can we compute $g(x)$? According to the definition, $g(x)$ is the least degree polynomial in the code, i.e., the least-degree polynomial satisfying $g(\alpha) = g(\alpha^3) = \cdots = g(\alpha^{2t-1}) = 0$. Now the coefficients of $g(x)$ are in $GF(2)$, but the various powers of α are in the larger field $GF(2^m)$. Thus (see Appendix C) $g(x)$ is the **minimal polynomial** over $GF(2)$ of the subset $A = \{\alpha, \alpha^3, \ldots, \alpha^{2t-1}\}$ of $GF(2^m)$. Hence if A^* is defined to be the set of all $GF(2)$-conjugates of elements in A, i.e. $A^* = \{\beta^{2^i} : \beta \in A, i \geqslant 0\}$, then

$$g(x) = \prod_{\beta \in A^*} (x - \beta). \tag{9.9}$$

We summarize these results in the following theorem.

Theorem 9.2 *If we define the t-error-correcting BCH code of length n by (9.7) or (9.8), then the code is cyclic, with generator polynomial given by (9.9). Thus the dimension of the code is given by $n - \deg(g)$, i.e., $k = n - |A^*|$, where A^* is the set of $GF(2)$-conjugates of $A = \{\alpha, \alpha^3, \ldots, \alpha^{2t-1}\}$ in $GF(2^m)$.* □

Example 9.1 Consider a three-error correcting BCH code of length 15. Let α be a primitive root in $GF(16)$; then by Theorem 9.2, the generator polynomial is the minimal polynomial of the set $A = \{\alpha, \alpha^3, \alpha^5\}$. The conjugates of α are $(\alpha, \alpha^2, \alpha^4, \alpha^8)$; of α^3, $(\alpha^3, \alpha^6, \alpha^{12}, \alpha^9)$; of α^5, (α^5, α^{10}). Hence

$$A^* = \{\alpha, \alpha^2, \alpha^3, \alpha^4, \alpha^5, \alpha^6, \alpha^8, \alpha^9, \alpha^{10}, \alpha^{12}\},$$

and so by Theorem 9.2, the dimension is $15 - 10 = 5$.

To actually compute $g(x)$ for this example, we need a concrete realization of $GF(16)$. Let's represent $GF(16)$ according to powers of a primitive root α that satisfies $\alpha^4 = \alpha + 1$. In Table 9.1 the element α^j is given as polynomial of degree $\leqslant 3$ in α; for example, $\alpha^{11} = \alpha^3 + \alpha^2 + \alpha$. The generator polynomial $g(x)$ is the product of the minimal polynomials of α, α^3, and α^5. The minimal polynomial of α is by definition $x^4 + x + 1$. The minimal polynomials of α^3—call it $g_3(x) = g_{30} + g_{31}x + g_{32}x^2 + g_{33}x^3 + g_{34}x^4$—must satisfy $g_3(\alpha^3) = 0$. From Table 9.1 this equivalent to $g_{30}[0001] + g_{31}[1000] + g_{32}[1100] + g_{33}[1010] + g_{34}[1111] = [0000]$. The only non-trivial solution to this set of 4 homogeneous equations in the 5 unknowns is $[g_{30}, g_{31}, g_{32}, g_{33}, g_{34}] = [11111]$, and so $g_3(x) = x^4 + x^3 + x^2 + x + 1$.

Table 9.1 The field $GF(16)$ represented as powers of α, where $\alpha^4 = \alpha + 1$.

i	α^i
0	0001
1	0010
2	0100
3	1000
4	0011
5	0110
6	1100
7	1011
8	0101
9	1010
10	0111
11	1110
12	1111
13	1101
14	1001

Similarily, $g_5(x) = g_{50} + g_{51}x + g_{52}x^2$ (we already know that α^5 has only two conjugates, α^5 and α^{10}) turns out to be $x^2 + x + 1$. Hence the generator polynomial of the three-error-correcting BCH code of length 15 is $g(x) = (x^4 + x + 1)(x^4 + x^3 + x^2 + x + 1)(x^2 + x + 1) = x^{10} + x^8 + x^5 + x^4 + x^2 + x + 1$. Similarly, the parity-check polynomial is $h(x) = (x^{15} + 1)/g(x) = x^5 + x^3 + x + 1$. (We emphasize, however, that $g(x)$ depends on the particular realization of $GF(16)$ given in Table 9.1. See Problem 9.6.) □

Let us summarize what we know about BCH codes so far: they can be designed to correct any desired number of errors up to about half the code's block length (Theorem 9.1), and they have a very nice algebraic characterization as cyclic codes. However, their practical importance is due almost wholly to the fact that they have a remarkably efficient *decoding* algorithm. We will begin our discussion of this algorithm in the following section.

9.3 Decoding BCH codes, Part one: the key equation

In this section, we will derive the so-called *key equation*, which is the basis for the BCH decoding algorithm. Before we get to the key equation, however, we must present some preliminary material. We shall present this material more generally than is strictly necessary, so that we can refer to it later, when we discuss the decoding *erasures* as well as errors, both for BCH codes and for *Reed–Solomon* codes.

Thus let F be a field which contains a primitive nth root of unity α.[3] We first note that

$$1 - x^n = \prod_{i=0}^{n-1}(1 - \alpha^i x). \tag{9.10}$$

This is because the polynomials on both sides of (9.10) have degree n, constant term 1, and roots α^{-i}, for $i = 0, 1, \ldots, n - 1$. Next, let

$$\mathbf{V} = (V_0, V_1, \ldots, V_{n-1})$$

be an n-dimensional vector over F, and let

$$\hat{\mathbf{V}} = (\hat{V}_0, \hat{V}_1, \ldots, \hat{V}_{n-1})$$

be its *discrete Fourier transform* (DFT), whose components are defined as follows.

$$\hat{V}_j = \sum_{i=0}^{n-1} V_i \alpha^{ij}, \qquad \text{for } j = 0, 1, \ldots, n - 1. \tag{9.11}$$

We sometimes call the V_i's the "time-domain" coordinates, and the $\hat{V}_j$'s the "frequency-domain" coordinates, of the vector $\mathbf{V}$. The time-domain components can be recovered from the frequency-domain components via the so-called "inverse DFT":

$$V_i = \frac{1}{n}\sum_{j=0}^{n-1} \hat{V}_j \alpha^{-ij}, \qquad \text{for } i = 0, 1, \ldots, n - 1. \tag{9.12}$$

In (9.12) the "$1/n$" factor in front of the sum must be interpreted with some care, in view of the possibly finite characteristic of F. The number "n" is the sum $1 + 1 + \cdots + 1$ (n terms), and "$1/n$" is the inverse of this number. For example, if F has characteristic 2 and n is odd, then $1/n = 1$. Apart from this small subtlety, however, the proof of (9.12) is identical to the usual proof of the inverse DFT formula, and we leave it as Problem 9.8. If we interpret the components of $\mathbf{V}$ and $\hat{\mathbf{V}}$ as the coefficients of polynomials, i.e., if we define generating functions $V(x)$ and $\hat{V}(x)$ by

$$V(x) = V_0 + V_1 x + \cdots + V_{n-1}x^{n-1} \tag{9.13}$$

and

$$\hat{V}(x) = \hat{V}_0 + \hat{V}_1 x + \cdots + \hat{V}_{n-1} x^{n-1}, \tag{9.14}$$

then the DFT and IDFT relationships (9.11) and (9.12) become

$$\hat{V}_j = V(\alpha^j) \tag{9.15}$$

and

$$V_i = \frac{1}{n} \hat{V}(\alpha^{-i}). \tag{9.16}$$

There are many interesting and useful relationships between the time-domain and frequency-domain coordinates of a given vector. One of them is that a "phase shift" in the time domain corresponds to a "time shift" in the frequency domain, in the following sense. It we multiply the ith component of $\mathbf{V}$ by $\alpha^{\mu i}$, i.e., if we define a new vector $\mathbf{V}_\mu$ as

$$\mathbf{V}_\mu = (V_0, \ V_1 \alpha^\mu, \ \ldots, \ V_{n-1} \alpha^{\mu(n-1)}), \tag{9.17}$$

then its DFT is

$$\hat{\mathbf{V}}_\mu = (\hat{V}_\mu, \ \hat{V}_{\mu+1}, \ \ldots, \ \hat{V}_{\mu+n-1}), \tag{9.18}$$

where in (9.18) the subscripts are taken mod n. We leave the proof of (9.18) as Problem 9.10.

As coding theorists, we are always interested in the *weight* of a vector. The following classical theorem tells us how to estimate the weight in the time domain if we know something about the vector in the frequency domain.

Theorem 9.3 *(the BCH argument) Suppose* $\mathbf{V}$ *is a nonzero vector with the property that* $\hat{\mathbf{V}}$ *has* m *consecutive* 0 *components, i.e.,* $\hat{V}_{j+1} = \hat{V}_{j+2} = \cdots = \hat{V}_{j+m} = 0$. *Then the weight of* $\mathbf{V}$ *is* $\geqslant m + 1$.

Proof Let $\hat{\mathbf{W}}$ be the vector obtained by cyclically shifting $\hat{\mathbf{V}}$ until its m consecutive 0's appear in positions $n - m, \ n - m + 1, \ \ldots, \ n - 1$, i.e.,

$$\hat{\mathbf{W}} = \left[* * \ \ldots \ * \overbrace{00 \ldots 0}^{m} \right].$$

By (9.17) and (9.18), $\hat{\mathbf{W}}$ is the DFT of a vector $\mathbf{W}$ whose weight is the same as the weight of $\mathbf{V}$. However, by (9.12), $W_i = \frac{1}{n} \hat{W}(\alpha^{-i})$, where $\hat{W}(x) = \hat{W}_0 + \hat{W}_1 x + \cdots + \hat{W}_{n-m-1} x^{n-m-1}$. Since $\hat{W}(x)$ is a nonzero polynomial of degree $\leqslant n - m - 1$, it follows that $W_i = 0$ for *at most* $n - m - 1$

values of i, and so $W_i \neq 0$ for *at least* $m + 1$ values of i. Thus $\text{wt}(\mathbf{V}) = \text{wt}(\mathbf{W}) \geqslant m + 1$. □

We are almost ready to introduce the key equation, but we need a few more definitions. With the vector $\mathbf{V}$ fixed, we define its *support set I* as follows:

$$I = \{i : 0 \leqslant i \leqslant n - 1 \text{ and } V_i \neq 0\}. \tag{9.19}$$

We now define several polynomials associated with $\mathbf{V}$, the locator polynomial, the punctured locator polynomials, and the evaluator polynomial. The *locator polynomial* for $\mathbf{V}$ is

$$\sigma_{\mathbf{V}}(x) = \prod_{i \in I} (1 - \alpha^i x). \tag{9.20}$$

For each value of $i \in I$ we also define the ith *punctured locator polynomial* $\sigma_{\mathbf{V}}^{(i)}(x)$:

$$\sigma_{\mathbf{V}}^{(i)}(x) = \sigma_{\mathbf{V}}(x)/(1 - \alpha^i x)$$

$$= \prod_{\substack{j \in I \\ j \neq i}} (1 - \alpha^j x). \tag{9.21}$$

Finally, we define the *evaluator polynomial* for $\mathbf{V}$ as

$$\omega_{\mathbf{V}}(x) = \sum_{i \in I} V_i \sigma_{\mathbf{V}}^{(i)}(x). \tag{9.22}$$

We will need the following lemma later on, for example, in Sections 9.5 and 9.7 when we discuss the RS/BCH decoding algorithms.

Lemma 1 $\gcd(\sigma_{\mathbf{V}}(x), \omega_{\mathbf{V}}(x)) = 1$.

Proof By (9.20), $\gcd(\sigma_{\mathbf{V}}(x), \omega_{\mathbf{V}}(x)) = \prod_{i \in J}(1 - \alpha^i x)$, where $J = \{i \in I : \omega_{\mathbf{V}}(\alpha^{-i}) = 0\}$. By (9.22), if $i \in I$, $\omega_{\mathbf{V}}(\alpha^{-i}) = V_i \sigma_{\mathbf{V}}^{(i)}(\alpha^{-i})$. But by the definition of I, if $i \in I$, $V_i \neq 0$, and by (9.21), $\sigma_{\mathbf{V}}^{(i)}(\alpha^{-i}) = \prod_{\substack{j \in I \\ j \neq i}}(1 - \alpha^{j-i}) \neq 0$. Hence the set J is empty, and so $\gcd(\sigma_{\mathbf{V}}(x), \omega_{\mathbf{V}}(x)) = 1$, as asserted. □

We now come to the promised "key equation."

Theorem 9.4 *(the key equation) For a fixed vector* $\mathbf{V}$, *the polynomials* $\hat{V}(x)$, $\sigma_{\mathbf{V}}(x)$, *and* $\omega_{\mathbf{V}}(x)$ *satisfy*

$$\sigma_{\mathbf{V}}(x)\hat{V}(x) = \omega_{\mathbf{V}}(x)(1 - x^n). \tag{9.23}$$

Proof Using the definitions (9.11), (9.14) and (9.22), we find that

$$\hat{V}(x) = \sum_{i \in I} V_i \sum_{j=0}^{n-1} x^j a^{ij}. \tag{9.24}$$

According to (9.21), $\sigma_V(x) = \sigma_V^{(i)}(x)(1 - a^i x)$ for all $i \in I$, and so from (9.24) we have

$$\sigma_V(x)\hat{V}(x) = \sum_{i \in I} V_i \sigma_V^{(i)}(x)(1 - a^i x) \sum_{j=0}^{n-1} x^j a^{ij}$$

$$= \sum_{i \in I} V_i \sigma_V^{(i)}(x)(1 - x^n)$$

$$= \omega_V(x)(1 - x^n). \qquad \square$$

The following corollary to Theorem 9.3 tells us how to reconstruct the nonzero components of **V** from $\sigma_V(x)$ and $\omega_V(x)$. It involves the *formal derivative* $\sigma'_V(x)$ of the polynomial $\sigma_V(x)$. (See Problem 9.18.)

Corollary 1 *For each $i \in I$, we have*

$$V_i = -a^i \frac{\omega_V(a^{-i})}{\sigma'_V(a^{-i})}. \tag{9.25}$$

Proof If we differentiate the key equation (9.23) we get

$$\sigma_V(x)\hat{V}'(x) + \sigma'_V(x)\hat{V}(x) = \omega_V(x)(-nx^{n-1}) + \omega'_V(x)(1 - x^n). \tag{9.26}$$

Note that if $x = a^{-i}$ with $i \in I$, from (9.20) and (9.10) we see that both $\sigma_V(x)$ and $1 - x^n$ vanish. Thus if $x = a^{-i}$, (9.26) becomes

$$\sigma'_V(a^{-i})\hat{V}(a^{-i}) = -na^i \omega_V(a^{-i}). \tag{9.27}$$

But from (9.16), $\hat{V}(a^{-i}) = nV_i$. This fact, combined with (9.27), completes the proof. $\qquad \square$

Corollary 1 says, in effect, that the time-domain coordinates of **V** can be recovered from $\sigma_V(x)$ and $\omega_V(x)$. The next corollary says that if the first few frequency-domain coordinates of **V** are known, the rest can be recovered from $\sigma_V(x)$ alone, via a simple recursion. In the statement of the corollary, we suppose that the coefficients of $\sigma_V(x)$ are given by

$$\sigma_{\mathbf{V}}(x) = 1 + \sigma_1 x + \cdots + \sigma_d x^d.$$

Corollary 2 *For all indices j, we have*

$$\hat{V}_j = -\sum_{i=1}^{d} \sigma_i \hat{V}_{j-i}, \tag{9.28}$$

where all subscripts are to be interpreted mod n.

Proof The key equation implies that

$$\sigma_{\mathbf{V}}(x)\hat{V}(x) \equiv 0 \quad (\bmod\ 1 - x^n). \tag{9.29}$$

What (9.29) says is that for each j in the range $0 \leqslant j \leqslant n - 1$, the coefficient of x^j in the polynomial $\sigma_{\mathbf{V}}(x)\hat{V}(x) \bmod(1 - x^n)$ is 0. But this coefficient is $\sum_{i=0}^{d} \sigma_i \hat{V}_{(j-i)\bmod n}$, so that for each j in the range $0 \leqslant j \leqslant n - 1$, we have

$$\sum_{i=0}^{d} \sigma_i \hat{V}_{j-i} = 0, \tag{9.30}$$

where subscripts are to be taken mod n and we have defined $\sigma_0 = 1$. But now equation (9.30) is equivalent to the equation (9.28). $\qquad\square$

Example 9.2 We illustrate this material using the field $GF(16)$, in which the nonzero elements are represented by the powers of a primitive root α satisfying the equation $\alpha^4 = \alpha + 1$. We consider the vector

$$\mathbf{V} = (0, 0, \alpha^2, 0, 0, 0, 0, \alpha^7, 0, 0, 0, 0, 0, 0, 0, 0).$$

Then the polynomial $V(x)$ defined in (9.13) is

$$V(x) = \alpha^2 x^2 + \alpha^7 x^7.$$

Using (9.11) or (9.15) we can calculate the DFT of $\mathbf{V}$:

$$\hat{\mathbf{V}} = (\alpha^{12}, \alpha^9, 0, \alpha^3, 1, 0, \alpha^9, \alpha^6, 0, 1, \alpha^{12}, 0, \alpha^6, \alpha^3, 0).$$

Thus $\hat{V}(x)$, as defined in (9.14), is

$$\hat{V}(x) = \alpha^{12} + \alpha^9 x + \alpha^3 x^3 + x^4 + \alpha^9 x^6 + \alpha^6 x^7 + x^9 + \alpha^{12} x^{10} + \alpha^6 x^{12} + \alpha^3 x^{13}$$

$$= (\alpha^{12} + \alpha^9 x)(1 + \alpha^6 x^3 + \alpha^{12} x^6 + \alpha^3 x^9 + \alpha^9 x^{12})$$

$$= (\alpha^{12} + \alpha^9 x)\frac{1 + x^{15}}{1 + \alpha^6 x^3}$$

$$= \alpha^{12}\frac{1 + x^{15}}{1 + \alpha^{12} x + \alpha^9 x^2}. \tag{9.31}$$

The support set of $\mathbf{V}$ is $I = \{2, 7\}$, and so the locator polynomial for $\mathbf{V}$ is

$$\sigma_{\mathbf{V}}(x) = (1 + \alpha^2 x)(1 + \alpha^7 x) = 1 + \alpha^{12} x + \alpha^9 x^2. \tag{9.32}$$

The polynomials $\sigma_{\mathbf{V}}^{(i)}(x)$ defined in (9.21) are in this case

$$\sigma_{\mathbf{V}}^{(2)} = (1 + \alpha^7 x), \quad \sigma_{\mathbf{V}}^{(7)} = (1 + \alpha^2 x).$$

The evaluator polynomial $\omega_{\mathbf{V}}(x)$ defined in (9.22) is

$$\omega_{\mathbf{V}}(x) = \alpha^2(1 + \sigma^7 x) + \alpha^7(1 + \alpha^2 x) = \alpha^{12}. \tag{9.33}$$

Combining (9.31), (9.32), and (9.33), we see that the key equation indeed holds in this case. To check Corollary 1, we note that from (9.32), $\sigma_{\mathbf{V}}'(x) = \alpha^{12} = \omega_{\mathbf{V}}(x)$, so that Corollary 1 becomes simply $V_i = \alpha^i$, for $i \in I$, which is true ($V_2 = \alpha^2$ and $V_7 = \alpha^7$). Finally, note that Corollary 2 says in this case that

$$\hat{V}_j = \alpha^{12}\hat{V}_{j-1} + \alpha^9 \hat{V}_{j-2} \qquad \text{for } j = 2, 3, \ldots, 14,$$

so that (using $\hat{V}_0 = \alpha^{12}$ and $\hat{V}_1 = \alpha^9$ as initial conditions)

$$\hat{V}_2 = \alpha^{12} \cdot \alpha^9 + \alpha^9 \cdot \alpha^{12} = 0,$$

$$\hat{V}_3 = \alpha^{12} \cdot 0 + \alpha^9 \cdot \alpha^9 = \alpha^3,$$

$$\hat{V}_4 = \alpha^{12} \cdot \alpha^3 + \alpha^9 \cdot 0 = 1,$$

$$\vdots$$

$$\hat{V}_{14} = \alpha^{12} \cdot \alpha^3 + \alpha^9 \cdot \alpha^6 = 0,$$

which agrees with our direct calculation of $\hat{\mathbf{V}}$.   □

With the preliminary material about the key equation out of the way, we can begin a serious discussion of the problem of decoding BCH codes. Suppose then that $\mathbf{C} = (C_0, C_1, \ldots, C_{n-1})$ is a codeword from the t-error-correcting BCH code of length n defined by (9.6), which is transmitted over a noisy channel, and that $\mathbf{R} = (R_0, R_1, \ldots, R_{n-1})$ is received. We assume that the components of $\mathbf{R}$ are 0's and 1's, i.e., are elements of $GF(2)$. We define the *error pattern* as the vector $\mathbf{E} = (E_0, E_1, \ldots, E_n) = \mathbf{R} - \mathbf{C}$. The decoder's first step is to compute the *syndromes* $S_1, S_2 \ldots, S_{2t}$, which are defined by

$$S_j = \sum_{i=0}^{n-1} R_i \alpha^{ij}, \qquad \text{for } j = 1, 2, \ldots, 2t. \tag{9.34}$$

Since $\mathbf{R} = \mathbf{C} + \mathbf{E}$, and $\mathbf{C}$ is a codeword, it follows that

$$S_j = \sum_{i=0}^{n-1} E_i \alpha^{ij}, \qquad \text{for } j = 1, 2, \ldots, 2t, \tag{9.35}$$

so that, as expected, the syndromes depend only on the error pattern and not on the transmitted codeword. Note also that on comparing (9.35) with (9.11), we see that S_j is the jth component of the DFT of the error pattern; in other words, the syndrome lets us see $2t$ consecutive components (the first, second, $\ldots$, $2t$th) of $\hat{\mathbf{E}}$. If we now define the *twisted* error patter $\mathbf{V}$ as

$$\mathbf{V} = (E_0, E_1\alpha, E_2\alpha^2, \ldots, E_{n-1}\alpha^{n-1}), \tag{9.36}$$

it follows from (9.17) and (9.18) that $(S_1, S_2, \ldots, S_{2t}) = (\hat{V}_0, \hat{V}_1, \ldots, \hat{V}_{2t-1})$.

The key equation applies to the vector $\mathbf{V}$ defined in (9.36); however, since we only know the first $2t$ coefficients of $\hat{V}(x)$ (i.e., $\hat{V}_0, \hat{V}_1, \ldots, \hat{V}_{2t-1}$), we focus instead on the key equation *reduced mod x^{2t}*:

$$\sigma(x)\hat{V}(x) = \omega(x) \pmod{x^{2t}}. \tag{9.37}$$

(In (9.37) we have dropped the subscript $\mathbf{V}$'s on $\sigma(x)$ and $\omega(x)$.) From (9.19) and (9.36) we see that the support set I for $\mathbf{V}$ is the set of indices such that $E_i \neq 0$, i.e., the set of *error locations*. For this reason, the polynomial $\sigma(x)$ in (9.37) is called the *error-locator polynomial*. Similarly, the polynomial $\omega(x)$ in (9.37) is called the *error-evaluator polynomial*. Equation (9.37) is called the *BCH key equation*.

Now observe that if, given the syndrome of the received word $\mathbf{R}$, or equivalently, $\hat{V}(x) \bmod x^{2t}$, we could somehow "solve" the BCH key equation (9.37) for the polynomials $\sigma(x)$ and $\omega(x)$, we could then easily recover the error pattern $\mathbf{E}$, and thus also the transmitted codeword $\mathbf{C} = \mathbf{R} - \mathbf{E}$. We could do this by first computing the n values $\sigma(\alpha^{-i})$, for $i = 0, 1, \ldots, n - 1$, which would identify the support set I of $\mathbf{V}$ defined in (9.19). Then the nonzero components of $\mathbf{V}$ could be computed by (9.25), and this would give us the complete vector **V**, or equivalently, **E** (see (9.36)). Alternatively, knowing $(\hat{V}_0, \hat{V}_1, \ldots, \hat{V}_{2t-1})$, we could complete the vector $\hat{\mathbf{V}}$ via (9.28), and then recover **V** via an inverse DFT. In the next section, we will see that there is a remarkably efficient algorithm for computing $\sigma(x)$ and $\omega(x)$ from the BCH equation, provided we make the additional assumption that the actual number of errors that occurred is at most t. (This assumption is necessary, since a t-error-correcting BCH code is not designed to correct more than t errors.)

9.4 Euclid's algorithm for polynomials

This section does not deal directly with the problem of decoding BCH codes. The reader should bear in mind, however, that our goal is to solve the BCH key equation (Eq. 9.37)) for $\sigma(x)$ and $\omega(x)$, given $\hat{V}(x) \bmod x^{2t}$.

Throughout this section $a(x)$ and $b(x)$ will be fixed polynomials over a field F, with $\deg a(x) \geqslant \deg b(x)$.[4] Later $a(x)$ will be replaced by x^{2t}, and $b(x)$ by the syndrome polynomial $S(x)$.

Euclid's algorithm is a recursive procedure for finding the greatest common divisor (gcd for short) $d(x)$ of $a(x)$) and $b(x)$, and for finding a linear combination of $a(x)$ and $b(x)$ equal to $d(x)$, i.e., an equation of the form

$$u(x)a(x) + v(x)b(x) = d(x). \qquad (9.38)$$

The algorithm involves four sequences of polynomials: $(u_i(x))$, $(v_i(x))$, $(r_i(x))$, $(q_i(x))$. The initial conditions are

$$u_{-1}(x) = 1, \quad v_{-1}(x) = 0, \quad r_{-1}(x) = a(x),$$
$$u_0(x) = 0, \quad v_0(x) = 1, \quad r_0(x) = b(x) \qquad (9.39)$$

($q_{-1}(x)$ and $q_0(x)$ are not defined). For $i \geqslant 1$, $q_i(x)$ and $r_i(x)$ are defined to be the *quotient* and *remainder*, respectively, when $r_{i-2}(x)$ is divided by $r_{i-1}(x)$:

$$r_{i-2}(x) = q_i(x)r_{i-1}(x) + r_i(x), \quad \deg r_i < \deg r_{i-1}. \qquad (9.40)$$

The polynomials $u_i(x)$ and $v_i(x)$ are then defined by

$$u_i(x) = u_{i-2}(x) - q_i(x)u_{i-1}(x), \qquad (9.41)$$

$$v_i(x) = v_{i-2}(x) - q_i(x)v_{i-1}(x). \qquad (9.42)$$

Since the degrees of the remainders r_i are strictly decreasing, there will be a last nonzero one; call it $r_n(x)$. It turns out that $r_n(x)$ is the gcd of $a(x)$ and $b(x)$, and furthermore that the desired equation expressing the gcd as a linear combination of the original two polynomials (cf. Eq. (9.38)) is

Table 9.2 Properties of Euclid's algorithm.

A	$v_i r_{i-1} - v_{i-1} r_i = (-1)^i a$	$0 \leq i \leq n+1$
B	$u_i r_{i-1} - u_{i-1} r_i = (-1)^{i+1} b$	$0 \leq i \leq n+1$
C	$u_i v_{i-1} - u_{i-1} v_i = (-1)^{i+1}$	$0 \leq i \leq n+1$
D	$u_i a + v_i b = r_i$	$-1 \leq i \leq n+1$
E	$\deg(u_i) + \deg(r_i - 1) = \deg(b)$	$1 \leq i \leq n+1$
F	$\deg(v_i) + \deg(r_i - 1) = \deg(a)$	$0 \leq i \leq n+1$

$$u_n(x)a(x) + v_n(x)b(x) = r_n(x). \tag{9.43}$$

Since this particular aspect of Euclud's algorithm is not our main concern, we leave the proof of these facts to Prob. 9.18(b).

What is more interesting to us at present is the list shown in Table 9.2 of intermediate relationships among the polynomials of Euclid's algorithm. It is not difficult to prove these properties by induction on i; see Prob. 9.19(a).

Example 9.3 Let $F = GF(2)$, $a(x) = x^8$, $b(x) = x^6 + x^4 + x^2 + x + 1$. The behavior of Euclid's algorithm is given in Table 9.3.

The $i = 4$ line of Table 9.3 shows that $\gcd(a(x), b(x)) = 1$ (which is obvious anyway), and with Property D from Table 9.2 yields the equation $(x^5 + x^4 + x^3 + x^2)a(x) + (x^7 + x^6 + x^3 + x + 1)b(x) = 1$. This example is continued in Example 9.4. □

We now focus our attention on Property D in Table 9.2, which can be rewritten as

$$v_i(x)b(x) \equiv r_i(x) \pmod{a(x)}. \tag{9.44}$$

Using Property F and the fact that deg $r_{i-1} >$ deg r_i, we get the estimate

$$\deg v_i + \deg r_i < \deg a. \tag{9.45}$$

The main result of this section (Theorem 9.5) is a kind of converse to (9.44) and (9.45). We begin with a lemma.

Lemma 2 *Suppose Euclid's algorithm, as described above, is applied to the two polynomials $a(x)$ and $b(x)$. Given two integers $\mu \geq 0$ and $v \geq 0$ with $\mu + v = \deg a - 1$, there exists a unique index j, $0 \leq j \leq n$, such that:*

Table 9.3 An example of Euclid's algorithm.

i	u_i	v_i	r_i	q_i
-1	1	0	x^8	$\cdots$
0	0	1	$x^6+x^4+x^2+x+1$	$\cdots$
1	1	x^2+1	x^3+x+1	x^2+1
2	x^3+1	$x^5+x^3+x^2$	x^2	x^3+1
3	x^4+x+1	$x^6+x^4+x^3+x^2+1$	$x+1$	x
4	$x^5+x^4+x^3+x^2$	$x^7+x^6+x^3+x+1$	1	$x+1$
5	$x^6+x^4+x^2+x+1$	x^8	0	$x+1$

$$\deg(v_j) \leq \mu, \tag{9.46}$$

$$\deg(r_j) \leq \nu. \tag{9.47}$$

Proof Recall that $\deg r_i$ is a strictly decreasing function of i until $r_n = \gcd(a, b)$, and define the index j uniquely by requiring

$$\deg r_{j-1} \geq \nu + 1, \tag{9.48}$$

$$\deg r_j \leq \nu. \tag{9.49}$$

Then by Property F we also have

$$\deg v_j \leq \mu, \tag{9.50}$$

$$\deg v_{j+1} \geq \mu + 1. \tag{9.51}$$

Equations (9.49) and (9.50) show the existence of an index j satisfying (9.46) and (9.47); Eqs. (9.48) and (9.51) show uniqueness. $\qquad\square$

The following theorem is the main result of this section.

Theorem 9.5 *Suppose* $a(x)$, $b(x)$, $v(x)$ *and* $r(x)$ *are nonzero polynomials satisfying*

$$v(x)b(x) \equiv r(x) \quad (\bmod\, a(x)), \tag{9.52}$$

$$\deg v(x) + \deg r(x) < \deg a(x). \tag{9.53}$$

Suppose further that $v_j(x)$ *and* $r_j(x)$, $j = -1, 0, \ldots, n+1$, *are the sequences of polynomials produced when Euclid's algorithm is applied to the pair* $(a(x), b(x))$. *Then there exist a unique index* j, $0 \leq j \leq n$, *and a polynomial* $\lambda(x)$ *such that*

$$v(x) = \lambda(x)v_j(x), \tag{9.54}$$

$$r(x) = \lambda(x)r_j(x). \tag{9.55}$$

Proof[5] Let j be the index satisfying (9.46) and (9.47) with $\nu = \deg r$, $\mu = \deg a - \deg r - 1$. Thus from (9.53) $\deg(v(x)) \leqslant \mu$. Then according to (9.51) and (9.48), $\deg v_{j+1} \geqslant \mu + 1 \geqslant \deg v + 1$, and $\deg r_{j-1} \geqslant \nu + 1 = \deg r + 1$. Hence if there is an index such that (9.54) and (9.55) hold, it must be unique.

Now rewrite Property D and Eq. (9.52) as follows:

$$u_j a + v_j b = r_j, \tag{9.56}$$

$$ua + vb = r, \tag{9.57}$$

where u is some unspecified polynomial. Multiply (9.56) by v and (9.57) by v_j:

$$u_j va + v_j vb = r_j v, \tag{9.58}$$

$$uv_j a + vv_j b = rv_j. \tag{9.59}$$

Together (9.58) and (9.59) imply $r_j v \equiv rv_j \pmod{a}$. But by (9.47) and (9.53), $\deg(r_j v) = \deg r_j + \deg v \leqslant \nu + \mu < \deg a$. Similarly, by (9.46) and (9.53), $\deg(rv_j) = \deg r + \deg v_j \leqslant \nu + \mu < \deg a$. It follows that $r_j v = rv_j$. This fact, combined with (9.58) and (9.59), implies that $u_j v = uv_j$. But since Property C guarantees that u_j and v_j are relatively prime, this means that

$$u(x) = \lambda(x)u_j(x),$$

$$v(x) = \lambda(x)v_j(x),$$

for some polynomial $\lambda(x)$. Then Equation (9.57) becomes $\lambda u_j a + \lambda v_j b = r$; comparing this with Eq. (9.58), we conclude that $r(x) = \lambda(x)r_j(x)$. $\qquad\square$

The results of Theorem 9.5 will be used constantly in our forthcoming discussion of decoding algorithms for BCH and Reed–Solomon codes. To facilitate these discussions, we now introduce the algorithmic procedure "Euclid($a(x)$, $b(x)$, μ, ν)".

Definition *If* $(a(x), b(x))$ *is a pair of nonzero polynomials with* $\deg a(x) \geqslant \deg b(x)$, *and if* (μ, ν) *is a pair of nonnegative integers such that* $\mu + \nu = \deg a(x) - 1$, Euclid($a(x)$, $b(x)$, μ, ν) *is the procedure that returns the unique pair of polynomials* $(v_j(x), r_j(x))$ *with* $\deg v_j(x) \leqslant \mu$ *and* $\deg r_j(x) \leqslant \nu$, *when Euclid's algorithm is applied to the pair* $(a(x), b(x))$.

The following theorem summarizes the results of this section.

Theorem 9.6 *Suppose $v(x)$ and $r(x)$ are nonzero polynomials satisfying*

$$v(x)b(x) \equiv r(x) \pmod{a(x)}, \tag{9.60}$$

$$\deg v(x) \leqslant \mu, \tag{9.61}$$

$$\deg r(x) \leqslant \nu, \tag{9.62}$$

where μ and ν are nonnegative integers such that $\mu + \nu = \deg r(x) - 1$. Then if $(v_j(x), r_j(x))$ is the pair of polynomials returned by Euclid$(a(x), b(x), \mu, \nu)$, *there is a polynomial $\lambda(x)$ such that*

$$v(x) = \lambda(x)v_j(x), \tag{9.63}$$

$$r(x) = \lambda(x)r_j(x). \tag{9.64}$$

Proof Theorem 9.4 guarantees that there exists a unique index j such that (9.63) and (9.64) hold. Furthermore the procedure Euclid$(a(x), b(x), \mu, \nu)$ must return this pair, since by (9.63) and (9.64), $\deg v_j(x) \leqslant \deg v(x) \leqslant \mu$ and $\deg r_j(x) \leqslant \deg r(x) \leqslant \nu$. $\qquad\square$

Example 9.4 Let $a(x) = x^8$, $b(x) = x^6 + x^4 + x^2 + x + 1$, $F = GF(2)$, as in Example 9.3. Using Table 9.2, we can tabulate the output of Euclid for the eight possible pairs (μ, ν):

(μ, ν)	Euclid$(x^8, x^6 + x^4 + x^2 + x + 1, \mu, \nu)$
$(0, 7)$	$(1, x^6 + x^4 + x^2 + x + 1)$
$(1, 6)$	$(1, x^6 + x^4 + x^2 + x + 1)$
$(2, 5)$	$(x^2 + 1, x^3 + x + 1)$
$(3, 4)$	$(x^2 + 1, x^3 + x + 1)$
$(4, 3)$	$(x^2 + 1, x^3 + x + 1)$
$(5, 2)$	$(x^5 + x^3 + x^2, x^2)$
$(6, 1)$	$(x^6 + x^4 + x^3 + x^2 + 1, x + 1)$
$(7, 0)$	$(x^7 + x^6 + x^3 + x + 1, 1)$

Now suppose we wished to "solve" the congruence $(x^6 + x^4 + x^2 + x + 1)\sigma(x) \equiv \omega(x) \pmod{x^8}$, subject to the restriction that $\deg \sigma(x) \leqslant 3$, $\deg \omega(x) \leqslant 4$. According to Theorem 9.5, we invoke

Euclid(x^8, $x^6 + x^4 + x^2 + x + 1$, 4, 3) which by the above table returns the pair ($x^2 + 1$, $x^3 + x + 1$), so that all solutions to the given problem are of the form $\sigma(x) = \lambda(x)(x^2 + 1)$, $\omega(x) = \lambda(x)(x^3 + x + 1)$, with deg $\lambda(x) \leqslant 1$. If we further required $\gcd(\sigma(x), \omega(x)) = 1$, then the *only* solution would be $\sigma(x) = x^2 + 1$, $\omega(x) = x^3 + x + 1$. $\qquad\square$

At this point the application of Theorem 9.4 to the problem of solving the key equation for BCH codes should be apparent. In any event we spell it out in the next section.

9.5 Decoding BCH codes, Part two: the algorithms

Let us recapitulate the BCH decoding problem, which we abandoned temporarily at the end of Section 9.3. We are given a received vector $\mathbf{R} = (R_0, R_1, \ldots, R_{n-1})$, which is a noisy version of an unknown codeword $\mathbf{C} = (C_0, C_1, \ldots, C_{n-1})$ from the t-error-correcting BCH code defined by (9.7), i.e., $\mathbf{R} = \mathbf{C} + \mathbf{E}$, where $\mathbf{E}$ is the error pattern. Our goal is to recover $\mathbf{C}$ from $\mathbf{R}$. The first step in the decoding process is to compute the *syndrome polynomial* $S(x)$, defined by

$$S(x) = S_1 + S_2 x + \cdots + S_{2t} x^{2t-1}, \tag{9.65}$$

where $S_j = \sum_{i=0}^{n-1} R_i \alpha^{ij}$, for $j = 1, 2, \ldots, 2t$. We saw at the end of Section 9.3 that $S(x) = \hat{V}(x) \bmod x^{2t}$, where $\hat{V}(x)$ is the generating function for the Fourier transform of the vector $\mathbf{V}$ defined in (9.36), so that the key equation (9.37) becomes

$$\sigma(x)S(x) \equiv \omega(x) \pmod{x^{2t}}, \tag{9.66}$$

where $\sigma(x)$ is the error-locator polynomial and $\omega(x)$ is the error-evaluator polynomial.

The next step in the decoding process is to use Euclid's algorithm, and in particular the procedure Euclid($a(x)$, $b(x)$, μ, ν) defined in Section 9.4, to solve the key equation for $\sigma(x)$ and $\omega(x)$. This is possible, since if the number of errors that actually occurred is $\leqslant t$, then by (9.20) and (9.22),

$$\deg \sigma(x) \leqslant t,$$

$$\deg \omega(x) \leqslant t - 1,$$

and by Lemma 1, $\gcd(\sigma(x), \omega(x)) = 1$. Thus the hypotheses of Theorem 9.5 are met with $a(x) = x^{2t}$, $b(x) = S(x)$, $v(x) = \sigma(x)$, $r(x) = \omega(x)$, $\mu = t$, $\nu = t - 1$, so that if the procedure Euclid(x^{2t}, $S(x)$, t, $t - 1$) is called it will return the polynomial pair $(v(x), r(x))$ where $v(x) = \lambda\sigma(x)$, $r(x)\lambda\omega(x)$, and λ is a nonzero scalar. The scalar λ can be determined by the fact that $\sigma(0) = 1$ (see (9.20)), i.e., $\lambda = v(0)^{-1}$, and so

$$\sigma(x) = v(x)/v(0),$$

$$\omega(x) = r(x)/v(0).$$

The final step in the decoding algorithm is to use $\sigma(x)$ and $\omega(x)$ to determine the error pattern $\mathbf{E} = (E_0, E_1, \ldots, E_{n-1})$, the hence the corrected codeword $\mathbf{C} = \mathbf{R} - \mathbf{E}$. As we observed at the end of Section 9.3, there are two ways to do this, which we shall call the *time-domain approach* and the *frequency-domain approach*.

The time-domain approach is based on the fact that

$$\sigma(x) = \prod_{i \in I}(1 - \alpha^i x)$$

where I is the *error-locator set*, i.e. $I = \{i : E_i \neq 0\}$ (see (9.20) and (9.36)). Thus in order to find the error locations, one needs to find the *reciprocals of the roots of the equation* $\sigma(x) = 0$. Since there are only n possibilities for the roots, viz., $1, \alpha^{-1}, \alpha^{-2}, \ldots, \alpha^{-(n-1)}$, a simple "trial and error" algorithm can be used to find $\mathbf{E}$. Thus the so-called "time-domain completion" can be described by the following pseudocode fragment. It takes as input $\sigma(x)$ and produces the error vector $(E_0, E_1, \ldots, E_{n-1})$.

```
/* Time-Domain Completion */
{
  for (i = 0 to n − 1)
  {
    if (σ(α⁻ⁱ) == 0)
      Eᵢ = 1;
    else
      Eᵢ = 0;
  }
}
```

A complete decoding algorithm using the time-domain completion is shown in Figure 9.1. Note that the error-evaluator polynomial $\omega(x)$ is not needed—its significance will become apparent only when we consider *Reed–Solomon codes* in the next section.

```
/* ``Time-Domain'' BCH Decoding Algorithm */
{
  for (j = 1 to 2t)
```
$$S_j = \sum_{i=0}^{n-1} R_i \alpha^{ij};$$
$$S(x) = S_1 + S_2 x + \cdots + S_{2t} x^{2t-1};$$

```
  if (S(x) ==0)
    print ``no errors occurred'';
  else
  {
```
  Euclid $(x^{2t}, S(x), t, t-1)$;
  $\sigma(x) = v(x)/v(0)$;
```
    for (i = 0 to n-1)
    {
      if (σ(α⁻ⁱ) == 0)
```
   if $(\sigma(\alpha^{-i}) == 0)$
    $E_i = 1$;
```
      else
```
    $E_i = 0$;
```
    }
    for (i = 0 to n-1)
```
  $\hat{C}_i = R_i + E_i$;
  print ``corrected codeword: $(\hat{C}_0, \hat{C}_1, \ldots, , \hat{C}_{n-1})$'';
```
  }
}
```

Figure 9.1 A time domain BCH decoding algorithm.

The *frequency-domain approach* is based on Corollary 2 to Theorem 9.4, which says that the components of $\hat{V} = (\hat{V}_0, \ldots, \hat{V}_{n-1})$ can be computed recursively, via the formula $\hat{V}_j = \sum_{i=1}^{d} \sigma_i \hat{V}_{j-i}$, where $\sigma(x) = 1 + \sigma_1 x + \cdots + \sigma_d x^d$, provided at least d "initial values" of the vector $\hat{V}$ are known. Since the syndrome provides $2t$ components of $\hat{V}$, viz. $\hat{V}_1, \hat{V}_2, \ldots, \hat{V}_{2t}$, and since $\texttt{Euclid}(x^{2t}, S(x), t, t-1)$ is guaranteed to return a polynomial $v(x)$ of degree $\leq t$, the syndrome values $S_1, S_2, \ldots, S_{2t}$ are more than enough to get the recursion started, so that the following "frequency-domain completion" will successfully calculate the error vector **E**:

```
/* Frequency-Domain Completion */
{
   for (j = 2t + 1 to n)
```
$$S_{j \bmod n} = \sum_{i=1}^{d} \sigma_i S_{j-i};$$
```
   for (i = 0 to n − 1)
```
$$E_i = \sum_{j=0}^{n-1} S_j \alpha^{-ij};$$
```
}
```

A complete decoding algorithm using the frequency-domain completion is shown in Figure 9.2.

Example 9.5 Consider the three-error correcting BCH code of length 15, with generator polynomial $g(x) = x^{10} + x^8 + x^5 + x^4 + x^2 + x + 1$ (see Example 9.1). Suppose the vector $\mathbf{R} = (110000110110101)$ is received. Then the syndrome components S_j are given by $S_j = 1 + \alpha^j + \alpha^{6j} + \alpha^{7j} + \alpha^{9j} + \alpha^{10j} + \alpha^{12j} + \alpha^{14j}$, where α is a primitive root in $GF(16)$. Using Table 9.1,

```
/* ``Frequency-Domain'' BCH Decoding Algorithm */
{
   for (j = 1 to 2t)
```
$$S_j = \sum_{i=0}^{n-1} R_i \alpha^{ij};$$
$$S(x) = S_1 + S_2 x + \cdots + S_{2t} x^{2t-1};$$
```
   if (S(x) ==0)
      print ``no errors occurred'';
   else
   {
      Euclid (x²ᵗ, S(x), t, t−1);
```
   $\sigma(x) = v(x)/v(0);$
```
      for (j = 2t +1 to n)
```
$$S_{j \bmod n} = \sum_{i=1}^{d} \sigma_i S_{j-i};$$
```
      for (i = 0 to n − 1)
```
$$E_i = \sum_{j=0}^{n-1} S_j \alpha^{-ij};$$
```
      for (i = 0 to n − 1)
```
   $\hat{C}_i = R_i + E_i;$
```
      print ``corrected codeword: (Ĉ₀, Ĉ₁, ..., Ĉₙ₋₁)'';
   }
}
```

Figure 9.2 A frequency-domain BCH decoding algorithm.

together with the fact that $S_{2j} = S_j^2$ (Problem 9.17), we find that $S_1 = \alpha^{12}$, $S_2 = \alpha^9$, $S_3 = 0$, $S_4 = \alpha^3$, $S_5 = 1$, $S_6 = 0$, and so $S(x) = x^4 + \alpha^3 x^3 + \alpha^9 x + \alpha^{12}$. Applying Euclid's algorithm to the pair $(x^6, S(x))$, we get the following table:

i	u_i	v_i	r_i	q_i
-1	1	0	x^6	—
0	0	1	$x^4 + \alpha^3 x^3 + \alpha^9 x + \alpha^{12}$	—
1	1	$x^2 + \alpha^3 x + \alpha^6$	α^3	$x^2 + \alpha^3 x + \alpha^6$

Thus the procedure Euclid(x^6, $S(x)$, 3, 2) returns the pair $(x^2 + \alpha^3 x + \alpha^6, \alpha^3)$. Multiplying both of these polynomials by α^{-6}, we therefore find that $\sigma(x) = 1 + \alpha^{12}x + \alpha^9 x^2$, and $\omega(x) = \alpha^{12}$. If we choose the time-domain completion, we find that $\sigma(\alpha^{-i}) = 0$ for $i = 2$ and 7, so that the error pattern is $\mathbf{E} = [00100001000000]$, and the corrected codeword is $\hat{C} = [111000100110101]$. On the other hand, if we choose the frequency-domain completion, we use the initial conditions $S_1 = \alpha^{12}$, $S_2 = \alpha^9$, $S_3 = 0$, $S_4 = \alpha^3$, $S_5 = 1$, $S_6 = 0$ and the recursion $S_j = \alpha^{12} S_{j-1} + \alpha^9 S_{j-2}$ to complete the syndrome vector, and find

$$\mathbf{S} = (S_0, S_1, \ldots, S_{15}) = (0, \alpha^{12}, \alpha^9, 0, \alpha^3, 1, 0, \alpha^9, \alpha^6, 0, 1, \alpha^{12}, 0, \alpha^6, \alpha^3).$$

Performing an inverse DFT on the vector $\mathbf{S}$ we find that $\mathbf{E} = [00100001000000]$, and $\hat{C} = [111000100110101]$ as before. □

The algorithms in Figures 9.1 and 9.2 will work perfectly if the number of errors that occurs is no more that t. If, however, *more* than t errors occur, certain problems can arise. For example, the procedure "Euclid(s^{2t}, $S(x)$, t, $t-1$)" could return a polynomial $v(x)$ with $(v)(0) = 0$, thereby causing a division by 0 in the step "$\sigma(x) = v(x)/v(0)$". Also, the decoder output $\hat{C} = (\hat{C}_0, \hat{C}_1, \ldots, \hat{C}_{n-1})$ may turn out not to be a codeword. Therefore in any practical implementation of the decoding algorithms, it will be necessary to test for these abnormal conditions, and print a warning, like "more than t errors" if they occur.

9.6 Reed–Solomon codes

In the first five sections of this chapter we have developed an elaborate theory for BCH codes. They are multiple-error-correcting linear codes over the binary field $GF(2)$, whose decoding algorithm requires computations in the larger field $GF(2^m)$. Thus for BCH codes there are *two* fields of interest: the codeword *symbol field* $GF(2)$, and the decoder's *computation field* $GF(2^m)$.

It turns out that almost the same theory can be used to develop another class of codes, the *Reed–Solomon* codes (RS codes for short). The main *theoretical* difference between RS codes and BCH codes is that for RS codes, the symbol field and the computation field are the same. The main *practical* difference between the two classes of codes is that RS codes lend themselves naturally to the transmission of information *characters*, rather than *bits*. In this section we will define and study Reed–Solomon codes.

Thus let F be any field which contains an element α of order n.[6] If r is a fixed integer between 1 and n, the set of all vectors $\mathbf{C} = (C_0, C_1, \ldots, C_{n-1})$ with components in F such that

$$\sum_{i=0}^{n-1} C_i \alpha^{ij} = 0, \qquad \text{for } j = 1, 2, \ldots, r, \tag{9.67}$$

is called a *Reed–Solomon code* of *length n* and *redundancy r* over *F*. The vectors $\mathbf{C}$ belonging to the code are called its *codewords*. The following theorem gives the basic facts about RS codes.

Theorem 9.7 *The code defined by (9.67) is an $(n, n - r)$ cyclic code over F with generator polynomial $g(x) = \prod_{j=1}^{r}(x - \alpha^j)$, and minimum distance $d_{\min} = r + 1$.*

Proof Let $\mathbf{C} = (C_0, C_1, \ldots, C_{n-1})$ be an arbitrary vector of length n over F and let $C(x) = C_0 + C_1 x + \cdots + C_{n-1} x^{n-1}$ be the corresponding generating function. Then (9.67) says that $\mathbf{C}$ is a codeword if and only if $C(\alpha^j) = 0$, for $j = 1, 2, \ldots, r$, which is the same as saying that $C(x)$ is a multiple of $g(x) = (x - \alpha)(x - \alpha^2) \ldots (x - \alpha^r)$. But since $x^n - 1 = \prod_{j=1}^{n}(x - \alpha^j)$, it follows that $g(x)$ is a divisor of $x^n - 1$, and so by Theorem 8.3(b) the code is an $(n, n - r)$ cyclic code with generator polynomial $g(x)$. To prove the assertion about $d_{\min}$, observe that (9.67) says that if $\hat{C} = (\hat{C}_0, \hat{C}_1, \ldots, \hat{C}_{n-1})$ is the DFT of a codeword, then $\hat{C}_1 = \hat{C}_2 = \cdots = \hat{C}_r = 0$ (cf. Eq. (9.11)). Thus by the BCH argument (Theorem 9.3), the weight of any nonzero codeword is $\geq r + 1$. On the other hand, the generator polynomial $g(x) = x^r + g_{r-1} x^{r-1} + \cdots + g_0$, when viewed as a codeword, has weight $\leq r + 1$. Thus $d_{\min} = r + 1$ as asserted. $\qquad\square$

Example 9.6 Consider the (7, 3) Reed–Solomon code over $GF(8)$. If α is a primitive root in $GF(8)$ satisfying $\alpha^3 = \alpha + 1$, the generator polynomial for the code is $g(x) = (x - \alpha)(x - \alpha^2)(x - \alpha^3)(x - \alpha^4) = x^4 + \alpha^3 x^3 + x^2 + \alpha x + \alpha^3$. If $g(x)$ is viewed as a codeword, it is $[\alpha^3, \alpha, 1, \alpha^3, 1, 0, 0]$, which is of weight 5, the minimum weight of the code. □

We note that the (7, 3) RS code over $GF(8)$ in Example 9.6 has $d_{min} = 5$, whereas the (7, 3) code over $GF(2)$ given in Example 8.2 (and elsewhere in Chapter 8) has only $d_{min} = 4$. The following theorem shows that for a given n and k, RS codes have the largest possible d_{min}, independent of the field F.

Theorem 9.8 *(the Singleton bound) If C is an (n, k) linear code over a field F, then $d_{min} \leq n - k + 1$.*

Proof We begin by recalling that if T is a linear transformation mapping a finite-dimensional vector space U to another vector space V, then

$$\text{rank}(T) + \text{nullity}(T) = \dim(U). \tag{9.68}$$

We apply this to the linear transformation T mapping the code C to the space F^{k-1} by projecting each codeword onto the first $k - 1$ coordinates:

$$T(C_0, C_1, \ldots, C_{n-1}) = (C_0, C_1, \ldots, C_{k-2}).$$

We know that $\text{rank}(T) \leq k - 1$, since the image F^{k-1} has dimension $k - 1$. Also, $\dim(C) = k$ by assumption. Thus (9.68) implies that $\text{nullity}(T) \geq 1$. Thus there exists at least one nonzero codeword $\mathbf{C}$ such that $T(\mathbf{C}) = 0$. Such a codeword has at least $k - 1$ zero components, and so has weight at most $n - k + 1$. □

Theorem 9.8 says that $d_{min} \leq n - k + 1$ for any (n, k) linear code. On the other hand, Theorem 9.7 says that $d_{min} = n - k + 1$ for any (n, k) Reed–Solomon code, and so Reed–Solomon codes are *optimal* in the sense of having the largest possible minimum distance for a given length and dimension. There is a special name give to linear codes with $d_{min} = n - k + 1$; they are called *maximum-distance separable (MDS)* codes. (Some other MDS codes are described in Problems 9.24–9.26.) All MDS codes share some very interesting mathematical properties; among the most interesting is the following, called the *interpolation property* of MDS codes.

Theorem 9.9 *Let C be an (n, k) MDS code over the field F, and let $I \subseteq \{0, 1, \ldots, n - 1\}$ be any subset of k coordinate positions. Then for any* set $\{\alpha_i : i \in I\}$ *of k elements from F, there exists a unique codeword* **C** *such that* $C_i = \alpha_i$ *for all* $i \in I$.

Proof We consider the linear transformation P_I mapping the code C to F^k by *projecting* each codeword onto the index set I; i.e., $P_I(C_0, C_1, \ldots, C_{n-1}) = (C_{i_1}, C_{i_2}, \ldots, C_{i_k})$, where $I = \{i_1, i_2, \ldots, i_k\}$. Applying (9.68), which in this case says that rank(P_I) + nullity(P_I) = dim(C) we see that dim$(C) = k$, since C is a k-dimensional code. Also, nullity$(P_I) = 0$, since if there were a nonzero codeword **C** with $P_I(\mathbf{C}) = 0$, that codeword would have weight at most $n - k$, contradicting the fact that **C** is an MDS code. Hence by (9.68) rank$(P_I) = k$, and so the mapping $P_I : C \to F^k$ is nonsingular, i.e. one-to-one and onto. Thus every vector in F^k appears exactly once as the projection of a codeword onto I, which is what the theorem promises. $\square$

We summarize the result of Theorem 9.9 by saying that any subset of k coordinate positions of a k-dimensional MDS code is an *information set* (see also Problem 7.13). The proof we have given is short but nonconstructive; however, for RS codes there is an efficient *interpolation algorithm*, which is closely related to the Lagrange interpolation formula of numerical analysis. The next theorem spells this out.

Theorem 9.10 *Consider the* (n, k) *Reed–Solomon code over the field F defined by (9.67), where* $k = n - r$. *There is a one-to-one correspondence between the codewords* $\mathbf{C} = (C_0, C_1, \ldots, C_{n-1})$ *of this code, and the set of all polynomials* $P(x) = P_0 + P_1 x + \cdots + P_{k-1} x^{k-1}$ *of degree* $k - 1$ *or less over F, given by*

$$C_i = \alpha^{-i(r+1)} P(\alpha^{-i}).$$

Thus apart from the scaling factors $\alpha^{-i(r+1)}$, *the components of a given RS codeword are the values of a certain* $(k - 1)$*st-degree polynomial.*

Proof Let $\mathbf{C} = [C_1, \ldots, C_{n-1}]$ be a fixed codeword. We define a "twisted" version of **C**, called $\mathbf{D} = [D_1, \ldots, D_{n-1}]$, by

$$D_i = \alpha^{-i(r+1)} C_i, \qquad \text{for } i = 0, 1, n - 1. \tag{9.69}$$

Since by (9.67) we have $\hat{C}_1 = \hat{C}_2 = \cdots = \hat{C}_r = 0$, it follows from (9.17) and (9.18) that $\hat{D}_{n-r} = \cdots = \hat{D}_{n-1} = 0$. Thus the DFT polynomial for **D**, denoted by $\hat{D}(x)$, is a polynomial of degree $n - r - 1 = k - 1$ or less:

$$\hat{D}(x) = \hat{D}_0 + \hat{D}_1 x + \cdots + \hat{D}_{k-1} x^{k-1}.$$

Let us define the polynomial $P(x)$ as follows:

$$P(x) = \frac{1}{n} \hat{D}(x).$$

Then by (9.16) we have $D_i = P(\alpha^{-i})$, for $i = 0, 1, \ldots, n - 1$. Combining this with (9.69), we obtain $C_i = \alpha^{-i(r+1)} P(\alpha^{-i})$, which is what we wanted. ☐

The following example illustrates Theorem 9.10.

Example 9.7 Consider the (7, 3) RS code described in Example 9.6. According to Theorem 9.9, there is a unique codeword **C** such that $C_1 = \alpha^3$, $C_4 = \alpha$, and $C_6 = \alpha^4$. Let us construct this codeword.

We begin by observing that if $I = \{1, 4, 6\}$, Theorem 9.9 guarantees, in essence, the existence of a 3×7 generator matrix for C of the form

$$G_{146} = \begin{matrix} & 0 & 1 & 2 & 3 & 4 & 5 & 6 \\ & \begin{pmatrix} * & 1 & * & * & 0 & * & 0 \\ * & 0 & * & * & 1 & * & 0 \\ * & 0 & * & * & 0 & * & 1 \end{pmatrix} \end{matrix}$$

where the $*$'s are unknown elements of $GF(8)$ which must be determined. Once G_{146} is known, the desired codeword **C** is given by $\mathbf{C} = [\alpha^3, \alpha, \alpha^4] \cdot G_{146}$. So let's construct the three rows of G_{146}, which we shall call C_1, C_4, and C_6.

By Theorem 9.9, any codeword **C** from the (7, 3) RS code can be represented as $C_i = \alpha^{-5i} P(\alpha^{-i})$, where $P(x) = P_0 + P_1 x + P_2 x^2$ is a polynomial of degree 2 or less. Thus for example, if $P_1(x)$ denotes the polynomial corresponding to the first row $\mathbf{C}_1$ of G_{146}, we have

$$P_1(\alpha^{-1}) = \alpha^5, \ P_1(\alpha^{-4}) = 0, \ P_1(\alpha^{-6}) = 0. \tag{9.70}$$

It follows from the conditions $P_1(\alpha^{-4}) = P_1(\alpha^{-6}) = 0$ in (9.70) that $P_1(x) = A(1 + \alpha^4 x)(1 + \alpha^6 x)$ for some constant A, which can be determined by the condition $P_1(\alpha^{-1}) = \alpha^5$. Indeed $P_1(\alpha^{-1}) = \alpha^5$ implies $A(1 + \alpha^3)(1 + a^5) = \alpha^5$, i.e., $A = \alpha^5/(1 + \alpha^3)(1 + \alpha^5) = 1$. Thus $P_1(x) = (1 + \alpha^4 x)(1 + \alpha^6 x)$, and so

$$\mathbf{C}_1 = [P_1(1), a^2 P_1(a^{-1}), a^4 P_1(a^{-2}), a^6 P_1(a^{-3}),$$

$$a^1 P_1(a^{-4}), a^3 P_1(a^{-5}), a^5 P_1(a^{-6})]$$

$$= [1, 1, a, a^3, 0, a, 0].$$

Similarly, if $P_4(x)$ and $P_6(x)$ denote the quadratic polynomials corresponding to the rows C_4 and C_6 of the generator matrix G_{146}, then we find that $P_4(x) = a^2(1 + ax)(1 + a^6 x)$ and $P_6(x) = a^6(1 + ax)(1 + a^4 x)$. Thus we compute

$$\mathbf{C}_4 = [1, 0, a^6, a^6, 1, a^2, 0],$$

$$\mathbf{C}_6 = [1, 0, a^4, a^5, 0, a^5, 1].$$

Combining $\mathbf{C}_1$, $\mathbf{C}_4$, and $\mathbf{C}_6$, we find that the generator matrix G_{146} is

$$G_{146} = \begin{pmatrix} 1 & 1 & a & a^3 & 0 & a & 0 \\ 1 & 0 & a^6 & a^6 & 1 & a^2 & 0 \\ 1 & 0 & a^4 & a^5 & 0 & a^5 & 1 \end{pmatrix},$$

and so, finally, the unique codeword $\mathbf{C}$ with $C_1 = a^3$, $C_4 = a$, $C_6 = a^4$ is

$$\mathbf{C} = [a^3, a, a^4] \cdot G_{146} = [a^5, a^3, a^6, 0, a, 1, a^4]. \qquad \square$$

This concludes our theoretical discussion of RS codes; now let's consider the practical issues of *encoding* and *decoding* them.

Since by Theorem 9.7, an (n, k) RS code is cyclic, it can be encoded using the shift-register techniques developed in Chapter 8. In particular, the general encoding circuit of Figure 8.5(a) can be used. However, since an RS code is defined over an arbitrary field F—which in practice will never be the binary field $GF(2)$ (Problem 9.27)—the three basic components (flip-flops, adders, and multipliers) will typically not be "off-the-shelf" items. Although the design of these components over the important fields $GF(2^m)$ is an important and interesting topic, it is beyond the scope of this book, and we will conclude our discussion of RS encoders with Figure 9.3, which shows a systematic shift-register encoder for the $(7, 3)$ RS code over $GF(8)$ with $g(x) = x^4 + a^3 x^3 + x^2 + ax + a^3$ (see Examples 9.6 and 9.7).

We turn now to the problem of *decoding* RS codes, which turns out to be quite similar to the decoding of BCH codes. In view of the similarity of their definitions (compare (9.7)) with (9.67)), this should not be surprising.

Let us begin by formally stating the RS decoding problem. We are given a received vector $\mathbf{R} = (R_0, R_1, \ldots, R_{n-1})$, which is a noisy version of an

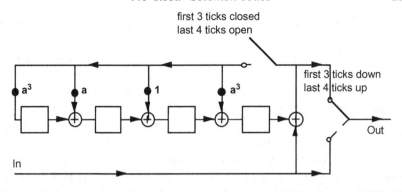

Figure 9.3 A systematic shift-register encoder for the (7, 3) RS code over $GF(8)$ with $g(x) = x^4 + a^3 x^3 + x^2 + ax + a^3$.

unknown codeword $C = (C_0, C_1, \ldots, C_{n-1})$ from the (n, k) RS code defined by (9.67), i.e., $R = C + E$, where E is the error pattern. Since by Theorem 9.7, $d_{min} = r + 1$, we cannot hope to correctly identify C unless $wt(E) \leq \lfloor r/2 \rfloor$, and so for the rest of the discussion we shall let $t = \lfloor r/2 \rfloor$, and assume that $wt(E) \leq t$.

The first step in the decoding process is to compute the *syndrome polynomial*

$$S(x) = S_1 + S_2 x + \cdots + S_r x^{r-1}, \tag{9.71}$$

where $S_j = \sum_{i=0}^{n-1} R_i a^{ij}$, for $j = 1, 2, \ldots, r$. By the results of Section 9.3, if we define the "twisted error pattern" by

$$V = (E_0, E_1 a, E_2 a^2, \ldots, E_{n-1} a^{n-1}),$$

then $S(x) = \hat{V}(x) \bmod x^r$, and the key equation (9.23), reduced $\bmod x^r$, becomes

$$\sigma(x) S(x) \equiv \omega(x) \pmod{x^r},$$

where $\sigma(x)$ is the locator polynomial, and $\omega(x)$ is the evaluator polynomial, for the vector V.

At this point the decoding problem is almost exactly the same as it was for BCH codes as described in Section 9.5. In particular, if the procedure $\texttt{Euclid}(x^r, S(x), t, t-1)$ is called, it will return the pair of polynomials $(v(x), r(x))$, where $v(x) = \lambda \sigma(x)$ and $r(x) = \lambda \omega(x)$ for some nonzero constant λ.

The final step in the decoding algorithm is to use $\sigma(x)$ and $\omega(x)$ to determine the error pattern $E = (E_0, E_1, \ldots, E_{n-1})$, and hence the original codeword

$\mathbf{C} = \mathbf{R} - \mathbf{E}$. As with BCH codes, there are two essentially different ways to do this, the *time-domain approach* and the *frequency-domain approach*.

The *time-domain approach* for RS decoding is similar to the time-domain approach for BCH decoding, with one important exception. For BCH codes, when the errors are located, their values are immediately known. This is because BCH codes are binary, so that $E_i = 0$ or 1 for all i. Thus if there is an error in position i, i.e., $E_i \neq 0$, then necessarily $E_i = 1$. However, for RS codes, the E_i's lie in the "big" field F, so that simply knowing that $E_i \neq 0$ is not enough to indentify E_i. In order to evaluate an error whose location is known, we use Corollary 1 to Theorem 9.4, which say that if $E_i \neq 0$, i.e., $\sigma(\alpha^{-i}) = 0$, then $V_i = \alpha^i E_i = -\alpha^i \omega(\alpha^{-i})/\sigma'(\alpha^{-i})$, i.e.,

$$E_i = -\frac{\omega(\alpha^{-i})}{\sigma'(\alpha^{-i})}. \tag{9.72}$$

```
/* ``Time-Domain'' RS Decoding Algorithm */
{
  for (j = 1 to r)
    S_j = ∑_{i=0}^{n-1} R_i α^{ij};
  S(x) = S_1 + S_2 x + ··· + S_r x^{r-1};
  if (S(x) == 0)
    print ``no errors occurred'';
  else
  {
    Euclid (x^r, S(x), t, t-1);
    σ(x) = v(x)/v(0);
    ω(x) = r(x)/v(0);
    for (i = 0 to n - 1)
    {
      if (σ(α^{-i}) == 0)
        E_i = -ω(α^{-i})/σ'(α^{-i});
      else
        E_i = 0;
    }
    for (i = 0 to n - 1)
      Ĉ_i = R_i - E_i;
    print ``corrected codeword: (Ĉ_0, Ĉ_1, ..., , Ĉ_{n-1})'';
  }
}
```

Figure 9.4 A time-domain RS decoding algorithm.

Thus the time-domain completion of the RS decoding algorithm can be written as follows:

```
/* Time-Domain Completion */
{
  for (i = 0 to n − 1)
  {
      if (σ(α⁻ⁱ) == 0)
      Eᵢ = −ω(α⁻ⁱ)/σ′(α⁻ⁱ);
      else
      Eᵢ = 0;
  }
}
```

A complete time-domain decoding algorithm for RS codes is shown in Figure 9.4.

The *frequency-domain approach* to RS decoding is nearly identical to the frequency-domain approach to BCH decoding, since the idea of recursive completion of the error vector works for an arbitrary field F. Here is a pseudocode listing for a frequency-domain completion.

```
/* Frequency-Domain Completion */
{
  for (j = r + 1 to n)
    Sⱼ mod n = −∑ᵈᵢ₌₁ Sⱼ₋ᵢ;
  for (i = 0 to n − 1)
    Eᵢ = ¹⁄ₙ ∑ⁿ⁻¹ⱼ₌₀ Sⱼα⁻ⁱʲ;
}
```

A complete RS decoding algorithm using the frequency-domain completion is given in Figure 9.5.

Example 9.8 Consider the $(7, 3)$ RS code over $GF(2^3)$ with $g(x) = (x - \alpha)(x - \alpha^2)(x - \alpha^3)(x - \alpha^4) = x^4 + \alpha^3 x^3 + x^2 + ax + \alpha^3$ already considered in Examples 9.6 and 9.7. Suppose the received vector is $\mathbf{R} = (\alpha^3, \alpha, 1, \alpha^2, 0, \alpha^3, 1)$. The syndromes $S_1 = \sum R_i \alpha^{ij}$ are $S_1 = \alpha^3$, $S_2 = \alpha^4$, $S_3 = \alpha^4$, $S_4 = 0$, so that $S(x) = \alpha^4 x^2 + \alpha^4 x + \alpha^3$. If we invoke the

```
/* ``Frequency-Domain'' RS Decoding Algorithm */
{
   for (j = 1 to r)
      S_j = ∑_{i=0}^{n-1} R_i α^{ij};
      S(x) = S_1 + S_2 x + ⋯ + S_r x^{r-1};
   if (S(x) == 0)
      print ``no errors occurred'';
   else
   {
      Euclid (x^r, S(x), t, t−1);
      σ(x) = v(x)/v(0);
      ω(x) = r(x)/v(0);
      for (j = r +1 to n)
         S_{j mod n} = −∑_{i=1}^{d} σ_i S_{j-i};
      for (i = 0 to n − 1)
         E_i = (1/n) ∑_{j=0}^{n-1} S_j α^{-ij};
      for (i = 0 to n − 1)
         Ĉ_i = R_i − E_i;
      print ``corrected codeword: [Ĉ_0, Ĉ_1, …, , Ĉ_{n-1}]'';
   }
}
```

Figure 9.5 A frequency-domain RS decoding algorithm.

procedure $\text{Euclid}(x^4, a^4 x^2 + a^4 x + a^3, 2, 1)$ we obtain the following table:

i	$v_i(x)$	$r_i(x)$	$q_i(x)$
-1	0	x^4	—
0	1	$a^4 x^2 + a^4 x + a^3$	—
1	$a^3 x^2 + a^3 x + a^5$	$x + a$	$a^3 x^2 + a^3 x + a^5$

Thus we conclude that $\sigma(x) = a^{-5}(a^5 + a^3 x + a^3 x^2) = 1 + a^5 x + a^5 x^2$, and $\omega(x) = a^{-5}(x + a) = a^2 x + a^3$.

With the time-domain approach, we find that $\sigma(a^{-3}) = \sigma(a^{-2}) = 0$, i.e., $\sigma(x) = (1 + a^2 x)(1 + a^3 x)$. Thus the error locations are $i = 2$ and $i = 3$. To evaluate these two errors, use the formula (9.72), together with the fact that $\sigma'(x) = a^5$, so that $\omega(x)/\sigma'(x) = a^4 x + a^5$, and find

Table 9.4 The field $GF(8)$ represented
as powers of α, where $\alpha^3 = \alpha + 1$.

i	α^i
0	001
1	010
2	100
3	011
4	110
5	111
6	101

$$E_2 = \frac{\omega(\alpha^{-2})}{\sigma'(\alpha^{-2})} = \alpha^4 \cdot \alpha^{-2} + \alpha^5 = \alpha^3,$$

$$E_3 = \frac{\omega(\alpha^{-3})}{\sigma'(\alpha^{-3})} = \alpha^4 \cdot \alpha^{-3} + \alpha^5 = \alpha^6.$$

Thus $\mathbf{E} = (0, 0, \alpha^3, \alpha^6, 0, 0, 0)$ and the decoder's output is $\hat{\mathbf{C}} = \mathbf{R} + \mathbf{E} = (\alpha^3, \alpha, \alpha, 1, 0, \alpha^3, 1)$.

With the frequency-domain approach, we use the initial conditions $S_1 = \alpha^3$, $S_2 = \alpha^4$, $S_3 = \alpha^4$, $S_4 = 0$ and the recursion (based on the coefficients of $\sigma(x)$) $S_j = \alpha^5 S_{j-1} + \alpha^5 S_{j-2}$ to find

$$S_5 = \alpha^5 \cdot 0 + \alpha^5 \cdot \alpha^4 = \alpha^2,$$

$$S_6 = \alpha^5 \cdot \alpha^2 + \alpha^5 \cdot 0 = 1,$$

$$S_7 = S_0 = \alpha^5 \cdot 1 + \alpha^5 \cdot \alpha^2 = \alpha^4.$$

Thus $\mathbf{S} = (S_0, S_1, S_2, S_3, S_4, S_5, S_6) = (\alpha^4, \alpha^3, \alpha^4, \alpha^4, 0, \alpha^2, 1)$. To obtain $\mathbf{E}$, we take an inverse DFT of $\mathbf{S}$, using (9.12):

$$\mathbf{E} = \hat{\mathbf{S}} = (0, 0, \alpha^3, \alpha^6, 0, 0, 0),$$

and now the decoding concludes as before. $\qquad\qquad\qquad\square$

We conclude this section with a brief discussion of two important applications of RS codes: *burst-error correction* and *concatenated coding*.

We can illustrate the application to burst-error correction by returning to Example 9.8. There we saw the $(7, 3)$ RS code over $GF(8)$ in action, correcting two symbol errors. But instead of viewing each codeword as a

7-dimensional vector over $GF(8)$, we can expand each element of $GF(8)$ into a 3-dimensional binary vector via Table 9.4 and thereby convert the codewords into 21-dimensional binary vectors. In other words, the (7, 3) RS code over $GF(8)$ can be viewed as a (21, 9) linear code over $GF(2)$. For example, the codeword

$$\mathbf{C} = (\alpha^3, \alpha, \alpha, 1, 0, \alpha^3, 1)$$

becomes the binary vector

$$\mathbf{C} = (011\ 010\ 010\ 001\ 000\ 011\ 001).$$

Now suppose this binary version of $\mathbf{C}$ was sent over a binary channel and suffered the following error burst of length 5:

$$\mathbf{E} = (000\ 000\ 0\overbrace{11\ 101}^{\substack{\text{error}\\\text{burst}}}\ 000\ 000\ 000)$$

Then the received vector would be

$$\mathbf{R} = (011\ 010\ 001\ 100\ 000\ 011\ 001),$$

which of course differs from $\mathbf{C}$ in four positions. Ordinarily it would be difficult or impossible to correct four errors in a (21, 9) binary linear code (see Problem 9.33), but we can take advantage of the fact that this particular set of four errors has occurred in a short burst by observing that when $\mathbf{E}$ is mapped into a 7-dimensional vector from $GF(8)$,

$$\mathbf{E} = (0, 0, \alpha^3, \alpha^6, 0, 0, 0),$$

it only has weight 2! Thus if we convert $\mathbf{R}$ into a vector from $GF(8)$,

$$\mathbf{R} = (\alpha^3, \alpha, 1, \alpha^2, 0, \alpha^3, 1),$$

we can (and we already did in Example 9.8) find the error pattern and correct the errors, via the decoding algorithm of Figure 9.4 or 9.5. In this way the original RS code has become a (21, 9) binary linear code which is capable of correcting many patterns of burst errors.

The generalization is this: *a t-error-correcting RS code of length n over $GF(2^m)$ can be implemented as an $(m(2^m - 1), m(2^m - 1 - 2t))$ linear code over $GF(2)$ which is capable of correcting any burst-error pattern that does not affect more than t of the symbols in the original $GF(2^m)$ version of the codeword.*

We come finally to the application of RS to *concatenated coding*, a subject already mentioned briefly in Chapter 6 (see p. 129). We illustrate with a numerical example.

Suppose the (7, 4) binary Hamming code is being used on a BSC with crossover probability $p = .025$, as shown in Figure 9.6. In the notation of Figure 9.6, $P\{u \neq v\} = \sum_{k=2}^{7} \binom{7}{k} p^k (1 - p)^{7-k} = .0121$. The idea of concatenation is to regard the "encoder–BSC–decoder" part of Figure 9.6 as one big noisy channel, called the *outer channel* (the BSC itself becomes the *inner channel*), and to design a code for it. In this example the outer channel is a DMC with 16 inputs; the results of this section suggest that we regard these inputs and outputs as elements from $GF(16)$ rather than as four-dimensional vectors over $GF(2)$. So let us now consider using a (15, 11) RS code over $GF(16)$ to reduce the noise in the outer channel, as illustrated in Figure 9.7.

The RS encoder in Figure 9.7 takes 11 information symbols $\alpha = (\alpha_0, \dots, \alpha_{10})$ from $GF(16)$ (which are really 44 bits from the original source) and produces an RS codeword $\mathbf{C} = (C_0, C_1, \dots, C_{14})$. The outer channel then garbles $\mathbf{C}$, and it is received as $\mathbf{R} = (R_0, \dots, R_{14})$. The RS decoder then produces an estimate $\beta = (\beta_0, \dots, \beta_{10})$ of α, which will be equal to α if the outer channel has caused no more than two symbol errors. Thus if $\epsilon \, (= 0.0121)$ denotes the probability of decoder error in Figure 9.6, the probability of decoder error in Figure 9.7 is not more that $\sum_{k=3}^{15} \binom{15}{k} \epsilon^k (1 - \epsilon)^{15-k} = 0.0007$. The overall rate of the coding system depicted in Figure 9.7 is $11/15 \times 4/7 = 0.42$; indeed, the system is really just a (105, 44) binary linear code which has been "factored" in a clever way. We might wish

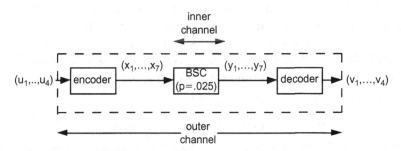

Figure 9.6 The (7, 4) Hamming code on a BSC with $p = .025$.

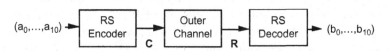

Figure 9.7 The (15, 11) Reed–Solomon code being used on the outer channel of Figure 9.6.

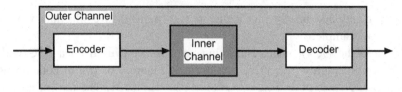

Figure 9.8 A general coded communication system, viewed as a noisy "outer" channel. (Compare with Fig. 9.6.)

to compare this with an approximately comparable *unfactored* system, say the 11-error-correcting binary BCH code of length 127 which is a (127, 57) code. Its rate (0.45) is slightly higher and its decoder error probability (.0004) is slightly lower, but its decoding complexity is considerably larger—for the BCH code, the error-locator polynomial will typically be an 11th-degree polynomial over $GF(128)$, whereas for the RS code it will be quadratic polynomial over $GF(16)$.

The preceding example illustrates both the general idea of concatenation and the reason why RS codes are so useful in concatenated systems. *Any* coded communication system can be regarded as a noisy outer channel, as in Figure 9.8. However, for this point of view to be useful, we must be able to design an outer code capable of correcting most of the errors caused by the outer channel, which is likely to be a very complex beast, since its errors are caused by inner decoder failures. When the inner decoder fails, that is when $(v_1, \ldots, v_k) \neq (u_1, \ldots, u_k)$ in Figure 9.8, the symbols $v_1, \ldots, v_k$ usually bear practically no resemblance to $u_1, \ldots, u_k$. This means that errors in the outer channel tend to occur in bursts of length k. And we have already seen that RS codes are well suited to burst-error correction. This is the reason why RS codes are in widespread use as outer codes in concatenated systems.

9.7 Decoding when erasures are present

We have seen that BCH and RS codes can correct multiple errors. In this section we will see that they can also correct another class of channel flaws, called *erasures*. An erasure is simply a channel symbol which is received illegibly. For example, consider the English word *BLOCK*. If the third letter is changed from *O* to *A*, we get *BLACK*; this is an *error* in the third position. However, if the same word suffers an *erasure* in the third position, the result is *BL*CK*, where "*" is the *erasure symbol*. In practice, erasures are quite common. They can be expected to occur when the channel noise becomes unusually severe for a short time. For example, if you are trying to talk at the airport and a low-flying jet passes overhead, your conversation is *erased*. Your listeners will not mistake what you are trying to say; they will simply not be able to understand you.

In this section, we will learn something about erasure correction. We will see that in principle, an erasure is only half as hard to correct as an error (Theorem 9.11); and we will see how to modify the BCH and RS decoding algorithms in order to correct both erasures and errors.

To model a channel which can produce erasures as well as errors, we simply enlarge the underlying symbol set F to $\bar{F} = F \cup \{*\}$, where "$*$" is as above a special erasure symbol. The only allowed *transmitted* symbols are the elements of F, but any element in $\bar{F}$ can be *received*. The main theoretical result about simultaneous erasure-and-error correction follows. (Compare with Theorem 7.2.)

Theorem 9.11 *Let C be a code over the alphabet F with minimum distance d. Then C is capable of correcting any pattern of e_0 erasures and e_1 errors if $e_0 + 2e_1 \leq d - 1$.*

Proof To prove the theorem, we first introduce the *extended Hamming distance* $\bar{d}_H(x, y)$ between symbols in $\bar{F}$:

$$\bar{d}_H(x, y) = \begin{cases} 0 & \text{if } x = y, \\ 1 & \text{if } x \neq y \text{ and neither } x \text{ nor } y \text{ is "}*\text{"}, \\ \frac{1}{2} & \text{if } x \neq y \text{ and one of } x \text{ and } y \text{ is "}*\text{"}. \end{cases}$$

Thus for example if $F = \{0, 1\}$ and $\bar{F} = \{0, 1, *\}$, then $\bar{d}_H(0, 1) = 1$, $\bar{d}_H(1, *) = 1/2$, $\bar{d}_H(1, 1) = 0$. We then extend the definition of $\bar{d}_H$ to vectors $\mathbf{x} = (x_1, \ldots, x_n)$ and $\mathbf{y} = (y_1, \ldots, y_n)$ with components in $\bar{F}$ as follows:

$$\bar{d}_H(\mathbf{x}, \mathbf{y}) = \sum_{i=1}^{n} \bar{d}_H(x_i, y_i).$$

With this definition, $\bar{d}_H$ becomes a *metric* on the set $\bar{F}^n$ of all n-dimensional vectors over $\bar{F}$. (See Problem 9.40). Indeed $\bar{d}_H(\mathbf{x}, \mathbf{y})$ is just the ordinary Hamming distance between $\mathbf{x}$ and $\mathbf{y}$, if no erasure symbols are involved in $\mathbf{x}$ or $\mathbf{y}$.

We next introduce a special decoding algorithm, called the *minimum-distance decoding* (MDD) algorithm for the code C. When the MDD algorithm is given as input a received word $R \in \bar{F}^n$, it produces as its output a codeword C_i for which the extended Hamming distance $\bar{d}_H(C_i, R)$ is smallest. We will prove Theorem 9.11 by showing that the MDD algorithm will correct e_0 erasures and e_1 errors, if $e_0 + 2e_1 \leq d - 1$.

Thus suppose that C_i is the transmitted codeword, and that in transmission it suffers e_0 erasures and e_1 errors, with $e_0 + 2e_1 \leq d - 1$. If R is the corresponding garbled version of C_i, then $\bar{d}_H(C_i, R) = \frac{1}{2}e_0 + e_1 \leq \frac{1}{2}(d - 1)$. There can be no other codeword this close to R, since if e.g. $\bar{d}_H(C_j, R) \leq \frac{1}{2}(d - 1)$ where $j \neq i$, then by the triangle inequality

$$\bar{d}_H(C_i, C_j) \leq \bar{d}_H(C_i, R) + \bar{d}_H(R, C_j)$$

$$\leq \frac{1}{2}(d - 1) + \frac{1}{2}(d - 1)$$

$$= d - 1,$$

which contradicts the fact that the code's minimum distance is d. Therefore the distance $\bar{d}_H(C_i, R)$ is uniquely smallest for $j = i$, and the MDD algorithm will correctly indentify C_i, the actual transmitted codeword.   □

Example 9.9 Let C be the $(7, 3)$ cyclic code from Example 8.2, with codewords

$$C_0 = 0000000,$$

$$C_1 = 1011100,$$

$$C_2 = 0101110,$$

$$C_3 = 0010111,$$

$$C_4 = 1001011,$$

$$C_5 = 1100101,$$

$$C_6 = 1110010,$$

$$C_7 = 0111001.$$

Since this code is linear, its minimum distance is the same as its minimum weight; thus $d = 4$. According to Theorem 9.11, then, this code is capable of correcting e_0 erasures and e_1 errors, provided $e_0 + 2e_1 \leq 3$. Here is a table of the allowed combinations of erasures and errors:

e_0	e_1
3	0
2	0
1	1
1	0
0	1
0	0

For example, suppose $R = [1\ 1\ 1\ 0 * 0\ 1]$ is received. The MDD algorithm would make the following computations:

i	$\bar{d}_H(C_i, R)$	Erasure positions	Error positions
0	4.5	{4}	{0, 1, 2, 6}
1	3.5	{4}	{1, 3, 6}
2	5.5	{4}	{0, 2, 3, 5, 6}
3	3.5	{4}	{0, 1, 5}
4	4.5	{4}	{1, 2, 3, 5}
5	1.5	{4}	{2}
6	2.5	{4}	{5, 6}
7	2.5	{4}	{0, 3}

Therefore the MDD would output C_5 and conclude that R had suffered an erasure in position 4 and an error in position 2, i.e. $e_0 = 1$ and $e_1 = 1$. On the other hand if $R = [* \ * \ * \ 1\ 0\ 1\ 0]$, the computation would run as follows:

i	$\bar{d}_H(C_i, R)$	Erasure positions	Error positions
0	3.5	{0, 1, 2}	{3, 5}
1	3.5	{0, 1, 2}	{4, 5}
2	2.5	{0, 1, 2}	{4}
3	4.5	{0, 1, 2}	{3, 4, 6}
4	2.5	{0, 1, 2}	{6}
5	5.5	{0, 1, 2}	{3, 4, 5, 6}
6	2.5	{0, 1, 2}	{3}
7	3.5	{0, 1, 2}	{5, 6}

Here the algorithm faces a three-way tie (between C_2, C_4, and C_6), but no matter which of these three it selects, it will conclude that the transmitted codeword has suffered 3 erasures and 1 error, which is beyond the code's guaranteed correction capabilities.   ☐

Theorem 9.11 gives the *theoretical* erasure-and-error correction capability of a code in terms of its minimum distance, but from a *practical* standpoint the MDD algorithm used in the proof leaves much to be desired, since it is plainly impractical to compare the received word with each of the codewords unless the code is very small. Fortunately, for BCH and RS codes, there is a simple modification of the basic "errors-only" decoding algorithms we have already presented in Section 9.6 (Figs. 9.4 and 9.5), which enables them to correct erasures as well as errors. In the remainder of this section, we will discuss this modification.

The erasures-and-errors decoding algorithms for BCH and RS codes, like their errors-only counterparts, are virtually identical, but for definiteness we'll consider in detail only RS codes. At the end of this section, we'll discuss the simple modifications required for BCH codes. By Theorem 9.7, the minimum distance of an (n, k) RS code is $r + 1$, where $r = n - k$, and so Theorem 9.11 implies the following.

Theorem 9.12 *Let C be an (n, k) RS code over a field F. Then C is capable of correcting any pattern of e_0 erasures and e_1 errors, if $e_0 + 2e_1 \leqslant r$, where $r = n - k$.*   ☐

Now let's begin our discussion of the erasures-and-errors decoding algorithm for RS codes. Suppose we are given a received vector $\mathbf{R} = (R_0, R_1, \ldots, R_{n-1})$, which is a noisy version of an unknown codeword $\mathbf{C} = (C_0, C_1, \ldots, C_{n-1})$, from an (n, k) RS code with generator polynomial $g(x) = (x - \alpha)(x - \alpha^2) \ldots (x - \alpha^r)$, with $r = n - k$. We assume $\mathbf{R}$ has suffered e_0 erasures and e_1 errors, where $e_0 + 2e_1 \leqslant r$. The first step in the decoding algorithm is to *store the locations of the erasures*. This is done by defining the *erasure set* I_0 as

$$I_0 = \{i : R_0 = *\}, \tag{9.73}$$

and then computing the *erasure-location polynomial* $\sigma_0(x)$:

$$\sigma_0(x) = \prod_{i \in I_0} (1 - \alpha^i x). \tag{9.74}$$

(If there are no erasures, $\sigma_0(x)$ is defined to be 1.)

Once the erasure locations have been "stored" in $\sigma_0(x)$, the algorithm replaces the *'s in R with 0's, i.e., a new received vector $\mathbf{R}' = (R_0', R_1', \ldots, R_{n-1}')$ is defined, as follows:

$$R_i' = \begin{cases} R_i & \text{if } R_i \neq *, \\ 0 & \text{if } R_i = *. \end{cases} \tag{9.75}$$

The advantage of replacing the *'s with 0's is that unlike *, 0 is an element of the field F, and so arithmetic operations can be performed on any component of $\mathbf{R}'$. The *disadvantage* of doing this is that when viewed as a garbled version of $\mathbf{C}$, $\mathbf{R}'$ will have suffered $e_0 + e_1$ errors,[7] which may exceed the code's errors-only correction capability. However, as we shall see, by using the "side information" provided by the erasure-locator polynomial $\sigma_0(x)$, the errors in $\mathbf{R}'$ can all be corrected.

With this preliminary "erasure management" completed, the decoding algorithm proceeds in a manner which is similar to the errors-only algorithm. In particular, the next step is to compute the syndrome polynomial $S(x) = S_1 + S_2 x + \cdots + S_r x^{r-1}$, where

$$S_j = \sum_{i=0}^{n-1} R_i' \alpha^{ij}, \qquad \text{for } j = 1, 2, \ldots, r.$$

If now we define the *errors-and-erasures vector* $E' = (E_0', E_1', \ldots, E_{n-1}')$ as $E' = \mathbf{R}' - \mathbf{C}$, and the "twisted" errors-and-erasures vector V by

$$V = (E_0', E_1'\alpha, \ldots, E_{n-1}'\alpha^{n-1}), \tag{9.76}$$

then it follows by the results of Section 9.3 that $S(x) = \hat{V}(x) \bmod x^r$, and the key equation (9.37) becomes

$$\sigma(x)S(x) \equiv \omega(x) \pmod{x^r}, \tag{9.77}$$

where $\sigma(x)$ is the locator polynomial, and $\omega(x)$ is the evaluator polynomial, for the vector V. From now on, we'll call $\sigma(x)$ the *errors-and-erasures-locator polynomial*, and $\omega(x)$ *errors-and-erasures-evaluator polynomial*.

Let's focus for a moment on $\sigma(x)$, the errors-and-erasures-locator polynomial. We have

$$\sigma(x) = \prod_{i \in I}(1 - \alpha^i x), \tag{9.78}$$

where I is the errors-and-erasures set, i.e.,

$$I = I_0 \cup I_1, \tag{9.79}$$

where I_0 is the erasure set defined in (9.73) and I_1 is the *error set* defined as follows:

$$I_1 = \{i : R_i \neq * \text{ and } R_i \neq C_i\}.$$

It thus follows from (9.78) and (9.79) that

$$\sigma(x) = \sigma_0(x)\sigma_1(x), \tag{9.80}$$

where $\sigma_0(x)$ is as defined in (9.74) and

$$\sigma_1(x) = \prod_{i \in I_1}(1 - \alpha^i x). \tag{9.81}$$

Naturally we call $\sigma_1(x)$ the *error-locator polynomial*.

Now we return to the key equation (9.77). In view of (9.80), we already know part of $\sigma(x)$, viz. $\sigma_0(x)$, and so the decoding algorithm's next step is to compute the *modified syndrome polynomial* $S_0(x)$, defined as follows:

$$S_0(x) = \sigma_0(x)S(x) \bmod x^r. \tag{9.82}$$

Combining (9.77), (9.80), and (9.82), the key equation becomes

$$\sigma_1(x)S_0(x) \equiv \omega(x) \pmod{x^r}. \tag{9.83}$$

At this point, the decoder will know $S_0(x)$, and wish to compute $\sigma_1(x)$ and $\omega(x)$, using Euclid's algorithm. Is this possible? Yes, because we have

$$\deg \sigma_1(x) = e_1,$$

$$\deg \omega(x) \leqslant e_0 + e_1 - 1$$

so that $\deg \sigma_1 + \deg \omega \leqslant e_0 + 2e_1 - 1 < r = \deg x^r$, since we have assumed $e_0 + 2e_1 \leqslant r$. Although it may no longer be true that $\gcd(\sigma(x), \omega(x)) = 1$, it will be true that $\gcd(\sigma_1(x), \omega(x)) = 1$ (see Prob. 9.45). It thus follows from Theorem 9.6 that the procedure $\texttt{Euclid}(x^r, S_0(x), \mu, \nu)$ will return $\sigma_1(x)$ and $\omega(x)$, if μ and ν are chosen properly. To chose μ and ν, we reason as follows. Since $e_0 + 2e_1 \leqslant r$, we have

$$\deg \sigma_1(x) = e_1 \leqslant \frac{r - e_0}{2},$$

so that $\deg \sigma_1(x) \leqslant \lfloor (r - e_0)/2 \rfloor$. Similarly,

$$\deg \omega(x) \leqslant e_0 + e_1 - 1 \leqslant e_0 + \left\lfloor \frac{r - e_0}{2} \right\rfloor - 1$$

$$= \left\lfloor \frac{r + e_0}{2} \right\rfloor - 1$$

$$\leqslant \left\lceil \frac{r + e_0}{2} \right\rceil - 1$$

It is an easy exercise to prove that $\lfloor (r - e_0)/2 \rfloor + \lceil (r + e_0)/2 \rceil = r$ (see Prob. 9.43), and so it follows that if we define

$$\mu = \left\lfloor \frac{r - e_0}{2} \right\rfloor,$$

$$\nu = \left\lceil \frac{r + e_0}{2} \right\rceil - 1, \tag{9.84}$$

then the procedure $\texttt{Euclid}(x^r, S_0(x), \mu, \nu)$ is guaranteed to return a pair of polynomials $(v(x), r(x))$ such that $\sigma_1(x) = \lambda v(x)$, $\omega(x) = \lambda r(x)$, where λ is a nonzero scalar. To find λ we recall that $\sigma_1(0) = 1$ (see (9.81)), and so we have

$$\sigma_1(x) = v(x)/v(0),$$

$$\omega(x) = r(x)/v(0).$$

Now, having computed the erasure-locator polynomial $\sigma_0(x)$ and the error-locator polynomial $\sigma_1(x)$, the algorithm computes the erasure-and-error-locator polynomial $\sigma(x)$ by polynomial multiplication—see (9.80).

At this stage, the algorithm has both the locator polynomial $\sigma(x)$ and the evaluator polynomial $\omega(x)$ for the errors-and-erasures vector E', and the decoding can be completed by either the "time-domain completion" or the "frequency-domain completion" described in Section 9.6. The errors-and-erasures decoding algorithm is thus summarized in Figure 9.9.

Example 9.10 We illustrate the erasures-and-errors RS decoding algorithm with the (7, 2) RS code over the field $GF(8)$, which has generator polynomial $g(x) = (x - \alpha)(x - \alpha^2)(x - \alpha^3)(x - \alpha^4)(x - \alpha^5) = x^5 + \alpha^2 x^4 + \alpha^3 x^3 + \alpha^6 x^2 + \alpha^4 x + \alpha$. (We are assuming that α, a primitive root in $GF(8)$, is a root of the $GF(2)$-primitive polynomial $x^3 + x + 1$.) The code's redundancy is $r = 5$ and so by Theorem 9.11, it can correct any pattern of e_0 erasures and e_1 errors, provided $e_0 + 2e_1 \leqslant 5$. Let us take the garbled codeword

$$\mathbf{R} = [\alpha^4, \alpha^3, \alpha^6, *, \alpha^2, \alpha^4, \alpha^2],$$

```
/*RS Errors-and-Erasures Decoding Algorithm*/
{
```
$\quad$ Input I_0; $e_0 = |I_0|$;

$\quad \sigma_0(x) = \prod_{i \in I_0} (1 - \alpha^i x)$;

$\quad$ for $(i \in I_0)$

$\quad\quad R_i = 0$;

$\quad$ for $(j = 1, 2, \ldots, r)$

$\quad\quad S_j = \sum_{i=0}^{n-1} R_i \alpha^{ij}$;

$\quad S(x) = S_1 + S_2 x + \cdots + S_r x^{r-1}$;

$\quad S_0(x) = \sigma_0(x) S(x) \bmod x^r$;

$\quad \mu = \lfloor (r - e_0)/2 \rfloor$; $v = \lceil (r + e_0)/2 \rceil - 1$;

$\quad$ Euclid$(x^r, S_0(x), \mu, v)$;

$\quad \sigma_1(x) = v(x)/v(0)$;

$\quad \omega(x) = r(x)/v(0)$;

$\quad \sigma(x) = \sigma_0(x)\sigma_1(x)$;

$\quad \vdots$

$\quad$ (Time-domain completion or frequency-domain
$\quad$ completion)
```
}
```

Figure 9.9 Decoding RS (or BCH) codes when erasures are present.

and try to decode it, using the algorithm in Figure 9.9.

The first phase of the decoding algorithm is the "erasure management," which in this case amounts simply to observing that the erasure set is $I_0 = \{3\}$, so that $e_0 = 1$, the erasure-locator polynomial is

$$\sigma_0(x) = 1 + \alpha^3 x,$$

and the modified received vector $\mathbf{R}'$ is

$$\mathbf{R}' = [\alpha^4, \alpha^3, \alpha^6, 0, \alpha^2, \alpha^4, \alpha^2].$$

The next step is to compute the syndrome values S_1, S_2, S_3, S_4, S_5, using $\mathbf{R}'$. We have

$$S_j = \alpha^4 + \alpha^{3+j} + \alpha^{6+2j} + \alpha^{2+4j} + \alpha^{4+5j} + \alpha^{2+6j},$$

so that a routine calculation gives

$$S_1 = 1, \ S_2 = 1, \ S_3 = \alpha^5, \ S_4 = \alpha^2, \ S_5 = \alpha^4.$$

Thus the modified syndrome polynomial $S_0(x)$ is

$$S_0(x) = (1 + x + \alpha^5 x^2 + \alpha^2 x^3 + \alpha^4 x^4)(1 + \alpha^3 x)(\bmod x^5)$$

$$= 1 + ax + \alpha^2 x^2 + \alpha^4 x^3 + x^4$$

Since $e_0 = 1$, $r = 5$, the parameters μ and ν are

$$\mu = \left\lfloor \frac{5-1}{2} \right\rfloor = 2,$$

$$\nu = \left\lceil \frac{5+1}{2} \right\rceil - 1 = 2.$$

Thus we are required to invoke $\texttt{Euclid}(x^5, S_0(x), 2, 2)$. Here is a summary of the work:

i	v_i	r_i	q_i
-1	0	x^5	—
0	1	$x^4 + \alpha^4 x^3 + \alpha^2 x^2 + ax + 1$	—
1	$x + \alpha^4$	$\alpha^4 x^3 + \alpha^5 x^2 + \alpha^4 x + \alpha^4$	$x + \alpha^4$
2	$\alpha^3 x^2 + \alpha^4 x + \alpha^6$	$\alpha^4 x^2 + \alpha^5 x + \alpha^6$	$\alpha^3 x + \alpha^5$

Thus $\texttt{Euclid}(x^5, S_0(x), 2, 2)$ returns $(v_2(x), r_2(x)) = (\alpha^3 x^2 + \alpha^4 x + \alpha^6, \alpha^4 x^2 + \alpha^5 x + \alpha^6)$, so that

$$\sigma_1(x) = \alpha v_2(x) = \alpha^4 x^2 + \alpha^5 x + 1,$$

$$\omega(x) = \alpha r_2(x) = \alpha^5 x^2 + \alpha^6 x + 1$$

and finally

$$\sigma(x) = \sigma_0(x)\sigma_1(x) = x^3 + \alpha^2 x^2 + \alpha^2 x + 1.$$

This completes the "erasure-specific" part of the decoding, i.e., the portion of the algorithm described in Figure 9.9. We will now finish the decoding, using both the time-domain and frequency-domain completions.

For the time-domain completion, we note that $\sigma'(x) = x^2 + \alpha^2$, and compute the following table:

i	$\sigma(\alpha^{-i})$	$\sigma'(\alpha^{-i})$	$\omega(\alpha^{-i})$	$E_i = \omega(\alpha^{-i})/\sigma'(\alpha^{-i})$
0	0	α^6	α^3	α^4
1	α^5			
2	α^4			
3	0	α^4	α^5	α
4	0	1	α^3	α^3
5	α^5			
6	α^5			

Thus the errors-and-erasures vector is $\mathbf{E}' = [\alpha^4, 0, 0, \alpha, \alpha^3, 0, 0]$ (which means that there are two errors, in positions 0 and 4, in addition to the erasure in position 3), and so the decoded codeword is $\hat{\mathbf{C}} = \mathbf{R}' + \mathbf{E}'$, i.e.,

$$\hat{\mathbf{C}} = [0, \alpha^3, \alpha^6, \alpha, \alpha^5, \alpha^4, \alpha^2].$$

For the frequency-domain completion, having already computed S_1, S_2, S_3, S_4, S_5, we compute S_6 and $S_7 (= S_0)$ via the recursion

$$S_j = \alpha^2 S_{j-1} + \alpha^2 S_{j-2} + S_{j-3}$$

(since $\sigma(x) = 1 + \alpha^2 x + \alpha^2 x^2 + \alpha^3$), and find that $S_6 = \alpha^2$, and $S_0 = \alpha_5$. Thus the complete syndrome vector S is

$$S = [\alpha^5, 1, 1, \alpha^5, \alpha^2, \alpha^4, \alpha^2].$$

we now compute the inverse DFT of S, i.e.,

$$E_i' = \alpha^5 + \alpha^{-i} + \alpha^{-2i} + \alpha^{5-3i} + \alpha^{2-4i} + \alpha^{4-5i} + \alpha^{2-6i}$$

$$= \alpha^5 + \alpha^{6i} + \alpha^{5i} + \alpha^{5+4i} + \alpha^{2+3i} + \alpha^{4+2i} + \alpha^{2+i}.$$

This gives

$$\mathbf{E}' = [\alpha^4, 0, 0, \alpha, \alpha^3, 0, 0]$$

just as in the time-domain completion, and so

$$\hat{\mathbf{C}} = [0, \alpha^3, \alpha^6, \alpha, \alpha^5, \alpha^4, \alpha^2]$$

as before. $\qquad\qquad\qquad\qquad\qquad\qquad\qquad\qquad\qquad\qquad\square$

Let's conclude this section with a brief discussion of how to decode BCH codes when erasures are present. The key difference between the (errors-only) decoding algorithm for BCH codes and RS codes is that BCH codes, once the errors have been *located*, there is no need to *evaluate* them, since the only possible error value is 1. What this means is that when erasures are present, the algorithm in Figure 9.9 still holds (with $2t$ replacing r); the only way in which the decoding of BCH codes is simpler is in the implementation of the time-domain completion. (Compare Figures 9.3 and 9.4.)

9.8 The (23, 12) Golay code

In this section we will discuss an extremely beautiful but alas! nongeneralizable code, the binary (23, 12) Golay code. It is arguably the single most important error-correcting code. (There is also an (11, 6) Golay code over $GF(3)$; see Probs. 9.64–9.67.)

We begin with a tantalizing number-theoretic fact. In the 23-dimensional vector space over $GF(2)$, which we call V_{23}, a Hamming sphere of radius 3 contains

$$1 + \binom{23}{1} + \binom{23}{2} + \binom{23}{3} = 2048 \text{ vectors.}$$

But $2048 = 2^{11}$ is an exact power of 2, and thus it is conceivable that we could pack V_{23} with $4096 = 2^{12}$ spheres of radius 3, exactly, with no overlap. If we could perform this combinatorial miracle, the centers of the spheres would constitute a code with 2^{12} codewords of length 23 (rate = $12/23 = 0.52$) capable of correcting any error pattern of weight $\leqslant 3$. In this section, not only will we prove that such a packing is possible; we will show that the centers of the spheres can be taken as the codewords in a (23, 12) binary cyclic code!

In coding-theoretic terms, then, we need to construct a binary cyclic (23, 12) three-error-correcting code, i.e., one with $d_{\min} \geqslant 7$. We base the construction on certain properties of the field $GF(2^{11})$. Since $2^{11} - 1 = 2047 = 23 \cdot 89$, $GF(2^{11})$ must contain a primitive 23rd root of unity, which we shall call β. The minimal polynomial of β over $GF(2)$ is $g(x) = \prod_{\gamma \in B}(x - \gamma)$, where $B = \{\beta^{2^i} : i = 0, 1, 2, \ldots\}$ is the set of conjugates of β. A simple computation shows that B contains only 11 elements; indeed,

$$g(x) = \prod_{\gamma \in B}(x - \gamma), \tag{9.85}$$

where

$$B = \{\beta^j : j = 1, 2, 4, 8, 16, 9, 18, 13, 3, 6, 12\}.$$

Similarly the minimal polynomial of $\beta^{-1} = \beta^{22}$ is

$$\tilde{g}(x) = \prod_{\gamma \in \tilde{B}}(x - \gamma) \tag{9.86}$$

where

$$\tilde{B} = \{\beta^j : j = 22, 21, 19, 15, 7, 14, 5, 10, 20, 17, 11\}.$$

Since every 23rd root of unity except 1 is a zero of either $g(x)$ or $\tilde{g}(x)$, it follows that the factorization of $x^{23} - 1$ into irreducible factors over $GF(2)$ is

$$x^{23} - 1 = (x - 1)g(x)\tilde{g}(x). \tag{9.87}$$

In fact, it can be shown that[8]

$$\begin{aligned}
g(x) &= x^{11} + x^9 + x^7 + x^6 + x^5 + x + 1, \\
\tilde{g}(x) &= x^{11} + x^{10} + x^6 + x^5 + x^4 + x^2 + 1,
\end{aligned} \tag{9.88}$$

but we will not need this explicit factorization in the rest of this section. We can now define the Golay code.

Definition *The (23, 12) Golay code is the binary cyclic code whose generator polynomial is g(x), as defined in (9.85) or (9.88).*

Now all (!) we have to do is show that the code's minimum weight is $\geqslant 7$. The first step in this direction is rather easy.

Lemma 3 *Each nonzero Golay codeword has weight $\geqslant 5$.*

Proof In view of the structure of the set B of zeros of $g(x)$ (cf. Eq. (9.85)), we see that for every codeword C,

$$C(\beta) = C(\beta^2) = C(\beta^3) = C(\beta^4) = 0. \tag{9.89}$$

It thus follows from the BCH argument (Theorem 9.3) that $d_{\min} \geqslant 5$. □

In view of Lemma 3, it remains to show that there can be no words of weight 5 or 6. The next lemma allows us to focus our attention on words of even weight.

Lemma 4 *If A_i denotes the number of Golay codewords of weight i, then, for $0 \leqslant i \leqslant 23$,*

$$A_i = A_{23-i}. \tag{9.90}$$

Proof By (9.87), $g(x)\tilde{g}(x) = (x^{23} - 1)/(x - 1) = 1 + x + x^2 + \cdots + x^{22}$, and so the constant vector $K = (11111 \ldots 111)$ is in the code. By adding K to a word of weight i, we get a word of weight $23 - i$, and conversely. Thus the correspondence $C \leftrightarrow C + K$ is one-to-one between words of weights i and $23 - i$. □

The next lemma eliminates words of weight 2, 6, 10, 14, 18, and 22; by Lemma 4 this also eliminates words of weight 1, 5, 9, 13, 17, and 21, and thus proves that the Golay code has minimum distance $\geqslant 7$.

Lemma 5 *If C is a Golay codeword of even weight w, then $w \equiv 0 \pmod 4$.*

Proof Let the generating function for C be denoted by $C(x)$, i.e.,

$$C(x) = x^{e_1} + x^{e_2} + \cdots + x^{e_w}, \tag{9.91}$$

where $0 \leqslant e_1 < e_2 < \cdots < e_w \leqslant 22$. Since C belongs to the Golay code, $C(\beta) = 0$, that is,

$$C(x) \equiv 0 \pmod{g(x)}. \tag{9.92}$$

Since C has even weight, $C(1) = 0$, that is

$$C(x) \equiv 0 \pmod{x - 1}. \tag{9.93}$$

Now if we define $\tilde{C}(x)$ by

$$\tilde{C}(x) = x^{-e_1} + x^{-e_2} + \cdots + x^{-e_w}, \tag{9.94}$$

with exponents taken modulo 23, it follows that $\tilde{C}(\beta^{-1}) = C(\beta) = 0$, that is,

$$\tilde{C}(x) \equiv 0 \pmod{\tilde{g}(x)}. \tag{9.95}$$

Combining (9.92), (9.93), (9.95) with (9.87), we have

$$C(x)\tilde{C}(x) \equiv 0 \pmod{x^{23} - 1}. \tag{9.96}$$

Now let us actually compute $C(x)\tilde{C}(x) \pmod{x^{23} - 1}$, using the defining equations (9.91) and (9.94):

$$C(x)\tilde{C}(x) \equiv \sum_{i,j=1}^{w} x^{e_j - e_j} \qquad (\text{mod } x^{23} - 1)$$

$$\equiv w + \sum_{\substack{i,j=1 \\ i \neq j}}^{w} x^{e_j - e_j} \quad (\text{mod } x^{23} - 1)$$

$$\equiv \sum_{\substack{i,j=1 \\ i \neq j}}^{w} x^{e_i - e_j} \qquad (\text{mod } x^{23} - 1) \qquad (9.97)$$

(the last congruence because w is even and all computations take place in $GF(2)$). Thus

$$C(x)\tilde{C}(x) \equiv \sum_{b=1}^{22} \mu_b x^b \quad (\text{mod } x^{23} - 1),$$

where μ_b is the number of ordered pairs (i, j) with $e_i - e_j \equiv b \pmod{23}$. By formula (9.96) each μ_b is even:

$$\mu_b \equiv 0 \pmod{2}, \quad b = 1, 2, \ldots, 22. \qquad (9.98)$$

Now, if $e_i - e_j \equiv b$, then also $e_j - e_i \equiv 23 - b \pmod{23}$. Thus

$$\mu_b = \mu_{23-b}, \quad b = 1, 2, \ldots, 11. \qquad (9.99)$$

Finally, since there are $w(w - 1)$ terms in the sum on the right side of (9.97),

$$\sum_{b=1}^{22} \mu_b = w(w - 1). \qquad (9.100)$$

Combining (9.98), (9.99), and (9.100), we have

$$w(w - 1) = \sum_{b=1}^{22} \mu_b$$

$$= 2 \sum_{b=1}^{11} \mu_b$$

$$\equiv 0 \pmod{4},$$

i.e., $w(w - 1)$ is a multiple of 4. But since $w - 1$ is *odd*, it follows that w itself is a multiple of 4, which completes the proof. ☐

Combining Lemmas 3, 4, and 5, we arrive at the following theorem.

Theorem 9.13 *The number of codewords of weight i in the* (23, 12) *Golay code is 0 unless i* = 0, 7, 8, 11, 12, 15, 16, *or* 23. *Thus the spheres of radius 3 around the codewords do indeed pack* V_{23} *perfectly.*

There is a simple but useful variant of the (23, 12) Golay code, that deserves mention here. If $C = (C_0, C_1, \ldots, C_{22})$ is a Golay codeword, let us extend C to length 24 by *appending an overall parity check*, i.e., by defining a 24th component C_{23} as follows:

$$C_{23} = C_0 + C_1 + \cdots + C_{22}.^9$$

If every Golay codeword is extended in this way, the resulting code is a binary linear (but no longer cyclic) (24, 12) code, which is called the (24, 12) *extended Golay code*. It is a simple matter then to prove the following theorem. (See Problem 9.59.)

Theorem 9.14 *The number of codewords of weight i in the* (24, 12) *extended Golay code is* 0 *unless i* = 0, 8, 12, 16, *or* 24.

The (24, 12) extended Golay code enjoys two small advantages over the original (23, 12) Golay code, which are however enough to make the extended code preferable in most applications. First, since 24 is a multiple of eight, the (24, 12) code is naturally suited to byte-oriented implementations. Second, since the minimum distance of the extended code is *eight*, if it is used to correct all patterns of three or fewer errors, all error patterns of weight 4, and many error patterns of higher weight, will still be *detectable*, whereas the original (23, 12) has no such extra detection capability. (See Problems 9.59, 9.62.)

We conclude with some remarks about the implementation of the Golay codes. Since the (23, 12) code is cyclic, it is clear that we could design an 11-sttage shift-register encoder for it (see Section 8.2, and do Problem 9.60). The design of an algebraic decoding algorithm is not so easy; we could easily modify the BCH decoding algorithm in Figures 9.1 and 9.2 to correct every pattern of two or fewer errors, but the code is "accidentally" capable of correcting three! Fortunately, however, the code is small enough so that the syndrome "table lookup" algorithm discussed in Section 7.2 is usually practical. (See Prob. 9.63.)

Problems

9.1 We saw in Section 9.1 that the function $\mathbf{f}(\mathbf{V}) = \mathbf{V}^3$ makes the matrix H_2 of Eq. (9.2) into the parity-check matrix of a two-error-correcting code. Investigate whether or not the following candidate $\mathbf{f}$'s work:

(a) $\mathbf{f}(\mathbf{V}) = T\mathbf{V}$, where T is a linear transformation of V_m.

(b) $\mathbf{f}(\mathbf{V}) = a_0 + a_1\mathbf{V} + a_2\mathbf{V}^2$, where $\mathbf{V}$ is regarded as an element of $GF(2)$.

(c) $\mathbf{f}(\mathbf{V}) = \mathbf{V}^{-1}$, where $\mathbf{V} \in GF(2^m)$.

9.2 Suppose F is a finite field with q elements.

(a) If a is an arbitrary element of F, define the $(q-1)$st-degree polynomial $f_a(x) = (x - a)^{q-1} - 1$. Find the value of $f_a(x)$ for each of the q elements of F.

(b) Using the results of part (a), or otherwise, show that every function $f : F \to F$ can be represented as a polynomial of degree at most $q - 1$.

9.3 (This problem gives a generalization of the Vandermonde determinant theorem, which is needed in the proof of Theorem 9.1.) Let $P_i(x)$ be a monic polynomial of degree i, for $i = 0, 1, \ldots, n - 1$, and let $x_1, x_2, \ldots, x_n$ be distinct indeterminantes. Show that

$$\det \begin{pmatrix} P_0(x_1) & \cdots & P_0(x_n) \\ P_1(x_1) & \cdots & P_1(x_n) \\ \vdots & & \vdots \\ P_{n-1}(x_1) & \cdots & P_{n-1}(x_n) \end{pmatrix} = \prod_{i=1}^{n-1} \prod_{j=i+1}^{n} (x_j - x_i).$$

[*Hint*: If $x_i = x_j$, the left side vanishes.] The Vandermonde determinant theorem is the special case $P_i(x) = x^i$.

9.4 Here is a pseudocode listing for an algorithm for computing the dimension of the t-error-correcting BCH code of length $2^m - 1$:

```
{
    S = {1, 3, ..., 2t − 1};
    k = 2^m − 1;
    while (S is not empty)
    {
        u_0 = least element in S;
        u = u_0;
        do
        {
            delete u from S;
            k = k − 1;
            u = 2u mod 2^m − 1;
        }
        while (u ≠ u_0)
    }
}
```

(a) Show that when the algorithm terminates, the integer k is equal to the dimension of the t-error-correcting BCH code of length $2^m - 1$.

(b) Use the algorithm to compute the dimension of the t-error-correcting BCH code of length 63, for $1 \leqslant t \leqslant 31$.

9.5 (a) Prove that the dimension of the two-error-correcting BCH code of length $n = 2^m - 1$ is $n - 2m$, for all $m \geqslant 3$.

(b) More generally show that for any fixed $t \geqslant 1$, the dimension of the t-error-correcting BCH code of length $n = 2^m - 1$ is $n - mt$ for all sufficiently large m.

(c) What is the smallest value of m_0 such that the *three*-error-correcting BCH code of length $n = 2^m - 1$ has dimension $n - 3m$ for all $n \geqslant m_0$?

9.6 (a) For each t in the range $1 \leqslant t \leqslant 7$, compute the dimension of the t-error-correcting BCH code of length 15.

(b) For each of the codes in part (a), calculate the generator polynomial, assuming a primitive root α in $GF(16)$ that satisfies the equation $\alpha^4 = \alpha + 1$. (Cf. Example 9.1.)

9.7 In Example 9.1 we computed the generator polynomial for the three-error-correcting BCH code of length 15, under the assumption that the primitive root α of $GF(16)$ satisfied $\alpha^4 = \alpha + 1$. If we had chosen a primitive root satisfying $\alpha^4 = \alpha^3 + 1$ instead, what would the generator polynomial have turned out to be?

9.8 Prove the inverse DFT formula, Eq. (9.12)

9.9 Consider the field $GF(7)$, as represented by the set of integers $\{0, 1, 2, 3, 4, 5, 6\}$, with all arithmetic done modulo 7.

(a) Show that 3 is a primitive 6th root of unity in $GF(7)$.

(b) Using 3 as the needed primitive 6th root of unity, find the DFT of the vectors $\mathbf{V}_1 = (1, 2, 3, 4, 5, 6)$ and $\mathbf{V}_2 = (1, 3, 2, 6, 4, 5)$.

(c) Why do you suppose the DFT of $\mathbf{V}_2$ is so much "simpler" than that of $\mathbf{V}_1$?

9.10 Prove that the DFT of the phase-shifted vector $\mathbf{V}_\mu$ in equation (9.17) is given by the formula (9.18).

9.11 (Generalized BCH codes). Let $g(x)$ be a polynomial with coefficients in $GF(q)$ which divides $x^n - 1$. Further assume that α is an nth root of unity in some extension field of $GF(q)$ and that

$$g(\alpha^i) = 0, \quad \text{for } i = m_0, m_0 + 1, \ldots, m_0 + d - 2$$

for integers m_0 and d. Let C be the cyclic code with generator polynomial $g(x)$. Show that the minimum distance of C is $\geqslant d$. [*Hint*: Use the BCH argument.]

9.12 Is the converse of the "BCH argument" (Theorem 9.3) true? That is , if $\mathbf{V}$ is a vector of weight w, does it necessarily follow that the DFT $\hat{V}$ must have $w - 1$ or more consecutive zero components? If your answer is *yes*, give a proof. If you answer is *no*, give an explicit counterexample.

9.13 Prove that $\gcd(\hat{V}(x), 1 - x^n) = \prod_{i \notin I}(1 - \alpha^i x)$, where $\hat{V}(x)$ is as defined in (9.14), and I is as defined in (9.19).

9.14 Show that if any d consecutive components of $\hat{V}$ are known, the rest can be calculated, provided we also know $\sigma_{\mathbf{V}}(x)$ (cf. Corollary 9.2).

9.15 The field $GF(16)$, as represented in Table 9.1, contains a a primitive 5th root of unity, namely α^3, which for the remainder of this problem we shall denote by β. Let $\mathbf{V} = (1, \beta^4, \beta^5, 0, \beta^7)$, a vector of length 5 over $GF(16)$. Using the definitions in Section 9.3, compute $\hat{\mathbf{V}}$, $\sigma_{\mathbf{V}}$, $\sigma_{\mathbf{V}}^{(i)}$ for $i = 1, 2, 4$, and $\omega_{\mathbf{V}}(x)$. (Cf. Example 9.2.)

9.16 In Example 9.2, the components of $\hat{\mathbf{V}}$ satisfy the recursion $\hat{V}_j = \alpha^6 \hat{V}_{j-3}$. Explain "why" this should be so. [*Hint*: Examine (9.31) carefully.]

9.17 If S_j represents the jth syndrome value in the decoding of a BCH code (cf. Eq. (9.34)), show that, for all j, $S_{2j} = S_j^2$.

9.18 If $f(x) = f_0 + f_1 x + \cdots + f_n x^n$ is a polynomial over a field F, its *formal derivative* $f'(x)$ is defined as follows.

$$f'(x) = f_1 + 2f_2 x + \cdots + nf_n x^{n-1}.$$

From this definition, without the use of limits, deduce the following facts:
(a) $(f + g)' = f' + g'$.
(b) $(fg)' = fg' + f'g$.
(c) $(f^m)' = mf^{m-1}f'$.
(d) If $f(x) = \prod_{i=1}^r (x - \beta_i)$, then

$$f'(x) = \sum_{i=1}^r \prod_{\substack{j=1 \\ j \neq i}}^r (x - \beta_j).$$

(e) If $f(x)$ is as given in part (d) and the β_i are distinct, then

$$\sum_{i=1}^r \frac{1}{x - \beta_i} = \frac{f'(x)}{f(x)}.$$

9.19 (a) Prove properties A–F of Euclid's algorithm given in Table 9.2.
(b) Prove that $r_n(x)$, the last nonzero remainder in Euclid's algorithm, is a greatest common divisor of $a(x)$ and $b(x)$, and that Eq. (9.43) holds.

9.20 Let $a(x) = x^8 - 1$, $b(x) = x^6 - 1$, polynomials over the field $GF(2)$.
(a) Apply Euclid's algorithm to the pair $(a(x), b(x))$, thus obtaining a table like that in Example 9.3.
(b) For each pair (μ, ν) with $\mu \geq 0$, $\nu \geq 0$, and $\mu + \nu = 7$, calculate Euclid(a, b, μ, ν). (Cf. Example 9.4.)

9.21 (Padé approximants). Let $A(x) = a_0 + a_1 x + a_2 x^2 + \cdots$ be a power series over a field F. If μ and ν are nonnegative integers, a (μ, ν) *Padé approximation* to $A(x)$ is a rational function $p(x)/q(x)$ such that

(a) $$q(x)A(x) \equiv p(x) \pmod{x^{\mu + \nu + 1}},$$

(b) $$\deg q(x) \leq \mu, \deg p(x) \leq \nu.$$

Using Theorem 9.5, show that for each (μ, ν) there is (apart from a scalar factor) a unique pair $(p_0(x), q_0(x))$ such that if (a) and (b) and hold, then $p(x) = \lambda p_0(x)$ and $q(x) = \lambda q_0(x)$ for a nonzero scalar λ. The pair $(p_0(x), q_0(x))$ is called *the* (μ, ν) *Padé approximant* to $A(x)$. Referring to Table 9.3, compute the Padé approximants to $A(x) = 1 + x + x^2 + x^4 + x^6 + \cdots$ over $GF(2)$ with $\mu + \nu = 7$.

9.22 With the same setup as Example 9.5, decode the following noisy codeword from the $(15, 5)$ three-error-correcting BCH code:

$$R = [R_0, \ldots, R_{14}] = [110101010010010].$$

9.23 Consider the three-error-correcting BCH code of length 31, defined in terms of a primitive root $\alpha \in GF(32)$ satisfying $\alpha^5 + \alpha^2 + 1 = 0$.
  (a) Compute the generator polynomial.
  (b) Decode the following received vector: [00000001111010111101110001 0000].
  (c) Decode the following received vector: [10110011111010100110001 0010 1001].

9.24 Let α be a primitive nth root of unity in the field F, and let $\mathcal{P}_k$ be the set of polynomials of degree $\leqslant k - 1$ over F. For each $P \in \mathcal{P}_k$, define the vector $\mathbf{C}(P) = (P(1), P(\alpha), \ldots, P(\alpha^{n-1}))$.
  (a) Show that the code consisting of all vectors $\mathbf{C}(P)$ is an MDS code, and find the corresponding n, k, and d.
  (b) Is the code cyclic? Explain.
  (c) What relationship, if any, does this code bear to the RS code as defined in (9.67)?

9.25 Let F be any field which contains a primitive nth root of unity α. If r and i are fixed integers between 0 and n, the set of all vectors $\mathbf{C} = (C_0, C_1, \ldots, C_{n-1})$ with components in F such that

$$\sum_{i=0}^{n-1} C_i \alpha^{ij} = 0, \quad \text{for } j = i + 1, i + 2, \ldots, i + r,$$

is called an *alternate* Reed–Solomon code.
  (a) Show that the code so defined is an $(n, n - r)$ cyclic code. Find the generator polynomial and $d_{\min}$.
  (b) Explicitly compute the generator polynomial $g(x)$ for the alternate RS code with $F = GF(8)$, $n = 7$, $r = 4$, and $i = 1$. (Cf. Example 9.6.)
  (c) Show that there exist n fixed nonzero elements from F, say $\gamma_0, \ldots, \gamma_{n-1}$, such that the alternate code just described can be obtained from the original RS code (defined in (9.67)) by mapping each original RS codeword $(C_0, \ldots, C_{n-1})$ to the vector $(\gamma_0 C_0, \ldots, \gamma_{n-1} C_{n-1})$.

9.26 Let α be an nth root of unity in a finite field F, and let C be the linear code of length n defined as follows: $C = (C_0, C_1, \ldots, C_{n-1})$ is a codeword if and only if

$$\sum_{i=0}^{n-1} \frac{C_i}{x - \alpha^i} \equiv 0 \pmod{x^r},$$

where x is an indeterminate.
  (a) Show that the code C is cyclic.
  (b) Find the dimension of the code, in terms of n and r.
  (c) Find the code's minimum distance.
  (d) What is the relationship (if any) betwen C and the (n, k) RS code over F with generator polynomial $g(x) = (x - \alpha) \ldots (x - \alpha^r)$?

9.27 Give a complete discussion of *all* MDS codes over $GF(2)$.

9.28 Show that over any field F, the following two codes are MDS codes.

(a) The $(n, 1)$ repetition code.

(b) The $(n, n - 1)$ parity-check code.

9.29 In Theorem 9.7, it is shown in a (n, k) Reed–Solomon code over a field F, the minimum weight codeword has weight $n - k + 1$. Question: If the field F has q elements, exactly how many words of weight $n - k + 1$ are there in the code? [*Hint*: Use Theorem 9.9.]

9.30 Using Table 9.1 for guidance, compute the generator polynomial $g(x)$ for a $(15, 7)$ RS code over $GF(16)$. (Cf. Example 9.6.)

9.31 Consider the $(7, 3)$ RS code of Example 9.6.

(a) Find the unique codeword $(C_0, \ldots, C_6)$ such that $C_0 = 1$ $C_1 = 0$, and $C_2 = 0$. (Cf. Example 9.7.)

(b) Is there a codeword with $C_0 = \alpha^3$, $C_1 = \alpha$, $C_2 = 1$, and $C_3 = 0$?

9.32 Decode the following garbled codeword from the $(7, 3)$ RS code from Example 9.8:

$$\mathbf{R} = [\alpha^3 \ 1 \ \alpha \ \alpha^2 \ \alpha^3 \ \alpha \ 1].$$

9.33 Do you think there is a $(21, 9)$ binary linear code which is capable of correcting any pattern of 4 or fewer errors?

Problems 9.34–9.39 are all related. They present an alternative approach to Reed–Solomon codes. More important, however, they culminate with the construction of the famous Justesen codes (see Ref. [15]). These codes are important because they (together with certain variants of them) are the only known explicity constructable family of linear codes which contain sequences of codes of lengths n_i, dimensions k_i, and minimum distances d_i such that:

$$\lim_{i \to \infty} n_i = \infty,$$

$$\lim_{i \to \infty} k_i/n_i > 0,$$

$$\lim_{i \to \infty} d_i/n_i > 0.$$

(See conclusions of Prob. 9.39. Also cf. Prob. 7.21, the Gilbert bound, which shows that such sequences of codes must exist, but does not explain how to construct them.)

9.34 (Cf. Theorem 9.10.) Denote by P_r the set of polynomials of degree $\leqslant r$ over the finite field $GF(q^m)$ and let $(\alpha_0, \alpha_1, \ldots, \alpha_{n-1})$ be a list of $n > r$ distinct elements from $GF(q^m)$. Corresponding to each $f(x) \in P_r$, let the vector $(C_0, C_1, \ldots, C_{n-1}) \in GF(q^m)^n$ be defined by $C_i = f(\alpha_i)$. Show that the set of vectors obtained from P_r in this way is a linear code over F_{q^m} with length n, dimension $r + 1$, and minimum distance $n - r$.

9.35 The setup being the same as in Prob. 9.34, corresponding to each $f \in P_r$ let

$$\mathbf{C} = (C_0, C'_0, C_1, C'_1, \ldots, C_{n-1}, C'_{n-1}) \in GF(q^m)^{2n}$$

be defined by $C_i = f(\alpha_i)$, $C'_i = \alpha_i f(\alpha_i)$. Show that the set of vectors thus obtained is a linear code over $GF(q^m)$ with length $2n$ and dimension $r + 1$, such that within each nonzero codeword there are at least $n - r$ distinct pairs (C_i, C'_i).

9.36 Let $\phi : GF(q^m) \to GF(q)^m$ be a one-to-one linear mapping from $GF(q^m)$ onto

$GF(q)^m$. Take the code over $GF(q^m)$ defined in Prob. 9.35, and make it into a code over $GF(q)$ by mapping the codeword $\mathbf{C}$ onto $(\phi(C_0), \phi(C_0'), \ldots, \phi(C_{n-1}), \phi(C_{n-1}'))$. Show that the resulting $GF(q)$ linear code has length $2mn$ and dimension $m(r+1)$, and that within each nonzero codeword, among the n subvectors $(\phi(C_i), \phi(C_i'))$, there are at least $n-r$ distinct ones.

9.37 (This problem is not out of place, despite appearances.) Let $\{\mathbf{x}_1, \ldots, \mathbf{x}_M\}$ be a set of M distinct vectors from $Vn(F_2)$, and let $w_i = w_H(x_i)$ denote the Hamming weight of x_i. Let $p = (w_1 + \cdots + w_M)/nM$. Prove that

$$\log M \leqslant nH_2(p),$$

where H_2 is the binary entropy function. [*Hint*: Let $\mathbf{X} = (X_1, X_2, \ldots, X_n)$ be a random vector which is equally likely to assume any of the values $\mathbf{x}_i$, and let p_j denote the fraction of the M vectors that have a 1 in their jth coordinate. Now verify the following string of equalities and inequalities:

$$\log M = H(\mathbf{X}) \leqslant \sum_{j=1}^{n} H(X_j) = \sum_{j=1}^{n} H(p_j) \leqslant nH(p).]$$

(This result due to James Massey [38].)

9.38 Returning to the codes defined in Prob. 9.35, specialize to $q = 2$, $n = 2^m$, and $(\alpha_0, \alpha_1 \ldots, \alpha_{n-1})$ any ordering of the elements of $GF(2^m)$. Let $r/2^m = \rho$, and show that the resulting codes have
   (i) length $= m2^{m+1}$,
   (ii) rate $= \frac{1}{2}\left(\rho + \frac{1}{2^m}\right)$,
   (iii) $\frac{d_{\min}}{n} \geqslant (1-\rho)H_2^{-1}\left[\frac{1}{2} + \frac{\log_2(1-\rho)}{2m}\right]$.
   These codes are the *Justesen codes* mentioned above. [*Hint*: To prove (iii), use the results of Prob. 9.37.]

9.39 Finally, show that for any $0 \leqslant R \leqslant \frac{1}{2}$, there is an infinite sequence of Justesen codes over $GF(2)$ with lengths n_i, dimensions k_i, and minimum distances d_i such that:

$$\lim_{i \to \infty} n_i = \infty,$$

$$\lim_{i \to \infty} k_i/n_i = R,$$

$$\limsup_{i \to \infty} d_i/n_i \geqslant H_2^{-1}(1/2) \cdot (1 - 2R)$$

$$= 0.110028(1 - 2R).$$

9.40 Show that $\bar{d}_H$, as defined in the proof of Theorem 9.11, is a bona fide metric (cf. Problem 7.4).

9.41 For the $(7, 3)$ code of Example 9.9, find a vector $\mathbf{y} \in \{0, 1, *\}^7$ for which $\min_i \bar{d}_H$ is as large as possible.

9.42 For a given value of d, how many pairs of nonnegative integers (e_0, e_1) are there such that $e_0 + 2e_1 \leqslant d - 1$?

9.43 If m and n are positive integers such that $m + n$ is even, show that

$$\left\lfloor \frac{m}{2} \right\rfloor + \left\lceil \frac{n}{2} \right\rceil = \frac{m+n}{2}.$$

(See the remarks immediately preceeding Eqs. (9.84).)

9.44 Consider the (7, 4) Hamming code of Example 7.3. The code has $d_{min} = 3$ and so by Theorem 9.11, it is capable of correcting any pattern of e_0 erasures and e_1 errors, provided $e_0 + 2e_1 \leqslant 2$. Bearing this in mind, decode (if possible) each of the words in parts (a), (b), and (c).
 (a) [1 1 1 0 0 * 0]
 (b) [0 * 1 1 1 0 1]
 (c) [0 1 * 1 0 * 1]
 (d) If **R** is a randomly chosen vector of length 7 containing exactly one erasure, what is the probability that it will be uniquely decoded by the MDD decoding algorithm introduced in the proof of Theorem 9.11?

9.45 Show that if $\sigma_1(x)$ is the error-locator polynomial, and $\omega(x)$ is the errors-and-erasures evaluator polynomial for RS errors-and-erasures decoding (see equations (9.77) and (9.81)), then $\gcd(\sigma_1, \omega) = 1$.

9.46 Investigate the probability of decoder error for a Reed–Solomon decoder under the following circumstances.
 (a) r erasures and 1 error.
 (b) $r - 1$ erasures and 1 error.
 (c) $r + 1$ erasures, no errors.

9.47 Consider the (15, 7) RS code over $GF(16)$ (generated by a primitive root satisfying $\alpha^4 = \alpha + 1$), and decode the following received word.

$$\mathbf{R} = [\alpha^{13} \ 1 * \alpha^{10} \ \alpha^{12} \ \alpha^6 * \alpha^5 \ \alpha^{13} * \alpha^1 \ \alpha^8 \ \alpha^7 \ \alpha^2 \ \alpha^9].$$

9.48 Using the suggestions at the end of Section 9.7, decode the following noisy vector from the (15, 5) BCH code with generator polynomial $g(x) = x^{10} + x^8 + x^5 + x^4 + x^2 + x^2 + 1$ (cf. Example 9.1).

$$\mathbf{R} = [1 \ 1 * 0 \ 0 \ 0 * 0 \ 0 \ 0 \ 1 * 1 \ 0 \ 1].$$

 (a) Use the time-domain completion.
 (b) Use the frequency-domain completion.

9.49 Consider an (n, k) linear code with $d_{min} = d$. If there are no errors, Theorem 9.11 guarantees that the code is capable of correcting any pattern of up to $d - 1$ erasures. Show that this result cannot be improved, by showing that there is at least one set of d erasures that the code isn't capable of correcting.

9.50 In Section 8.4, we considered codes capable of correcting single bursts of errors. It turns out that is is very much easier to correct single bursts of *erasures*. After doing this problem, you will agree that this is so.
 (a) If C is an (n, k) linear code which is capable of correcting any erasure burst of length b or less, show that $n - k \geqslant b$. (Cf. the Reiger bound, corollary to Theorem 8.10.)
 (b) Show that any (n, k) cyclic code is capable of correcting any erasure burst of length $n - k$ or less.
 (c) Consider the (7, 3) cyclic code of Example 8.2. Correct the following

codewords, which have suffered erasure bursts of length 4: $(10****0)$, $(****101), (*101***)$.

9.51 When a linear code has suffered erasures *but no errors*, there is a very simple general technique for correcting the erasures, which we shall develop in this problem. The idea is to replace each of the erased symbols with a distinct indeterminate, and then to solve for the indeterminates, using the parity-check matrix. For example, consider the (7, 4) binary Hamming code of Example 7.3. The code has $d_{min} = 3$ and so by Theorem 9.11, it is capable of correcting any pattern of two or fewer erasures, provided there are no errors. If say the received word is $R = (1 * 1 * 1\ 0\ 1)$, we replace the two erased symbols with indeterminates x and y, thus obtaining $R = (1\ x\ 1\ y\ 1\ 0\ 1)$.

    (a) Use the fact that every codeword in the (7, 4) Hamming code satisfies the equation $HC^T = 0$, where H is the parity-check matrix given in Section 7.4, to obtain three simultaneous linear equations in the indeterminates x and y, and solve these equations, thus correcting the erasures.

    (b) If there were *three* erased positions instead of only two, we could use the same technique to obtain three equations in the three indeterminates representing the erasures. We could then solve these equations for the indeterminates, thus correcting three erasures. Yet Theorem 9.11 only guarantees that the code is capable of correcting two erasures. What's wrong here?

9.52 In this problem, we consider an alternative approach to correcting erasures and errors, which involves the idea of "guessing" the values of the erasures.

    (a) Assume first that the code is binary, i.e., the field F in Theorem 9.11 is $GF(2)$. Suppose we have a code C with minimum distance d, and that we have received a noisy codeword containing e_0 erasures and e_1 errors, with $e_0 + 2e_1 \leq d - 1$. Suppose that we change all the erased positions to 0, and then try to decode the word, using an errors-only decoding algorithm capable of correcting any pattern of up to $(d-1)/2$ errors. If the decoder succeeds, we stop. Otherwise, we try again, this time assuming that all the erased positions are 1. Show that this procedure, i.e., guessing that the erasures are all 0's, and then guessing that they are all 1's, will always succeed in correcting the errors and erasures.

    (b) Illustrate the decoding technique suggested in part (a) by decoding the word $[1 * *0 * 01]$ from the (7, 3) binary cyclic code. (Cf. Example 9.9.)

    (c) Does this "guessing" procedure work for nonbinary fields? In particular, how could you modify it to work over the ternary field $GF(3)$?

9.53 Consider the (7, 3) binary cyclic code with generator polynomial $g(x) = x^4 + x^3 + x^2 + 1$, as described, e.g., in Example 9.9. It has $d_{min} = 4$, and so by Theorem 9.10, it can correct any pattern of up to 3 erasures (if no attempt is made to correct errors as well). It can also, however, correct some, though not all, patterns of *four* erasures. In this problem, you are aksed to investigate the $\binom{7}{4} = 35$ possible erasure patterns of size four, and to determine which or them are correctable. In particular, please find the *number* of weight four erasure patterns that are correctable.

9.54 This problem concerns the probability that a randomly selected vector from $GF(q)^n$ will be decoded by a decoder for a Reed–Solomon code.

(a) Derive the following formula for the fraction of the total "volume" of $GF(q)^n$ occupied by nonoverlapping Hamming spheres of radius t around the codewords in an (n, k) code over $GF(q)$:

$$\frac{q^k \sum_{i=0}^{t} \binom{n}{i}(q-1)^i}{q^n}.$$

(b) Use the formula from part (a) to compute the limit, for a fixed value of t, as $q \to \infty$, of the probability that a randomly selected word of length $q - 1$ will fall within Hamming distance t or less of some codeword in a t-error-correcting RS code of length $q - 1$ over $GF(q)$. (Assume that the code's redundancy is $r = 2t$.)

9.55 Let C be an (n, k) binary cyclic code, with generator polynomial $g(x)$ and parity-check polynomial $h(x)$.

(a) Show that if $h(1) \neq 0$, then every codeword has even weight.

(b) Show that if there is no pair (θ_1, θ_2) of roots of $h(x)$ such that $\theta_1 \theta_2 = 1$, then every codeword's weight is divisible by four. [*Hint*: This is a generalization of the result in Lemma 5.]

9.56 In the text we proved that the $(23, 12)$ Golay code had $d_{\min} \geq 7$. Show that in fact, $d_{\min} = 7$ for this code. Do this in two ways:

(a) By examining the generator polynomial $g(x)$.

(b) By showing that *any* $(23, 12)$ binary linear code must have $d_{\min} \leq 7$.

9.57 Show that there is no $(90, 78)$ binary linear code with $d_{\min} = 5$, i.e., a perfect two-error-correcting code of length 90, despite the fact that $1 + \binom{90}{1} + \binom{90}{2} = 2^{12}$. [*Hint*: Let r denote the number of 12-bit syndromes of odd weight corresponding to one-bit errors. Show that the number of odd-weight syndromes corresponding to two-bit errors is $r(90 - r)$, and attempt to determine r.]

9.58 The $(23, 12)$ binary Golay code defined in Section 9.8, when combined with syndrome table lookup decoding, has the property that every error pattern of weight ≤ 3 will be corrected.

(a) Describe in detail what the decoder will do if the error pattern has weight 4.

(b) What if the error pattern has weight 5?

(c) Generalize the result of parts (a) and (b). For each integer t in the range $4 \leq t \leq 23$, what will the decoder do, if the error pattern has weight t?

9.59 Prove Theorem 9.14.

9.60 This problem concerns the number of codewords of weight 8 in the $(24, 12)$ extended Golay code.

(a) Show that the number of codewords of weight 8 is not zero. Do this in two ways: (1) By examining the generator polynomial $g(x)$, for the original $(23, 12)$ Golay code; (2) By showing that *any* $(24, 12)$ binary linear code must have $d_{\min} \leq 8$.

(b) Given [10] that the code contains exactly 759 words of weight 8, show that for any subset $\{i_1, \ldots, i_5\}$ of five elements from $\{0, 1, \ldots, 23\}$ there is exactly one codeword of weight 8 which is 1 at these five coordinate positions.

9.61 This problem concerns the error-*detecting* capabilities of the extended (24, 12) Golay code. (In parts (a) and (b), we assume that the code is being used to correct all error patterns of weigh three or less.)

(a) Show that the code can detect all error patterns of weight 4.

(b) Give that the weight enumerator of the code is

$$1 + 759x^8 + 2576x^{12} + 759x^{16} + x^{24},$$

for each e in the range $4 \leqslant e \leqslant 24$, compute the number of error patterns of weight e that the code will detect.

(c) Now assume the decoder only attempts to correct error patterns of weight *two* or less, and repeat part (b).

(d) Now assume the decoder only attempts to correct error patterns of weight *one* or less, and repeat part (b).

(e) Finally, assume the decoder is used in a *detect-only mode*, i.e., if the syndrome is zero, the received word is accepted as correct, but otherwise it is rejected. Repeat part (b).

9.62 In this problem, we will briefly consider encoders for the (23, 12) and (24, 12) Golay codes.

(a) Design a shift-register encoder for the (23, 12) Golay code.

(b) By modifying your design in part (a), or otherwise, come up with a design for an encoder for the (24, 12) extended Golay code.

9.63 Discuss the size and complexity of a syndrome table lookup decoder for the (24, 12) Golay code.

In Probs. 9.64–9.67 we will investigate the ternary Golay code. Observe that in the vector space $GF(3^{11})$ a Hamming sphere of radius 2 contains

$$1 + 2\binom{11}{1} + 4\binom{11}{2} = 243 = 3^5$$

vectors. This suggests that it might be possible to perfectly pack $GF(3^{11})$ with $729 = 3^6$ spheres of radius 2. The ternary Golay code does this. It is an (11, 6) linear code over $GF(3)$ whose codewords, when taken as sphere centers, produce such a packing. The code is defined as follows. Since $3^5 - 1 = 11 \cdot 22$, it follows that $GF(3^5)$ contains a primitive 11th root of unity, which we shall call β. Since over $GF(3)$ the factorization of $x^{11} - 1$ is $x^{11} - 1 = (x - 1)g(x)\tilde{g}(x)$, where $g(x) = x^5 + x^4 - x^3 + x^2 - 1^8$ and $\tilde{g}(x) = x^5 - x^3 + x^2 - x - 1$, we may assume that β is a zero of $g(x)$. The (11, 6) ternary Golay code is then defined to be the cyclic code with generator polynomial $g(x)$. To show that the spheres of radius 2 around the 729 codewords are disjoint, we must show that the minimum Hamming distance between codewords is $\geqslant 5$, that is, every nonzero codeword has weight $\geqslant 5$. The following problems contain a proof of this fact.[10]

9.64 Show that the code's minimum weight is $\geqslant 4$. [*Hint*: Use the BCH argument, Theorem 9.3.]

9.65 Show that if $C_0 + C_1 + \cdots + C_{10} = 0$, the codeword $\mathbf{C} = (C_0, C_1, \ldots, C_{10})$ has Hamming weight divisible by 3. [*Hint*: See Lemma 4, Section 9.8.]

9.66 If, on the other hand, $C_0 + C_1 + \cdots + C_{10} = \alpha \neq 0$, show that $(C_0 + \alpha, C_1 + \alpha, \ldots, C_{10} + \alpha)$ is a codeword and has weight divisible by 3.

9.67 Use the preceding results to show that the code contains no codewords of weight 4, 7, or 10. [*Hint*: If the weight is 4, by appropriate scalar multiplication the nonzero components can be transformed to either $(1, 1, 1, 1)$ or $(1, 1, 1, -1)$.]

Notes

1 (p. 230). The approach taken in this section is largely that of Berlekamp [14], Section 4.

2 (p. 230). The codes that are now universally called BCH codes were discovered in 1959 by the French mathematician A. Hocquenghem and independently in 1960 by R. C. Bose and D. K. Ray-Chaudhuri. However, Hocquenghem's work went unnoticed for several years and the new codes were called Bose-Chaudhuri codes (not Bose-Ray-Chaudhuri!) for a while. When it was belatedly discovered that in fact Hocquenghem had anticipated Bose and Ray-Chaudhuri the codes were rechristened Bose-Chaudhuri-Hocquenghem or *BCH codes*. It is important to remember that only the codes, not the decoding algorithms, were discovered by these early writers. For the history of the decoding algorithms see pp. 354–355.

3 (p. 237). If the characteristic of F is finite, we assume that the characteristic does not divide n.

4 (p. 244). By convention the degree of the zero polynomial is $-\infty$. This is done so that basic facts like $\deg(ab) = \deg(a) + \deg(b)$, $\deg(a + b) \leq \max(\deg(a), \deg(b))$, etc., will hold even if one of a or b is the zero polynomial.

5 (p. 247). This proof is due to J. B. Shearer.

6 (p. 254). In almost, but not all, applications, the field F will be $GF(2^m)$ for some $m \geq 1$. However, the underlying theory goes through equally well for any field, finite or not, and we shall make no unnecessary restrictions in F.

7 (p. 271). Unless, of course, some of the erased components of **C** were actually 0, in which cases **C** and **R**$'$ would differ in fewer than $e_0 + e_1$ positions.

8 (p. 278, 291). These factorizations are difficult to come by without further study. We recommend that the interested reader consult Berlekamp [14], Chapter 6.

9 (p. 281). See Problem 7.17(b).

10 (p. 290, 291). The weight enumerators for the Golay codes are as follows (reference: MacWilliams and Sloane [19], Chapter 20):

Golay (23, 12)		Golay (24, 12)		Golay (11, 6)	
i	A_i	i	A_i	i	A_i
0	1	0	1	0	1
7	253	8	759	5	132
8	506	12	2576	6	132
11	1288	16	759	8	330
12	1288	24	1	9	110
15	506			11	24
16	253				
23	1				

10

Convolutional codes

10.1 Introduction

In this chapter we study the class of *convolutional*[1] codes. Throughout we assume for simplicity that our codes are for use on a binary input channel, i.e., that the channel input alphabet can be identified with the finite field F_2. (It is quite easy, however, to generalize everything to the case $A_X \cong F_q$, where q is a prime power.)

Convolutional codes can be studied from many different points of view. In this introductory section we present three approaches, which we have called the *polynomial matrix* approach, the *scalar matrix* approach, and the *shift-register* approach. (Three other approaches, the *state-diagram* approach, the *trellis* approach, and the *tree* approach, will be given later in the chapter.[2])

•*The polynomial matrix approach.* Recall from Chapter 7 that a binary (n, k) linear block[3] code can be characterized by a $k \times n$ generator matrix $G = (g_{ij})$ over F_2. An (n, k) *convolutional code* (CC) is also characterized by a $k \times n$ generator matrix $\mathbf{G}$; the difference is that for a convolutional code the entries g_{ij} are *polynomials* over F_2. For example, the matrix

$$\mathbf{G} = [x^2 + 1, x^2 + x + 1]$$

is the generator matrix for a $(2, 1)$ CC, which we label CC 1 for future reference. Similarly,

$$\mathbf{G} = \begin{bmatrix} 1 & 0 & x+1 \\ 0 & 1 & x \end{bmatrix}$$

is the generator matrix for a $(3, 2)$ CC, which we call CC 2.

We now define three important numbers that are associated with a CC:
The *memory*:

$$M = \max_{i,j} [\deg(g_{ij})]. \qquad (10.1)$$

The *constraint length*:[4]

$$K = M + 1. \qquad (10.2)$$

The *rate*:

$$R = k/n. \qquad (10.3)$$

Thus CC 1 has $M = 2$, $K = 3$, $R = 1/2$, and CC 2 has $M = 1$, $K = 2$, $R = 2/3$. The physical significance of these quantities will emerge as the theory develops. For now, simply observe that since the entries in the generator matrix of a block code are polynomials of degree 0 (i.e., scalrs), and (n, k) block code can be viewed as a CC with $M = 0$, $K = 1$, and $R = k/n$.

In order to use the polynomial matrix $\mathbf{G}$ to encode scalar information, the information bits must be mapped into the coefficients of a k-tuple of polynomials $\mathbf{I} = (I_0(x), \ldots, I_{k-1}(x))$. Then the "codeword" $\mathbf{C} = (C_0(x), \ldots, C_{n-1}(x))$, which is an n-tuple of polynomials, is defined by

$$\mathbf{C} = \mathbf{I} \cdot \mathbf{G}, \qquad (10.4)$$

where the dot denotes vector-matrix multiplication. Thus in the polynomial matrix approach the CC with generator matrix $\mathbf{G}$ is the rowspace of $\mathbf{G}$.

Example 10.1 In CC 1, the polynomial information $\mathbf{I} = (x^3 + x + 1)$ would be encoded via Eq. (10.4) into the polynomial codeword $\mathbf{C} = (x^5 + x^2 + x + 1, x^5 + x^4 + 1)$. $\qquad \square$

Example 10.2 In CC 2, the polynomial information $\mathbf{I} = (x^2 + x, x^3 + 1)$ would be encoded via Eq. (10.4) into the polynomial codeword $\mathbf{C} = (x^2 + x, x^3 + 1, x^4 + x^3)$. $\qquad \square$

We have not yet specified the exact correspondence between k- and n-tuples of polynomials and bit patterns. Consideration of this correspondence leads naturally to the scalar matrix approach to CC's.

● *The scalar matrix approach.* The most natural bitwise (scalar) representation of a polynomial codeword $\mathbf{C} = (C_0(x), \ldots, C_{n-1}(x))$ is obtained by interleaving the polynomials' coefficients. Thus if the jth polynomial is $C_j(x) = C_{j0} + C_{j1}x + \ldots$, the scalar representation of $\mathbf{C}$ is

$$\mathbf{C} = (C_{00}, C_{10}, \ldots, C_{n-1,0}, C_{01}, \ldots, C_{n-1,1}, \ldots). \qquad (10.5)$$

Recalling that the CC with generator matrix $\mathbf{G}$ is the rowspace of $\mathbf{G}$, it is now easy to construct a scalar version of **G** with the property that the scalar codewords defined by Eq. (10.5) form the rowspace of the scalar **G**. Let

$$\mathbf{G} = \sum_{\nu=0}^{M} G_\nu x^\nu \tag{10.6}$$

be the expansion of **G** as a polynomial of degree M (cf. Eq. (10.1)), with coefficients G_ν that are $k \times n$ scalar matrices. Then the scalar version of **G** that we want is described in Fig. 10.1.

Notice that is scalar generator matrix has an infinite number of rows and columns. This corresponds to the fact that the information and codeword polynomials can have arbitrarily large degree.

Example 10.3 The polynomial generator matrix for CC 1, expanded as in Eq. (10.6), is

$$\mathbf{G} = [1, 1] + [0, 1]x + [1, 1]x^2.$$

Hence by Fig. 10.1 the scalar generator matrix for CC 1 is

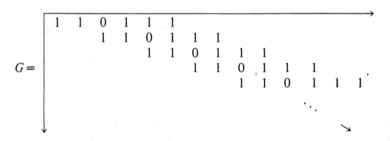

The scalar information corresponding to the polynomial information $\mathbf{I} = (x^3 + x + 1)$ is (1101) (not (1011)) and the scalar codeword correspond-

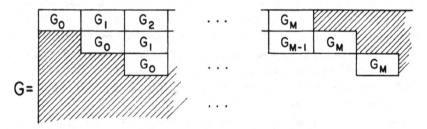

Figure 10.1 The scalar generator matrix of a CC with polynomial generator matrix given by Eq. 10.6) (shaded area = all 0's).

ing to the polynomial codeword $C = (x^5 + x^2 + x + 1, x^5 + x^4 + 1)$ is (111010000111)(cf. Example 10.1). □

Example 10.4 The polynomial generator matrix for CC 2, expanded as in Eq. (10.6), is

$$G = \begin{bmatrix} 1 & 0 & 1 \\ 0 & 1 & 0 \end{bmatrix} + \begin{bmatrix} 0 & 0 & 1 \\ 0 & 0 & 1 \end{bmatrix} x.$$

Hence by Fig. 10.1 the scalar generator matrix for CC 2 is

$$G = \begin{bmatrix}
1 & 0 & 1 & 0 & 0 & 1 \\
0 & 1 & 0 & 0 & 0 & 1 \\
  &   & 1 & 0 & 1 & 0 & 0 & 1 \\
  &   & 0 & 1 & 0 & 0 & 0 & 1 \\
  &   &   &   & 1 & 0 & 1 & 0 & 0 & 1 \\
  &   &   &   & 0 & 1 & 0 & 0 & 0 & 1 \\
  &   &   &   &   &   &   & \ddots
\end{bmatrix} .$$

The scalar information corresponding to the polynomial information $I = (x^2 + x, x^3 + 1)$ is (01101001). The scalar codeword corresponding to the polynomial codeword $C = (x^2 + x, x^3 + 1, x^4 + x^3)$ is (010100100011001) (cf. Example 10.2). □

We mentioned above in principle the degrees of the information polynomials (and so also the codeword polynomials) can be arbitrarily large, but of course in any practical application there will be a maximum allowable degree. This leads us to define the *Lth truncation* of a CC as follows.

First we require that $\deg[I_i(x)] \le L - 1$, $i = 0, 1, \ldots, k - 1$. Then, by Eqs. (10.1) and (10.4), in the corresponding polynomial codeword $C = (C_0(x), \ldots, C_{n-1}(x))$, each component has degree $\le M + L - 1$. Thus the information $I = (I_0(x), \ldots, I_{k-1}(x))$ can be represented by kL bits, and the codeword C by $n(M + L)$ bits. The encoding mapping from I to C can then be represented in scalar notation as $C = I \cdot G_L$, where the scalar matrix G_L, which is a truncation of the matrix of G of Fig. 10.1, is depicted in Fig. 10.2. (The shaded areas denote blocks of 0's.)

Thus the Lth truncation of an (n, k) CC can be viewed as an $(n(M + L), kL)$ linear block code, and so in this sense a convolutional code is a special kind of block code. The rate of the truncated code is given by

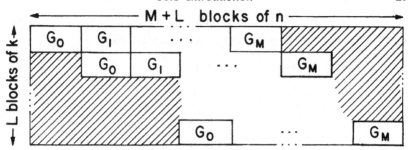

Figure 10.2 The matrix G_L.

$$R_L = \frac{kL}{n(M + L)}$$

$$= R\left(1 - \frac{M}{M + L}\right),$$

where $R = k/n$ is the rate of the untruncated CC (see Eq. (10.3)). In most practical situations L is taken to be much larger than M, in which case the rate R_L will be very close to R. This is one reason R is called the rate of the code. (For another reason, see p. 298.)

Example 10.5 If we take $L = 6$ in CC 1, we obtain a (16, 6) linear block code with generator matrix

$$G_6 = \begin{bmatrix} 1 & 1 & 0 & 1 & 1 & 1 & & & & \\ & 1 & 1 & 0 & 1 & 1 & 1 & & & \\ & & 1 & 1 & 0 & 1 & 1 & 1 & & \\ & & & 1 & 1 & 0 & 1 & 1 & 1 & \\ & & & & 1 & 1 & 0 & 1 & 1 & 1 \\ & & & & & 1 & 1 & 0 & 1 & 1 & 1 \end{bmatrix}. \quad \square$$

Example 10.6 If we take $L = 2$ in CC 2, we obtain a (9, 4) linear code with generator matrix

$$G_2 = \begin{bmatrix} 1 & 0 & 1 & 0 & 0 & 1 & & & \\ 0 & 1 & 0 & 0 & 0 & 1 & & & \\ & & 1 & 0 & 1 & 0 & 0 & 1 \\ & & 0 & 1 & 0 & 0 & 0 & 1 \end{bmatrix}. \quad \square$$

•*The shift-register approach.* Suppose now we wished to design an encoder for CC 1, that is, a device which would accept a stream of information bits

$\mathbf{I} = (I_0, I_1, \ldots)$ as input and produce an encoded stream $\mathbf{C} = (C_{00},$ $C_{10}, C_{01}, C_{11}, \ldots)$ as output, where $\mathbf{I}$ and $\mathbf{C}$ are related by

$$C_0(x) = C_{00} + C_{01}x + \cdots$$

$$= (x^2 + 1)(I_0 + I_1x + \cdots)$$

$$= (x^2 + 1)I(x)$$

and

$$C_1(x) = C_{10} + C_{11}x + \cdots$$

$$= (x^2 + x + 1)(I_0 + I_1x + \cdots)$$

$$= (x^2 + x + 1)I(x).$$

Such an encoder must be capable of multiplying the input stream by the two polynomials $x^2 + 1$ and $x^2 + x + 1$. But we have already seen in Chapter 8 (cf. Fig. 8.1) how to do this. The circuit

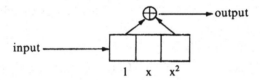

will multiply by $x^2 + 1$, and the circuit

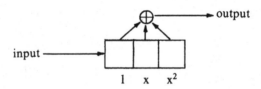

will multiply by $x^2 + x + 1$. Hence the circuit depicted in Fig. 10.3 will be an encoder for CC 1.[5] From Fig. 10.3 we see that the input and output streams are related by

$$\begin{aligned} C_{0j} &= I_j & +I_{j-2} \\ C_{1j} &= I_j + I_{j-1} & +I_{j-1} \end{aligned} \Bigg\}, \qquad j = 0, 1, \ldots \qquad (10.7)$$

Thus the stream $\mathbf{C}_i$ is obtained by *convolving* the input stream $\mathbf{I}$ with the appropriate polynomial.[1]

Notice from Fig. 10.3 that the jth output bits C_{0j} and C_{1j} depend not only on the jth input bit I_j but also on the preceding two bits I_{j-1} and I_{j-2}. The encoder must remember two input bits in addition to the current bit; this is why the code is said to have memory $M = 2$. Also, for every input bit there are two output bits; thus the code's rate is $1/2$. Finally, note that the encoder of Fig. 10.3 has three flip-flops; this is why the code is said to have constraint length 3.

The shift-register encoder for a general (n, k) convolutional code is a straightforward generalization of the one in Fig. 10.3. There will be k shift registers, one corresponding to each of the k input streams $\mathbf{I}_0, \mathbf{I}_1, \ldots, \mathbf{I}_{k-1}$. The ith shift register will convolve the ith input stream with each of the n polynomials $g_{i0}(x), \ldots, g_{i,n-1}(x)$. The jth output stream will then be formed by summing the jth output stream from each of the shift registers.

Example 10.7 A shift-register encoder for CC 2 is illustrated in Fig. 10.4. □

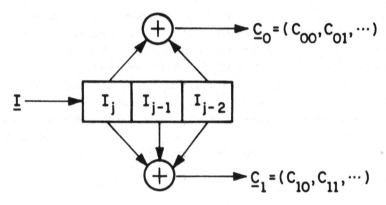

Figure 10.3 A shift-register encoder for CC 1.

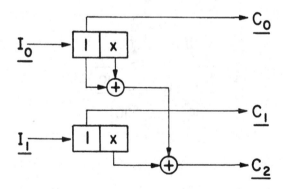

Figure 10.4 A shift-register encoder for CC 2.

From the shift-register viewpoint, a CC is the collection of all possible output streams from a particular encoder. At first glance this approach is not very appealing. And yet we will see in the next section that the shift-register approach leads directly to what is perhaps the most powerful known approach, the *state-diagram* approach.

10.2 State diagrams, trellises, and Viterbi decoding

This section is in two parts. In the first part we illustrate all the concepts for a particular convolutional code, CC 1. After we have completed this study of CC 1, we will discuss (see pp. 304–305) the generalization to other CC's.

Let us define the *state* of the encoder of Fig. 10.3 at given instant as the contents of the rightmost two flip-flops in the shift-register, that is, as the pair (I_{j-1}, I_{j-2}). There are four possible states: 00, 01, 10, 11. At every tick of the clock the encoder accepts an input bit (I_j), in response moves to another state (I_j, I_{j-1}), and emits two output bits C_{0j} and C_{1j} (see Eq. (10.7)). Thus the behavior of the encoder of Fig. 10.3 can be completely described by the *state diagram* of Fig. 10.5 in which the four boxes represent the four states; a transition from one state to another corresponding to an input of "0" is represented by a solid edge, a transition corresponding to "1" by a dashed edge, and the label on an edge represents the encoder's output as it moves from one state to another. For example, consider the edge going from state $c(10)$ to state $d(11)$. It represents the encoder's behavior when the shift-register in Fig. 10.3 contains $(I_j, I_{j-1}, I_{j-2}) = (110)$. The edge is dashed

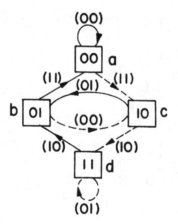

Figure 10.5 State diagram for the encoder of Fig. 10.3.

because $I_j = 1$. Using Eq. (10.7), we calculate $C_{0j} = 1$, $C_{1j} = 0$, and so the label on the edge is (10).

With the help of the state diagram of Fig. 10.5, it is easy to calculate the output of the encoder of Fig. 10.3 corresponding to any given input sequence. We simply start in state a and walk through the state diagram, following a solid edge if the input is "0," a dashed edge if it is a "1," and output the label of each edge traversed. For example if the input stream is 110100, we follow the path *acdbcba* and output 111010000111. [Note. This is the same output sequence calculated in Example 10.3.]

As we simulate the encoder's behavior by wandering through the state diagram, we may travel the same edge many times; this makes it difficult to keep track of where we have been, and it becomes desireable to modify the state diagram by including the dimension of time. Perhaps the best way to do this is to have a different copy of the state diagram for each tick, and to join them together into a *trellis diagram* as shown in Fig. 10.6, in which each column of four dots represents the four states a, b, c, d. There is one such column for each value of $j = 0, 1, 2, \ldots$. The index j will be called the *depth*, and the depth j appearance of a state will be denoted by a subscript j. A depth j state is joined by an edge to a depth $(j + 1)$ state iff there is an edge joining the two states in the state diagram. (An exception occurs for $j = 0$ and 1 because at $j = 0$ the shift register of Fig. 10.3 contains $[I_0, 0, 0]$ and so must be in state $a = \boxed{00}$.) The possible encoder outputs can be found by tracing the appropriate path through the trellis. For example, the output stream (111010000111...) corresponding to the input stream 110100... can be found by tracing the path $a_0 c_1 d_2 b_3 c_4 b_5 a_6 \ldots$ in the trellis of Fig. 10.6.

As we mentioned in Section 10.1, in any practical situation it wil be necessary to work with a truncated version of the code. Corresponding to the

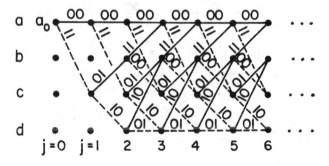

Figure 10.6 The trellis diagram corresponding to the state diagram of Fig. 10.5.

Lth truncation of the code, there is a truncated trellis, which yields a graphical representation of all possible encoder output streams when the input stream is L bits long. The $L = 6$ truncated trellis for CC 1 is illustrated in Fig. 10.7. (The reader may wonder why the 6 input bits I_0, I_1, ..., I_5 give rise to an 8-edge path through the trellis. This is due to the code's 2-bit memory. The 8 edges in a path in Fig. 10.7 correspond to the encoder's output when the contents of the shift register of Fig. 10.3 are $[I_0, 0, 0]$, $[I_1, I_0, 0]$, ..., $[I_5, I_4, I_3]$, $[0, I_5, I_4]$, $[0, 0, I_5]$. This also explains why the final two edges in any path through the trellis of Fig. 10.7 are solid.)

We will now explain how the trellis diagram can be used to decode a CC. Suppose for example that the $L = 6$ truncation of CC 1 is being used on a BSC with raw bit error probability $p < \frac{1}{2}$, and that $\mathbf{R} = [1011001110111100]$ is received. (This sequence has been written over the trellis of Fig. 10.7 for reference.) We have already noted (cf. Example 10.5) that this code is a (16, 6) linear block code; a maximum likelihood decoder must find that one of the 64 codewords which is closest in Hamming distance to $\mathbf{R}$ (cf. Prob. 2.13 and Section 7.3). A brute-force approach to this problem would be to compare, bit by bit, each of the 64 codewords to $\mathbf{R}$. However, there is a much simpler method which takes advantage of the fact that the 64 codewords correspond to the 64 paths from a_0 to a_8 in the trellis of Fig. 10.7.

Let us draw a new version of Fig. 10.7 in which each trellis edge is labeled with the Hamming distance between its Fig. 10.7 label and the corresponding two bits of $\mathbf{R}$. For example, the $b_3 \rightarrow c_4$ edge gets the label $d_H(00, 11) = 2$. The result is the trellis of Fig. 10.8. If we think of the edge labels in Fig. 10.8 as lengths, the total Hamming distance between $\mathbf{R}$ and a given codeword is just the total length of the trellis path corresponding to the codeword.

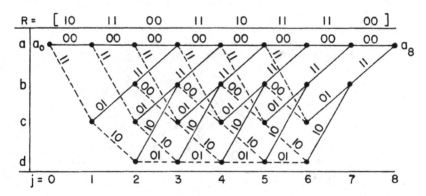

Figure 10.7 The $L = 6$ truncated trellis for CC 1. (See Example 10.5.)

For example, the Hamming distance between **R** and the codeword (0000110100101011) corresponding to the path $a_0a_1a_2c_3b_4c_5d_6b_7a_8$ is $1+2+2+1+1+1+1+2 = 11$. Thus the problem of finding the codeword closest to **R** becomes the problem of finding the shortest path from a_0 to a_8 in the trellis of Fig. 10.8, and we now focus on this shortest-path problem.

Suppose the shortest path P from a_0 to a_8 passes through some intermediate vertex, labelled x in Fig. 10.9. Let us denote the segment of P joining a_0 to x by P_1, and the segment joining x to a_8 by P_2. Clearly the path P_1 is the shortest path from a_0 to x, for if another path, say Q, were shorter, the path QP_2 would be shorter than $P = P_1P_2$, contradicting the minimality of P. This observation is the key to the *Viterbi decoding algorithm*, which works by constructing, for each j, a list of the shortest paths from a_0 to the vertices at depth j. The $(j+1)$st list is easily constructable from the jth list since if $a_0 \dots s_j t_{j+1}$ is a shortest path to t_{j+1}, then $a_0 \dots s_j$ must be a shortest path to s_j, and so the shortest depth $(j+1)$ paths can be obtained by extending the shortest depth j paths by one edge.

We are almost ready to state Viterbi's algorithm formally, but first we need some more notation. Denote by S the set of states $\{a, b, c, d\}$. Next if $s, t \in S$ and there is an edge going from s to t in the state diagram, we define

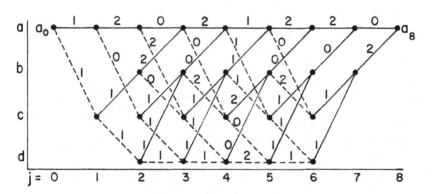

Figure 10.8 Another version of the trellis of Fig. 10.7.

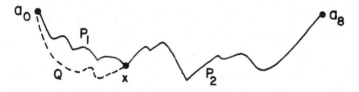

Figure 10.9 Paths between a_0 and a_8.

$B(s, t)$ to be 0 or 1 according to whether the transition from s to t corresponds to an 0 or a 1 input. If there is no such edge, $B(s, t)$ is not defined (See Fig. 10.10). Also, for $s, t \in S$, we define $l_{j-1,j}(s, t)$ to be the label on the trellis edge joining s_{j-1} to t_j. If there is no such edge, we set $l_{j-1,j}(s, t) = +\infty$. For example, $l_{0,1}(a, c) = 1$, $l_{2,3}(d, b) = 1$, $l_{7,8}(a, c) = +\infty$.

We are prepared to state Viterbi's algorithm. It computes two things, *metrics* and *survivors*. The metric $\mu_j(s)$, $s \in S$, represents the length of the shortest path from a_0 to s_j; the survivor $B_j(s)$ is a binary string of length j which represents a shortest path from a_0 to s_j. Thus $B_4(b) = 1010$ means that the shortest path from a_0 to b_4 is $a_0 c_1 b_2 c_3 b_4$. Viterbi's algorithm is given in Fig. 10.11. [Note. In step 2 the operation $*$ denotes concatenation; e.g., $1101 * 0 = 11010$.] The reader should now have no difficulty verifying (by induction on j) that $\mu_j(s)$ as computed by Viterbi's algorithm is in fact the length of a shortest path from a_0 to s_j, and that $B_j(s)$ describes such a path.

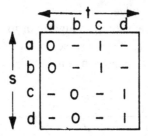

Figure 10.10 The function $B(s, t)$ for CC 1.

---

1. Initially set $\mu_0(a) \leftarrow 0$, and $\mu_0(s) = +\infty$ for all $s \neq a$. Also, set $B_0(a) = \phi$, and $j = 1$.

2. For each $s \in S$, find a $t \in S$ for which $\mu_{j-1}(t) + l_{j-1,j}(t,s)$ is a minimum. Then set

$$\mu_j(s) \leftarrow \mu_{j-1}(t) + l_{j-1,j}(t,s),$$
$$B_j(s) \leftarrow B_{j-1}(t) * B(t,s).$$

3. If $j = L + M$, output the first L bits of $B_j(a)$ and stop; otherwise set $j \leftarrow j + 1$ and go to Step 2.

Figure 10.11 Viterbi's decoding algorithm. Notation: S is the set of states, a is the all-zero state, $l_{j-1,j}(t, s)$ is the label of the trellis edge joining t_{j-1} to s_j, and $B(t, s)$ is 0 or 1 according to whether the transition from s to t corresponds to a 0 or a 1 input.

We can describe the performance of Viterbi's algorithm on the trellis[6] of Fig. 10.8 graphically, as in Fig. 10.12, where the metric $\mu_j(s)$ appears above the vertex s_j and the survivor $B_j(s)$ is represented by the unique path from a_0 to s_j. For example, $\mu_4(a) = 2$ and $B_4(a) = 0100$. The shortest path from a_0 to a_8 is seen to be $a_0 a_1 c_2 b_3 a_4 c_5 b_6 a_7 a_8$; it has length 4. The decoder's output is therefore 010010, and this is the maximum likelihood estimate of the information sequence $\mathbf{I} = (I_0, \ldots, I_5)$ that gave rise to $\mathbf{R}$.

This completes our discussion of this particular code (CC 1) on this particular channel (a BSC). We now discuss generalizations.

First we consider more general CC's. We restrict out attention to $(n, 1)$ codes, leaving the case of larger values of k to Prob. 10.5. The encoder (see Fig. 10.3) will contain one shift register of length $K = M + 1$; the state of the encoder is defined to be the length M vector consisting of the rightmost M flip-flops. Thus there will be 2^M states, and in the state diagram there will be two edges leading into, and two leading out from, each state. For example, the state diagram for an $(n, 1)M = 3$ CC would have the form shown in Fig. 10.13.[7] The edge labels are missing from Fig. 10.13. They would in general be length n binary vectors and would depend on the polynomials defining the code. As M becomes larger, the state diagram becomes harder to draw; but more important, as M approaches 10 or so, the structure of the state diagram becomes so complex that it is difficult to handle, even in a computer.

The Viterbi algorithm of Fig. 10.11 remains valid without change in this more general situation (assuming the channel is a BSC with $p < \frac{1}{2}$). The only problem is that the complexity of the algorithm is an exponentially increasing function of M (though it is only linear in L), and so it is only a practical method of decoding CC's with relatively small values of M. Indeed, with

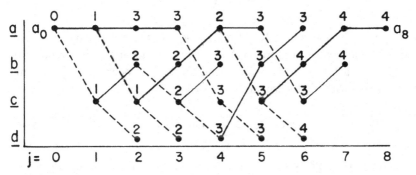

Figure 10.12 Viterbi's algorithm, applied to the trellis of Fig. 10.8.

Convolutional codes

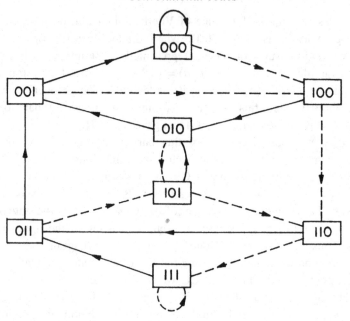

Figure 10.13 State diagram for $M = 3(K = 4)(n, 1)$ convolutional code.

today's available digital logic $M = 7$ or 8 appears to be the limit even for codes with $k = 1$.

Now we consider more general channels, and restrict ourselves to two input DMC's (but see Prob. 10.7). Recall from Prob. 2.13 that on a general DMC a maximum likelihood decoder must locate a codeword $\mathbf{C}_i = (C_{i0}, \ldots, C_{i,n-1})$ for which the probability

$$P\{\mathbf{R}|\mathbf{C}_i\} = \prod_{j=0}^{n-1} p(R_j|C_{ij})$$

is as large as possible, where $\mathbf{R} = (R_0, \ldots, R_{n-1})$ is received and the $p(y|x)$'s are the channel's transition probabilities. Since the logarithm is a monotonically increasing function of its argument, the ML decoder may equivalently look for a codeword $\mathbf{C}_i$ for which

$$L(\mathbf{C}_i, \mathbf{R}) = \sum_{j=0}^{n-1} \log p(R_j|C_{ij}) \tag{10.8}$$

is as large as possible. What this means is that the Viterbi algorithm of Fig. 10.11 works for an arbitrary DMC, except that the edge labels $l_{j-1,j}(s, t)$ must be redefined. To give the appropriate definition, assume that the Lth truncation of the code is being used and that $\mathbf{R} = (\mathbf{R}_0, \mathbf{R}_1, \ldots, \mathbf{R}_{L+M-1})$ is the received codeword broken up into $L + M$ n-bit vectors. If there is an edge going from state s to state t in the state diagram, denote by $\mathbf{C}_{s,t}$ the n-bit encoder output corresponding to this transition, that is, the label on the $s \to t$ edge (see Fig. 10.5). Then define

$$l_{j-1,j}(s, t) = L(\mathbf{C}_{s,t}, \mathbf{R}_{j-1}), \tag{10.9}$$

where L is defined by Eq. (10.8). If s and t are not connected, set $l_{j-1,j}(s, t) = +\infty$ as before.

10.3 Path enumerators and error bounds

We saw in Section 10.2 that there is a one-to-one correspondence between the possible output streams from a convolutional encoder and the paths through the state diagram beginning and ending in the all-zero state. In this section we apply combinatorial techniques to the problem of enumerating all such paths, and apply the results to obtain performance estimates for specific convolutional codes. As usual, we will illustrate the concepts with CC 1, and discuss generalizations later.

Consider again the state diagram of Fig. 10.5. Let us define the (Hamming) *weight* of a path through this graph as the number of 1's on the labels of the edges comprising the path. (For example, the path *acbaacddba* has weight 12.) We would perhaps like to count the number of paths from a to a of weight i for some fixed integer i, but this number will always be either 0 or infinite because of the loop of weight 0 at a. One way around this problem is to observe that every path from a to a can be uniquely decomposed into a sequence of paths from a to a with no intermediate returns to a. (For example, the path *acbaacddba* can be decomposed into the paths *acba*, *aa*, and *acddba*.) Let us call a path (other than the trivial path *aa*) beginning and ending at a with no intermediate returns a *fundamental path*, and for each i denote by A_i the number of fundamental paths of weight i. By trial and error one can calculate $A_0 = A_1 = A_2 = A_3 = A_4 = 0$, $A_5 = 1$, etc. We shall now describe a powerful technique that will allow us to compute all the A_i's simultaneously.

Figure 10.14 depicts the state diagram of Fig. 10.5 in a modified form that will be convenient for the study of the A_i's. Note that in Fig. 10.14 the initial state a has been split into two states a_0 and a_1, and that the loop at a has been removed. This means that there is a one-to-one correspondence between the set of fundamental paths in the original state diagram and the set of all paths

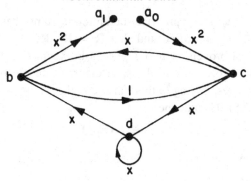

Figure 10.14 A modified state diagram for CC 1.

joining a_0 to a_1 in the modified state diagram of Fig. 10.14. The edges in Fig. 10.14 are labeled with powers of the indeterminate x corresponding to the Hamming weight of the edge label in Fig. 10.5. For example, edge cb is labeled x in Fig. 10.14 because in Fig. 10.5 its label (01) has weight 1. This labeling is a convenient bookkeeping device for enumerating paths from a_0 to a_1 by weight, for if we define the label of a path P to be the product of the labels of its edges, then clearly the exponent on the label of P will equal the weight of P. For example, the label of the path $P = a_0cbcdba_1$ is x^7, and so the weight of P is 7.

Recalling that our goal is the computation of the numbers A_i (where now A_i equals the number of paths from a_0 to a_1 of weight i), we define the *path weight enumerator*[8] for CC 1 as the generating function

$$A(x) = A_0 + A_1x + A_2x^2 + \cdots \qquad (10.10)$$

In terms of the graph of Fig. 10.14, $A(x)$ is the sum of the labels of all paths from a_0 to a_1. There is a standard combinatorial technique for computing this kind of generating function, which we have outlined in Appendix D. When this technique is applied to the graph of Fig. 10.14, the result is

$$A(x) = x^5/(1 - 2x)$$

$$= x^5 + 2x^6 + 4x^7 + \cdots + 2^ix^{5+i} + \cdots, \qquad (10.11)$$

and so there are $A_i = 2^{i-5}$ fundamental paths of weight i in CC 1 for $i \geqslant 5$. (See Eq. (D.3) in Appendix D.)

For future use we will also need a more refined generating function, which enumerates the paths not only according to their weight, but also according to their length and the number of 1's in the corresponding input sequence. To do this we need a state diagram with more elaborate labels, as in Fig. 10.15, where the exponents of x describe the Hamming weight of the encoder output corresponding to the edge, and the exponents of z describe the Hamming weight of the corresponding input. The exponent of y is always 1, corresponding to the fact that the length of each edge is 1. Once again we define the label of a path to be the product of the labels of its edges. For example, the path $P = a_0 cbcdba_1$ has label $x^7 y^6 z^3$. This means that the Hamming weight of the corresponding encoder output (110100101011) is 7, the path has length 6, and the input stream that gives rise to this path (101100) has Hamming weight 3. We now define the *complete path enumerator* $A(x, y, z)$ to be the sum of the labels of all paths in Fig. 10.15 from a_0 to a_1:

$$A(x, y, z) = \sum_{i,j,k} A_{i,j,k} x^i y^j z^k, \tag{10.12}$$

where $A_{i,j,k}$ denotes the number of paths from a_0 to a_1 with label $x^i y^j z^k$. Again referring to Appendix D, we find that for CC 1 this turns out to be (see Eq. (D.4))

$$A(x, y, z) = \frac{x^5 y^3 z}{1 - xyz(1 + y)}. \tag{10.13}$$

Of course, if we set $y = z = 1$, Eq. (10.13) reduces to Eq. (10.11).

We shall now see how path enumerators can be used to obtain bounds on the performance of convolutional codes.

For a given convolutional code and a given channel we are naturally interested in the resulting "error probability." There are, however, several possible definitions of error probability, and each must be handled slightly

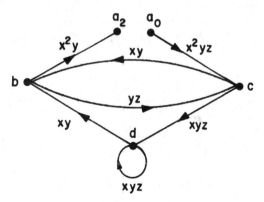

Figure 10.15 A more elaborately labeled version of Fig. 10.14.

differently. So before discussing techniques for bounding the error probability, we shall describe what we mean by an "error."

Suppose we are using a particular (n, k) convolutional code on a particular discrete memoryless channel (but see Prob. 10.14), and that the truncation depth L is large. We assume that a maximum likelihood decoding algorithm (e.g., Viterbi) is being used, and that the all-zero codeword is transmitted.[9] This means that the path followed by the encoder is the horizontal path at the top of the trellis diagram. The decoder does not know which path the encoder has followed, but will make a guess based on the received (noisy) version of the encoder's path labels. Let us call the path actually taken by the encoder the *correct path*, and the path postulated by the decoder the *decoder's path* (see Fig. 10.16). The decoder's path can clearly be partitioned into a (possibly empty) set of correct path segments separated by a set of paths which lie entirely below the correct path except for their end points. These incorrect path segments we call *error events* (in Fig. 10.16 there are 5 error events). Note that each possible error event corresponds to a fundamental path in the encoder's state diagram. The various decoder error probabilities we shall now discuss are all probabilities related to error events.

One obvious thing we could try is to compute the probability that there are no error events. But since we have assumed L to be very large, this probability will be near 0 unless the channel is noiseless (see Prob. 10.12). A more interesting thing we might try is to estimate the *first error probability* $P_{E,1}$, which is the probability that the first error event begins at depth 0, i.e., that the decoder's path has the shape shown in Fig. 10.17. Now let $\mathbf{0} = (0, 0, \dots)$ denote the correct (transmitted) path, $\mathbf{x}^{\text{inc}} = (x_1, x_2, \dots)$ any incorrect path beginning with the error event E, and $\mathbf{y} = (y_1, y_2, \dots)$ the received sequence, which is a noisy version of $\mathbf{0}$. Let $\mathbf{0}_l$, $\mathbf{x}_l^{\text{inc}}$, $\mathbf{y}_l$ denote the first ln-bit segments of each of these sequences. A ML decoder will choose a path $\mathbf{x}$ through the trellis for which $P(\mathbf{y}|\mathbf{x})$ is as large as possible. This means that it cannot choose $\mathbf{x}^{\text{inc}}$ unless $P(\mathbf{y}_l|\mathbf{x}_l^{\text{inc}}) \geqslant P(\mathbf{y}_l|\mathbf{0}_l)$, since otherwise the path obtained

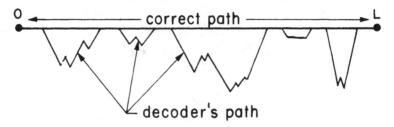

Figure 10.16 Some trellis paths.

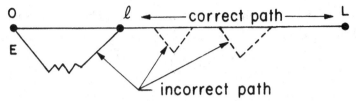

Figure 10.17 A depth 0 error event.

from $\mathbf{x}^{\text{inc}}$ by replacing the error event E with $\mathbf{O}_l$ will be preferred to $\mathbf{x}^{\text{inc}}$. By reasoning identical with that given in the proof of Theorem 7.5 (cf. Eq. (7.13)) the probability that E will be preferred to $\mathbf{0}_l$ by a ML decoder is bounded by $\gamma^{w_H(E)}$, where γ is given by Eq. (7.8) and $w_H(E)$ is the Hamming weight of E. Thus the first error probability $P_{E,1}$ is bounded by

$$P_{E,1} \leq \sum_E \gamma^{w_H(E)} \qquad (10.14)$$

the summation in (10.14) being extended over all error events E. We have already observed, however, that the error events are identical to the fundamental paths through the state diagram; and $A(x)$ (Eq. (10.10)) enumerates these paths by Hamming weight. Hence the bound (10.14) is equivalent to

$$P_{E,1} \leq A(\gamma). \qquad (10.15)$$

Although we defined $P_{E,1}$ as the probability that the decoder's path departs from the correct path at depth 0, it is clearly also the probability that the decoder is off the correct path at depth j, given that it is on the correct path at depth $j - 1$. This conditioning is somewhat unnatural, and so we now define the *error event probability* P_E as the probability that the decoder is off the correct path at a given depth j. This is just the probability that there is an error event hanging below the correct path at depth j (see Fig. 10.18). The situation is much the same as it was before, except that now a given error event E of

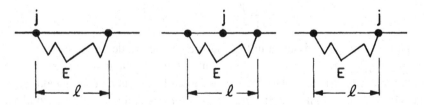

Figure 10.18 Some depth j error events.

length l can appear in any one of l positions. Hence the bound corresponding to (10.14) for P_E is

$$P_E \leqslant \sum_E \text{length } (E) \cdot \gamma^{w_H(E)}$$

$$= \sum_{i,j,k} j A_{i,j,k} \gamma^i, \tag{10.16}$$

where the $A_{i,j,k}$ are the coefficients of the complete path enumerator (Eq. (10.12)). But this sum can be obtained by partial differentiation of $A(x, y, z)$:

$$P_E \leqslant \left. \frac{\partial A(x, y, z)}{\partial y} \right|_{x=\gamma, y=z=1}. \tag{10.17}$$

Let us observe finally that the decoder outputs the information bits corresponding to its postulated path, and that even if it is on the wrong path, some of the individual bits will "accidentally" be right. Thus the *bit error probability* P_e will in general be less than the error event probability. Using reasoning similar to that leading to Eqs. (10.15) and (10.17), we can obtain the following bound on P_e (see Prob. 10.13):

$$P_e \leqslant \frac{1}{k} \left. \frac{\partial A(x, y, z)}{\partial z} \right|_{x=\gamma, y=z=1}. \tag{10.18}$$

Example 10.8 We illustrate these three bounds on CC 1, since we already know the path enumerators $A(x)$ and $A(x, y, z)$ for this code (Eqs. (10.11) and (10.13)). The results are (assuming $0 < \gamma < \frac{1}{2}$):

$$P_{E,1} \leqslant \frac{\gamma^5}{1 - 2\gamma} = \gamma^5 + O(\gamma^6),$$

$$P_E \leqslant \frac{3\gamma^5 - 3\gamma^6}{(1 - 2\gamma)^2} = 3\gamma^5 + O(\gamma^6),$$

$$P_e \leqslant \frac{\gamma^5}{(1 - 2\gamma)^2} = \gamma^5 + O(\gamma^6). \qquad \square$$

Note that all three bounds in the above example are dominated for small γ by order γ^5 terms. Examining the definitions of $A(x)$ and $A(x, y, z)$, we see this is so because 5 is the smallest weight of any fundamental path. In a general convolutional code the *free distance* d_f is defined to be the smallest weight of any fundamental path. It is not hard to show (see Prob. 10.15) for example, $P_{E,1} = a\gamma^{d_f} + O(\gamma^{d_f+1})$, for a certain constant a. The free distance is usually considered to be the most important single measure of a convolutional code's ability to combat noise, and much effort has been expended on the problem of finding convolutional codes with large free distances (see Section 12.3 and also Prob. 10.16).

10.4 Sequential decoding[10]

(In this section we restrict ourselves to convolutional codes with $k = 1$, and leave consideration of the general case as Prob. 10.20.)

We saw in Section 10.2 that Viterbi decoding of a convolutional codes is maximum likelihood, and so for any particular code it cannot be improved upon. We also saw (p. 304) that the complexity of Viterbi's algorithm is an exponential function of the code's memory M; unfortunately, the larger M is, the better the code is likely to be. For example, consider the class of (2,1) convolutional codes. We saw in Section 10.3 that the performance of a convolutional code can be measured by its free distance d_f: the larger d_f, the better the code, in general. But in Prob. 10.16 it is shown, for example, that for a (2, 1) code $d_f \le M[1 + o(1)]$, and so such a code with a free distance of say 30 would have such a large M that Viterbi decoding would be hopelessly complex. Thus we are motivated to find a decoding algorithm that will work on convolutional codes with very large values of M. There is a class of such algorithms, called *sequential decoding algorithms*. They are not quite as good as ML algorithms for a fixed code, but that defect is largely compensated for by the fact that they can be used to decode some codes with very large M.

The key to understanding sequential decoding algorithms is the *tree diagram*. In a shift-register encoder for an $(n, 1)$ convolutional code, one information bit entering the encoder causes n encoded bits to leave (cf. Fig. 10.3)). Thus it is possible to view the encoding process conceptually as a walk (climb?) through a *binary tree*, as depicted in Fig. 10.19.

The encoder begins at the START vertex, and after d inputs it will be at some vertex in the tree at depth d. If the next input is "0," it moves to depth $d + 1$ along the branch in the upward direction; if it is "1," in the downward direction. For example, if the input is 0100..., the encoder will follow the path indicated in Fig. 10.19. Although this has not been done in Fig. 10.19, in general each branch of the tree will be labeled with the n bits the encoder will output as it travels the branch.

The tree in Fig. 10.19 apparently extends to infinite depth, but of course for any concrete realization of a code the tree will be finite. If the Lth truncation of the code is being used, the tree will terminate at depth $L + M$; and beyond

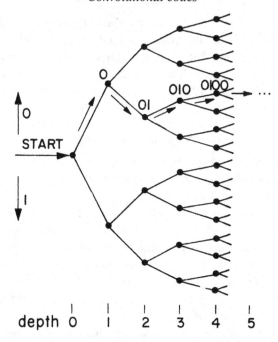

depth 0   1   2   3   4   5

Figure 10.19 A binary tree for an $(n, 1)$ convolutional code.

depth $L - 1$ there will be no bifurcation of the paths, since the last M inputs to the encoder will always be "0" (see discussion on p. 301). In Fig. 10.20 we give the completely labeled tree corresponding to the $L = 3$ truncation of CC 1.

From this new viewpoint each transmitted sequence corresponds to a path through the tree, beginning at the START vertex and ending at depth $L + M$. The received sequence will be a noisy version of this path, and the decoder's job will be to guess which path the encoder actually took. Since there will be a total of 2^L possible paths, if L is large it will not in general be possible to compare the received message to each of these paths. The approach taken to this problem by *sequential decoding algorithms* is to explore a very small subset of the possible paths. If at a given stage a certain partially explored path looks promising, it is explored further; if it does not, it is abandoned and another path is tried.

We shall now describe the two best known sequential decoding algorithms, the *stack algorithm* and *Fano's algorithm*. To illustrate the ideas, we begin with a "thought experiment."

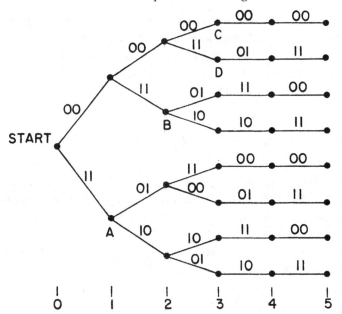

Figure 10.20 The binary tree corresonding to the $L = 3$ truncation of CC 1.

Example 10.9 Imagine that we are using the tree in Fig. 10.20 to decode CC 1, and that $\mathbf{y} = (y_1, y_2, \ldots, y_{10})$ is the received sequence. Imagine further that we are using a sequential decoding algorithm that has (by unexplained means) already explored the four paths ending at vertices A, B, C, D. Note that whatever path the encoder actually took, it must have passed through one and only one of these four vertices. As a next step in our decoding algorithm we would like to extend the most promising of these four paths. But which one is most promising? To answer this question, we pause to consider briefly the problem of decoding a code whose codewords have *different lengths*.

Let $\{\mathbf{x}_0, \mathbf{x}_1, \ldots, \mathbf{x}_{M-1}\}$ be a code with M different codewords, let $\mathbf{x}_i$ have length n_i, that is, $\mathbf{x}_i = (x_{i1}, x_{i2}, \ldots, x_{in_i})$, and let $n = \max n_i$. Suppose this code is being used to transmit information over a DMC in the following way. Codeword $\mathbf{x}_i$ is transmitted with probability p_i, $\sum_{i=0}^{M-1} p_i = 1$. When the n_i components of $\mathbf{x}_i$ have been transmitted, a "random tail" of $n - n_i$ channel input symbols is transmitted. These symbols are selected indepenently according to a fixed probability distribution $P(x)$ on the channel input alphabet A_X. In this way the receiver will always receive an n-symbol sequence $\mathbf{y} = (y_1, y_2, \ldots, y_n)$, in spite of the fact that the codewords are of different lengths.

Under these assumptions, if **y** is received, the decoding error probability will be minimized if the decoder selects the codeword $\mathbf{x}_i$ for which the conditional probability $P\{\mathbf{x}_i \text{ sent}|\mathbf{y} \text{ received}\}$ is as large as possible (cf. Prob. 2.13a). But since **y** is fixed and $P\{\mathbf{x}_i|\mathbf{y}\} = P\{\mathbf{x}_i, \mathbf{y}\}/P\{\mathbf{y}\}$, this is equivalent to maximizing $P\{\mathbf{x}_i, \mathbf{y}\} = P\{\mathbf{x}_i\}P\{\mathbf{y}|\mathbf{x}_i\}$, which is given by the following formula:

$$P\{\mathbf{x}_i \text{ sent, } \mathbf{y} \text{ received}\} = p_i \prod_{j=1}^{n_i} p(y_j|x_{ij}) \prod_{j=n_i+1}^{n} p(y_j), \qquad (10.19)$$

where $p_i = P\{\mathbf{x}_i \text{ sent}\}$, $p(y|x)$ are the channel transition probabilities, and the probability distribution $p(y)$ on the channel output alphabet A_Y is the one induced by the probability distribution $p(x)$ governing the random tail. Recall that the object is to find the $i \in \{0, 1, \ldots, M-1\}$ that maximizes (10.19); to do this we can divide by the positive constant $\prod_{j=1}^{n} p(y_j)$, take logarithms, and maximize the resulting expression, which is called a *metric*:

$$\mu(\mathbf{x}_i) = \sum_{j=1}^{n_i} \left[\log \frac{p(y_j|x_{ij})}{p(y_j)} - \frac{1}{n_i} \log \frac{1}{p_i} \right]. \qquad (10.20)$$

In summary, if **y** is received, the best way for the decoder to guess which codeword was sent is to compute $\mu(\mathbf{x}_i)$ for each $i = 0, 1, \ldots, M-1$; the codeword for which this metric is largest is the codeword most likely to have been sent.

Now let us return to Example 10.9; recall that the problem was to decide which of the vertices A, B, C, D the encoder was most likely to have passed through, on the basis of the received vector $\mathbf{y} = (y_1, y_2, \ldots, y_{10})$. If it passed through A, the first two transmitted symbols must have been 11; the a priori probability of this event is $\frac{1}{2}$, assuming that "0" and "1" inputs are equally likely. Similarly, if it passed through B the first four transmitted symbols were 0011, and this event has a priori probability $\frac{1}{4}$. Continuing in this way, we arrive at the following table:

Vertex	Codeword	Probability
A	11	$\frac{1}{2}$
B	0011	$\frac{1}{4}$
C	000000	$\frac{1}{8}$
D	000011	$\frac{1}{8}$

The situation is practically the same as the one that led to Eq. (10.20): we are asked to choose between codewords of different lengths and different a priori transmission probabilities on the basis of a noisy received sequence. The only way in which the model we used to derive (10.20) does not quite fit the present situation is in the selection of the "random tail": previously we assumed the tail symbols were selected independently according to a fixed probability distribution, but here the tail symbols correspond to the encoder's path after it passes the vertex being examined. Without actually exploring the tree further, however, it is best to assume that further symbols have been selected randomly. Since for any fixed convolutional code it is easy to show that the jth transmitted symbol is equally likely to be 0 or 1 (see Prob. 10.22), the appropriate choice for the distribution governing the random tail is $p(0) = p(1) = \frac{1}{2}$.

Hence to choose between the four paths A, B, C, D we should compute the corresponding four metrics, using Eq. (10.20). And for this application there is an important simplification: the term $n_i^{-1} \log p_i^{-1}$ is always equal to $\frac{1}{2}$. This is of course no accident: for a general $(n, 1)$ convolutional code a path of depth $d \leq L - 1$ into the code tree will yield a codeword of length nd with a priori transmission probability 2^{-d}, and so the term $n_i^{-1} \log p_i^{-1}$ of Eq. (10.20) will be $1/n$, the rate R of the code (this term is also equal to R for a general (n, k) convolutional code; see Prob. 10.20b). Hence the appropriate metric, called the *Fano metric* in honor of its discoverer, for evaluating paths through the code tree is

$$\mu_F(\mathbf{x}_i) = \sum_{j=1}^{n_i} \left[\log \frac{p(y_j | x_{ij})}{p(y_j)} - R \right]. \tag{10.21}$$

In summary: given a set $\{\mathbf{x}_0, \mathbf{x}_1, \ldots\}$ of partial paths through the tree with the property that any path through the tree must be an extension of exactly one of them, the most likely path is the one for which the Fano metric $\mu_F(\mathbf{x}_i)$ is largest. [Note. In Eq. (10.21) we have assumed that the channel is a DMC—but see Prob. 10.21—and that $p(y|x)$ are the channel transition probabilities. Also, $p(y)$ is the probability distribution on the channel output alphabet A_Y induced by the distribution $(\frac{1}{2}, \frac{1}{2})$ on the channel input alphabet $A_X = \{0, 1\}$. Thus $p(y) = \frac{1}{2}[p(y|0) + p(y|1)]$ for all $y \in A_Y$.] ☐

The above considerations lead us naturally to the simplest sequential decoding algorithm, the *stack algorithm*. The stack[11] decoder works with a finite set $S = \{\mathbf{x}_0, \mathbf{x}_1, \ldots\}$ of paths through the tree, ordered by the Fano metric: $\mu_F(\mathbf{x}_0) \geq \mu_F(\mathbf{x}_1) \geq \ldots$. The best path $\mathbf{x}_0$ is at the top of the stack, the next best is in the second position, and so on. Initially S contains only the trivial path of length 0 that begins and ends at START; its metric is defined to be 0. At each step the path $\mathbf{x}_0$ at the top of the stack is deleted and replaced by the possible extensions of $\mathbf{x}_0$, which are one branch longer. (If $\mathbf{x}_0$ extends to depth L or beyond, only one extension will be possible; see Fig. 10.20.) The metrics for these new paths are computed via Eq. (10.21), and the new paths are inserted into their proper place on the stack. Note that the calculation of the metric for a new path will only require the computation of the last n terms in the summation of Eq. (10.21), since the sum of the remaining terms equals $\mu_F(\mathbf{x}_0)$, which is already known. The decoder's decision is simply the first full path (i.e., of length $L + M$) to reach the top of the stack.

Example 10.10 Consider the code tree of Fig. 10.21, which corresponds to a convolutional code with $L = 2$, $M = 2$. We have simplified things in this example by giving the Fano metric of each path directly above the path's terminal vertex. Of course in practice the metrics would have to be computed using Eq. (10.21), the branch labels, and the received sequence $\mathbf{y} = (y_1, y_2, \ldots)$. For example, the path ABE has metric -2. The operation of the stack algorithm is described in the following table; the paths in each stack are represented by their terminal vertices, and the corresponding metrics are given in parentheses.

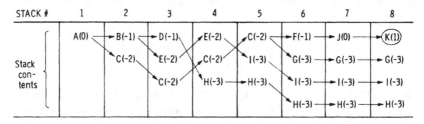

STACK #	1	2	3	4	5	6	7	8
Stack con-tents	A(0)	B(−1)	D(−1)	E(−2)	C(−2)	F(−1)	J(0)	K(1)
		C(−2)	E(−2)	C(−2)	I(−3)	G(−3)	G(−3)	G(−3)
			C(−2)	H(−3)	H(−3)	I(−3)	I(−3)	I(−3)
						H(−3)	H(−3)	H(−3)

For example, stack 6 is obtained from stack 5 by deleting C and replacing it

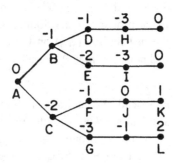

Figure 10.21 An $L = 2$, $M = 2$ tree for sequential decoding.

with F and G. F goes to the top of the stack because its metric (-1) is larger than that of G, I, or H. The decoder's final decision is the vertex K; the corresponding information bits are 10. Note, however, that K is not the terminal vertex with the largest metric; that honor goes to L. The stack decoder is prevented from reaching L because of the very low metric of its predecessor G. For a simple characterization of the path chosen by the stack algorithm see Prob. 10.23. □

The stack algorithm is quite an efficient way to find a good path through the tree, but it has one unpleasant feature: it requires a large and variable amount of storage, since at every stage the decoder must have a list of each of the paths visited so far. There is another sequential decoding algorithm, the *Fano algorithm*, that nearly always finds the same path through the tree using very little storage.

Essentially what Fano's algorithm does is this. At every stage the decoder is located at some vertex in the tree. From this vertex the decoder looks forward (i.e., deeper) into the tree. If it sees a vertex it likes, it moves forward to that vertex. If not, it moves backward and then tries to move forward along another branch. It decides whether or not it likes a given vertex by comparing the metric μ of that vertex (computed by Eq. (10.21)) to a running threshold T. The decoder stops as soon as it reaches a terminal vertex (i.e., one at depth $L + M$) and outputs the information sequence corresponding to the branch ending at the vertex.

The details of Fano's algorithm are given in the flowchart of Fig. 10.22. Here are some explanatory notes for the flowchart.

1. The decoder begins at the START vertex wth $T = 0$.

2. The number T (the *threshold*) changes throughout the algorithm by multiples of Δ, a preselected constant. At all times T will be $\leqslant$ the metric of the current vertex.

3. F. means FORWARD (increasing depth) into the tree; B. means BACK-WARD. To LOOK F. is to compute the metric μ of a vertex one branch F. in the tree. To MOVE F. is to move the decoder to a vertex one branch F. LOOK B. and MOVE B. have corresponding meanings. If the algorithm LOOKS B. from the START vertex, we assume the metric μ to be $-\infty$ there. This is to prevent a backward move from START.

4. V. stands for VERTEX. From a given V., the BEST V. is that vertex one branch deeper into the tree with the largest metric; the WORST V. is the one with the smallest metric.

5. The threshold T is *tight* at a given vertex if increasing T by Δ would

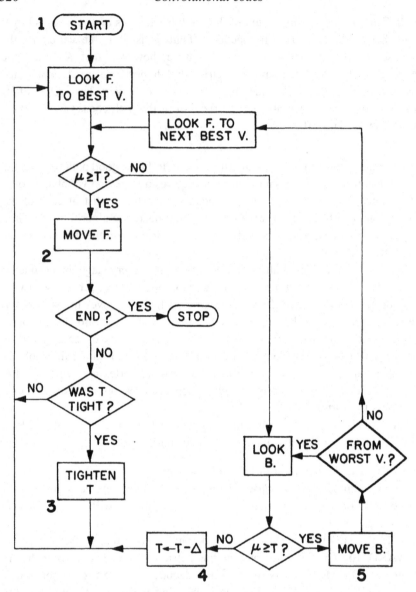

Figure 10.22 A flowchart description of Fano's sequential decoding algorithm.

make it larger than the metric of that vertex. To *tighten the threshold* is to increase it in increments of Δ until it is tight. The WAS T TIGHT? query occurs only after MOVE F. It refers to the state of affairs just before the move.

6. The END? query asks whether or not the current vertex is a terminal vertex; if it is, the algorithm stops.

Example 10.11 We illustrate Fano's algorithm on the tree of Fig. 10.21. (We set $\Delta = 1$ for this example.) The following table summarizes the algorithm's behavior by listing each of the changes in either the vertex being visited or the threshold, together with the location in the flowchart (see labels 1, 2, 3, 4, 5 in Fig. 10.22) where the change occurred.

Step	Vertex	T	Location	Step	Vertex	T	Location
1	A	0	1	11	E	-2	2
2	A	-1	4	12	B	-2	5
3	B	-1	2	13	A	-2	5
4	D	-1	2	14	C	-2	2
5	B	-1	5	15	F	-2	2
6	A	-1	5	16	F	-1	3
7	A	-2	4	17	J	-1	2
8	B	-2	2	18	J	0	3
9	D	-2	2	19	K	0	2
10	B	-2	5	20		-STOP-	

The reader is encouraged to work through each step of this table. We single out, however, three transitions as being especially interesting.

$6 \rightarrow 7$: At step 6 the decoder is at vertex A with $T = -1$, having just moved backward from B. The answer to the query FROM WORST V.? is NO, since B is the BEST V. relative to A. We then LOOK F. TO NEXT BEST V., which is C, with $\mu = -2$. We then take the NO branch from the $\mu \geqslant$ T? query, and LOOK B.; this results in setting $\mu = -\infty$ (see explanatory note 3 above). This means we take the NO branch from the next $\mu \geqslant$ T? query, and this results in reducing T to -2; thus we have arrived at step 7 in the table.

$12 \rightarrow 13$: At step 12 the decoder is at vertex B with $T = -2$, having just moved backward from E. The answer to the FROM WORST V.? query is YES, since E is the WORST V. relative to B. A backward look to $A(\mu = 0)$ results in a YES answer to the $\mu \geqslant$ T? query, and so the decoder moves backward to A, thus arriving at step 13.

$15 \rightarrow 16$: At step 15 the decoder is at vertex F, with $T = -2$, having just moved forward from C. Since F is not a terminal vertex, we take the NO branch from END?. The answer to WAS T TIGHT? is YES, since the threshold $T = -2$ was tight at the previous vertex C. Thus T is tightened up to $T = -1$, and we are at step 16. □

Note that in Example 10.11 the Fano decoder arrived at the same answer (vertex K) as the stack algorithm did in Example 10.10. This is no accident; provided that certain details are arranged properly, the Fano and stack algorithms practically always yield the same result. For a proof of this fact see Probs. 10.25–10.31.

Finally we remark that, although these sequential decoding algorithms perform almost as well as would Viterbi's algorithm on codes that have much more memory than can be accommodated by the Viterbi algorithm, the amount of computation required to complete a search through the code tree is highly variable, and this leads to certain problems; see Section 12.3 for details.

Problems

10.1 (This problem develops some advanced algebraic properties of convolutional codes and is related to the polynomial matrix approach. To do this problem it will be helpful to know something of the theory of modules over principal ideal rings; consult, e.g., Lang [9], Chapter XV, Section 2.) Let F be a field, and let $F[x]$ denote the ring of polynomials in the indeterminate x over F. Let $F[x]^n$ denote the set of all n-tuples $\mathbf{f} = (f_1(x), \ldots, f_n(x))$ of polynomials from $F[x]$. A *convolutional code* C is defined to be a subset of $F[x]^n$, which is (a) closed under componentwise addition and subtraction, and (b) closed under multiplication by elements of $F[x]$; that is, if $\mathbf{f} = (f_1, \ldots, f_n) \in C$, then so is $a\mathbf{f} = (af_1, \ldots, af_n)$ for all $a \in F[x]$. A *basis* for C is a set $\{\mathbf{g}_1, \ldots, \mathbf{g}_k\} \subseteq F[x]^n$ such that every $\mathbf{f} \in C$ can be expressed uniquely as $\mathbf{f} = \sum_{i=1}^{k} a_i \mathbf{g}_i$ for some $a_1, \ldots, a_k \in F[x]$. The number k is called the dimension (or rank) of the code.

(a) Show that every convolutional code C has such a basis and that every basis for C has the same number of elements; also, find a basis for $F[x]^n$ itself.

(b) Show that the encoding described by Eq. (10.4) is one-to-one iff the rows $\mathbf{g}_1, \ldots, \mathbf{g}_k$ of the matrix G form a basis for the code. Show that this is equivalent to saying that $\mathbf{g}_1, \ldots, \mathbf{g}_k$ are *linearly independent*, that is $\sum_{i=1}^{k} a_i \mathbf{g}_i = 0$ implies $a_1 = \ldots = a_k = 0$. Show that the matrix

$$G = \begin{bmatrix} x & x^3 + x & x + 1 \\ 0 & x^4 & x^2 \\ x^2 & x^2 & x \end{bmatrix}$$

does not meet this condition, and find a two-rowed generator matrix for the code that does. [Note. $F = F_2$ in this example and all further examples in this problem.]

(c) Let C be a convolutional code of dimension k. Show that there exist a basis $\{\mathbf{g}_1, \ldots, \mathbf{g}_k\}$ for C, a basis $\{\mathbf{f}_1, \ldots, \mathbf{f}_n\}$ of $F[x]^n$, and monic polynomials $a_1, \ldots, a_k$ with $a_i | a_{i+1}$, $i = 1, 2, \ldots, k-1$, such that $\mathbf{g}_i = a_i \mathbf{f}_i$, $i = 1, 2, \ldots, k$. The polynomials a_i are unique and are called the *invariant factors* of the code. Find such bases for CC's 1 and 2 (p. 292) and for the (2, 1) code with polynomial generator matrix $G = (x + 1, x^2 + 1)$.

(d) Define the *dot product* $\mathbf{f} \cdot \mathbf{g} = \sum_{i=1}^{n} f_i g_i \in F[x]$ between two elements of $F[x]^n$, and show that $\mathbf{f} \cdot \mathbf{g} = 0$ for all $\mathbf{g} \in F[x]^n$ iff $\mathbf{f} = \mathbf{0}$.

(e) If $\{\mathbf{f}_1, \ldots, \mathbf{f}_n\}$ is a basis for $F[x]^n$, a *dual basis* is a basis $\{\mathbf{f}_1^*, \ldots, \mathbf{f}_n^*\}$ such that $\mathbf{f}_i \cdot \mathbf{f}_j = 0$ if $i \neq j$, $= 1$ if $i = j$. Show that a dual basis exists and is unique.

(f) If C is a convolutional code, its *dual code* $C^\perp$ is defined as $C = \{\mathbf{h} \in F[x]^n : \mathbf{h} \cdot \mathbf{g} = 0 \text{ for all } \mathbf{g} \in C\}$. Notation being as in parts (c), (d), (e), show that $\{\mathbf{f}_{k+1}^*, \ldots, \mathbf{f}_n^*\}$ forms a basis for $C^\perp$, and hence that C has dimension $n - k$. Find a basis for the dual codes of each of the three codes cited in part (c).

(g) Let C be a convolutional code with $k \times n$ polynomial generator matrix G. C is said to be invertible if there exists a $k \times n$ polynomial matrix H such that $GH^T = I_k$, the $k \times k$ identity matrix. Show that C is invertible iff the invariant factors $a_1, \ldots, a_k$ (see part (c)) are all equal to 1. Find explicit inverses for CC's 1 and 2 of this chapter.

(h) More generally, a matrix H is called a $\psi(x)$ *inverse* of G if $GH^T = \psi I_k$. Show that G has a $\psi(x)$ inverse iff $a_k | \psi$. Find a $\psi(x)$ inverse for $G = [x + 1, x^2 + 1]$, with $\psi(x) = x + 1$.

10.2 We mentioned on p. 293 that an (n, k) block code could be viewed as an (n, k) convolutional code with $M = 0$. From this viewpoint what is the Lth truncation of a block code?

10.3 It is possible that a given pair of states in the state diagram of a convolutional code will be joined by more than one edge. When will this occur?

10.4 Consider a fixed $(n, 1)$ convolutional code, truncated at length L. Show that the amount of work involved in decoding by the brute-force "compare the received sequence to each of the 2^L possible codewords" method is about 2^{L-K} times as much as that involved in using the Viterbi algorithm. (Here K = the code's constraint length; see Eq. (10.2).)

10.5 We described Viterbi's decoding algorithm in the text only for codes with $k = 1$. Generalize to general k. (Watch out for multiple edges between states—see Prob. 10.3; make sure your algorithm will work for block codes, since they are $M = 0$ convolutional codes.)

10.6 Let C be the $(3, 2)M = 1$ convolutional code described by the generator matrix

$$G = \begin{bmatrix} 1 + x & 1 & 1 + x \\ x & 1 + x & 0 \end{bmatrix}.$$

(a) Design an encoding circuit (cf. Figs 10.3 and 10.4).
(b) Draw a state diagram (cf. Fig. 10.5).

(c) Assuming that the $L = 6$ truncation of the code is being used on a BSC, decode the received sequence 011 011 111 100 101 001 101, using Viterbi's algorithm.

10.7 Consider using Viterbi's algorithm on a Gaussian channel (cf. Chapter 4). Show that an appropriate definition of the label $l_{j-1,j}(s, t)$ is

$$l_{j-1,j}(s, t) = \|\mathbf{y} - \mathbf{x}\|^2,$$

where $\mathbf{x}$ is the n-dimensional transmitted vector corresponding to the $s \to t$ transition in the state diagram, and $\mathbf{y}$ is the corresponding n-dimensional received vector. [Note. $\|\mathbf{z}\|^2 = \sum_{i=1}^{n} z_i^2$ is the Euclidean norm of $\mathbf{z} = (z_1, \ldots, z_n)$.]

10.8 Referring to the state diagram of Fig. 10.5, let B_i denote the total number of paths of weight i from a to a that do not traverse the loop at a (but may have intermediate returns to a). Give a general formula for B_i.

10.9 Consider the following labeled, directed graph:

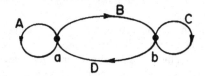

Compute the transmission between vertices a and b in terms of the edge labels A, B, C, D. (Consult Appendix D for definitions.)

10.10 Compute the path enumerator $A(x)$ and the complete path enumerator $A(x, y, z)$ for CC 2.

10.11 Let C be the $(3, 1)M = 2$ convolutional code described by the generator matrix $(x^2 + 1, x^2 + x + 1, x^2 + x + 1)$. Compute $A(x)$ and $A(x, y, z)$ for C. Now suppose the code is to be used on the following binary erasure channel:

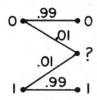

Give upper bounds on the first error probability $P_{E,1}$, the error event probability P_E, and the bit error probability P_e.

10.12 Consider using a binary (n, k) convolutional code on a fixed two-input DMC which is not noiseless, that is, whose capacity is less than $\log 2$. Let $P_{\text{err}}^{(L)}$ denote the probability that at least one error event will occur in the decoding of a codeword from the Lth truncation of the code. Show that $\lim_{L \to \infty} P_{\text{err}}^{(L)} = 1$ (see p. 309).

10.13 Prove that the bound on P_e given in Eq. (10.18) is valid. [*Advice:* Do the case $k = 1$ first.]

10.14 Extend the bounds of Section 10.3 (Eqs. (10.15), (10.17), (10.18)) to cover a Gaussian channel. [*Hint:* The bounds will have the same form, but γ will be equal to $e^{-\beta/2\sigma^2}$. See Prob. 7.20.]

10.15 Show that for a fixed convolutional code with free distance d_f, the error event probability satisfies $P_E = a\gamma^{d_f} + O(\gamma^{d_f+1})$ for small γ, and identify the constant a. Derive analogous results for $P_{E,1}$ and P_e (cf. Example 10.8.)

10.16 Apply Plotkin's bound (Prob. 7.24) to the Lth truncation of an (n, k) convolutional code to show that its free distance (see p. 311) d_f satisfies

$$d_f \leqslant \min_{L=1,2,\ldots} \left\lfloor n(M+L)\frac{2^{kL-1}}{2^{kL}-1} \right\rfloor.$$

10.17 This problem deals with a certain algorithm on the state diagram of a convolutional code. The algorithm outputs a nonnegative integer; you are supposed to figure out what the integer represents. It is assumed that each edge of the state diagram is labeled with the number of 1's ouptut by the encoder when the state transition corresponding to that edge is made. For example, for CC 1 the state diagram would look like this (cf. Fig. 10.5):

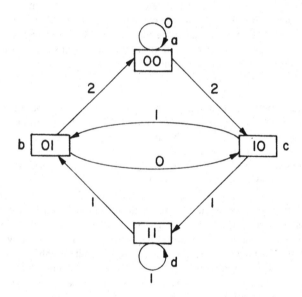

Some notational conventions: (1) the set of all states is denoted by V; (2) the label of an edge connecting state v_i to state v_j is $l(v_i, v_j)$; (3) the states reachable in one step from a state v are called the *successors* of v—for example, the successors of state d in the above state diagram are $\{d, b\}$; (4) the all-zero state is denoted by v_0. Now here is the algorithm (it involves an auxiliary function $d(v)$, $v \in V$, and a subset S of V):

Algorithm X

---

1. Set $d(v) = +\infty$, all $v \in V$; set $S \leftarrow \varnothing$, the empty set.
2. For all successors of v of v_0 except v_0, set $d(v) \leftarrow l(v_0, v)$.
3. Choose $v \in V - S$ with $d(v)$ minimum. If $v = v_0$, output $d(v_0)$ and stop.
4. Add v to S, and set $d(v') \leftarrow \min(d(v'), d(v) + l(v, v'))$ for all successors v' of v which are not in S. Got to step 3.

Now here are the problems:

(a) What does algorithm X do?

(b) Apply algorithm X to the convolutional codes with polynomial generator matrices

$$[x^3 + x^2 + 1, \ x^3 + x^2 + x + 1], \ [x^3 + x + 1, \ x^3 + x^2 + 1, \ x^3 + x^2 + x + 1],$$

$$\begin{bmatrix} 1+x & 1 & 1+x \\ x & 1+x & 0 \end{bmatrix}.$$

The following two problems deal with a serious problem that can arise if the wrong kind of convolutional code is chosen, *catastrophic error propagation*. In Prob. 10.18 we present a concrete example of a code that suffers from this flaw, and in Prob. 10.19 we give several equivalent definitions of a *catastrophic* convolutional code. For simplicity, in both problems we assume that the code is truncated, that is, both the input and output sequences are infinite.

10.18 Consider the $(2, 1)M = 2$ convolutional code with polynomial generator matrix $\mathbf{G} = [x + 1, \ x^2 + 1]$.

(a) Draw the state diagram.

(b) Attempt to calculate the path enumerator $A(x)$. Why does the attempt fail?

(c) Assume the code is being used on a BSC with raw bit error probability $0 < p < \frac{1}{2}$. Suppose that the input stream is $0000000\ldots$, which yields the transmitted stream $00 \ 00 \ 00 \ 00\ldots$. Suppose further that the received stream is $11 \ 00 \ 00 \ 00\ldots$. Use Viterbi's algorithm to decode. Are you disturbed by the results? (You should be.)

(d) Let $(I_1, I_2 \ldots)$ be the input stream, and $(\hat{I}_1, \hat{I}_2, \ldots)$ the Viterbi decoder's estimate of the input stream. Again assuming that a BSC is being used, show that $\lim_{j \to \infty} P\{\hat{I}_j = I_j\} = \frac{1}{2}$. [*Hint:* Use the results of part (c) and observe that two consecutive bit errors are certain to occur eventually.]

10.19 In Prob. 10.18 we saw that for a certain code a finite number of channel errors could cause an infinite number of decoder errors. A code for which this can happen is called a *catastrophic* code. In this probelm we shall give several equivalent conditions for testing whether or not a given code is catastrophic. Thus let C be an (n, k) convolutional code with polynomial generator matrix $\mathbf{G}$, and show that the following conditions are equivalent (also observe that the code of Prob. 10.18 satisfies each of them):

(a) There exists an input stream containing an infinite number of 1's such that the corresponding output stream has only a finite number of 1's.

(b) There exist k rational functions $I_j(x) = p_j(x)/q_i(x)$, $j = 1, 2, \ldots, k$ (none of the $q_j(x)$'s being divisible by (x), such that the vector

$C = (C_1(x), \ldots, C_n(x)) = (I_1(x), \ldots, I_k(x)) \cdot G$ has polynomial components.

(c) There is a closed path in the state diagram (other than the little loop at the all-zero state) in which every edge has label $00\ldots0$ (edge labels as in Fig. 10.5).

(d) Assuming the code is being used on a BSC with $0 < p < \frac{1}{2}$, there exists a pattern of a finite number of channel errors that will cause an infinite number of decoder errors. [*Hint*: Let the channel error pattern be the vector C of part (b).]

(d) The kth invariant factor a_k of the code is *not* a power of x (cf. Prob. 10.1c). [*Note*: From the theory of modules over principal ideal domains, this is equivalent to asserting that the greatest common divisor of the $k \times k$ subdeterminants of G is not a power of x. In particular, if $k = 1$ as in Prob. 10.18, this condition is that the g.c.d. of the n polynomials in $G = (g_1(x), \ldots, g_n(x))$ is not a power of x.]

10.20 (a) How does the tree diagram for a general (n, k) convolutional code differ from the one for $k = 1$ (Fig. 10.19)?

(b) Show that the term $1/n_i \log 1/p_i$ in Eq. (10.20) is equal to k/n for an (n, k) convolutional code (see discussion on p.316).

10.21 How should the Fano metric (cf. Eq. (10.21)) be defined for a Gaussian channel?

10.22 Consider a fixed truncation of a fixed convolutional code C. If no column of the polynomial generator matrix G is identically 0, show that exactly half of the codewords $(x_1, x_2, \ldots, x_{n(M+L)})$ have $x_i = 0$ and half have $x_i = 1$.

10.23 In this problem we shall give a characterization of the path through the code tree that will be chosen by the stack algorithm (or rather a characterization of the paths that will not be chosen). Assume we have a tree such as the one depicted in Fig. 10.20, and that associated with each vertex v there is a metric $\mu(v)$. Assume further that the stack algorithm described on p. 316 is used to choose a path through the tree. Let P and P' be two paths through the tree and assume they diverge at depth d:

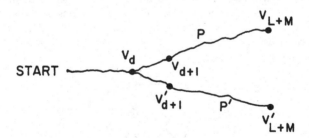

Show that, if $\min\{\mu(v_{d+1}), \ldots, \mu(v_{L+M})\} > \min\{\mu(v'_{d+1}), \ldots, \mu(v'_{L+M})\}$, then P' will not be chosen by the stack algorithm. (See Prob. 10.30 for a similar description of the path chosen by Fano's algorithm.)

10.24 Assign new metrics to the tree of Fig. 10.21 in such a way that (a) the path of length 4 with the largest metric, (b) the path chosen by the stack algorithm, and (c) the path chosen by Fano's algorithm will all be different.

In the next 7 problems we will develop several of the most important properties of Fano's algorithm (Fig. 10.22). In particular, we shall obtain a description of the path chosen by Fano's algorithm (Prob. 10.30) which when combined with Prob. 10.23 will show that the stack algorithm and Fano's algorithm practically always choose the same path. Throughout, we assume that the decoder's behavior is described by a sequence of pairs $(v_1, T_1) (v_2, T_2), \ldots$, where v_i is a vertex and T_i is the value of the threshold as v_i is being visited. The metric of a vertex v is denoted by $\mu(v)$; we assume that the metrics have been rounded off so that they are all multiples of the quantity Δ. If two vertices v and v' are on the same path through the tree, and the depth of v' is greater than that of v, we say v' is a *successor* of v, v is a *predecessor* of v'. Also, v' is an *immediate* successor if its depth is just one more than the depth of v.

10.25 Suppose the Fano algorithm, somewhere in its operation, produces the sequence $(v, T), (v_1, T_1), \ldots, (v_n, T_n)$, where v_1 is a successor of v, $v_n = v$, but $v_i \neq v$ for $i = 1, 2, \ldots, n - 1$. Show that $T_n = T$.

10.26 Under the same assumptions as in Prob. 10.25, except that v_1 is a predecessor of v, show that $T_n = T - \Delta$.

10.27 Show that at the first arrival at any vertex (except START) the threshold will be tight before the next LOOK F., that is, the threshold T will be equal to the metric of the vertex.

10.28 Show that after an unsuccessful forward look from a vertex v with threshold T (i.e., when the NO branch is taken from the upper $\mu \geq T$? query) every path forward from v must contain some vertex whose metric is $< T$.

10.29 If the Fano algorithm is at (v, T), show that every path forward from every predecessor of v must contain a vertex with metric $\leq T$. As a corollary show that every immediate successor of v with metric $> T$ must already have been visited.

10.30 Refer to the sketch for Prob. 10.23. Show that, if $\min\{\mu(v_d), \mu(v_{d+1}), \ldots, \mu(v_{L+M})\} > \min\{\mu(v_d), \mu(v'_{d+1}), \ldots, \mu(v'_{L+M})\}$, then the Fano algorithm will not choose P'.

10.31 The result of Prob. 10.30 shows that, given any two paths P and P', the Fano algorithm will not choose one of them. It is conceivable, however, that the algorithm will fail to choose any path at all by getting stuck in a loop. Show that this cannot happen. [*Hint*: Use the results of Prob. 10.26.]

10.32 Consider using CC 1 (see p. 292) on a binary erasure channel with erasure probability ϵ:

Decode the following received sequence: ??????11000000... (you may assume all subsequent received bits are 0), using (a) the Viterbi algorithm, (b) a sequential decoding algorithm (if you use Fano, take $\Delta = \frac{1}{2}$). What do you conclude about the relative performances of the two algorithms in the presence of a burst of errors?

Notes

1 (pp. 293, 298). A word about the term *convolutional* is in order here. If $\{a_n\}$ and $\{b_n\}$ are two sequenes, their *convolution* is the sequence $\{c_n\}$, defined by

(i) $c_n = \sum_v a_v b_{n-v}$.

Now if we define the generating functions $a(x) = \sum_n a_n x^n$, $b(x) = \sum_n b_n x^n$, etc., Eq. (i) becomes

(ii) $c(x) = a(x)b(x)$.

Thus multiplication of the generating functions corresponds to the convolution of the sequences. In the present application the data to be encoded are represented by generating functions, and the encoding process (see Eq. (10.4)) involves multiplying these generating functions by a fixed family of polynomials.

2 (p. 293). There is at least one other important approach, the *linear sequential circuit* approach, which we shall not present (consult [15], pp. 205–232).

3 (p. 293). We call the linear codes of Chapters 7–9 linear *block* codes to distinguish them from the linear *convolutional* codes of the present chapter. The terminology stems from the fact that block codes encode the data into independent blocks of length n, whereas in a convolutional code adjacent blocks of size n are interdependent. See Note 2, Chapter 7.

4 (p. 294). Other authors use different definitions of constraint length. Forney [15, pp. 213–232] defines the constraint length for the i-th input by $v_i = \max_j [\deg(g_{ij})]$, and the overall constraint length by $v = \sum v_i$. Massey [21] defines the constraint length to be $(M + 1)n$. Our definition agrees with that of Viterbi and Omura [26].

5 (p. 298). Missing from Fig. 10.3, but an important component of a practical encoder, is a mulitplexing switch that would send alternate bits from C_0 and C_1 down the channel.

6 (p. 305). The trellis diagram is conceptually convenient, but not strictly necessary for the implementation of a Viterbi encoder. The state diagram alone can easily serve as a dynamic model for the algorithm's behavior.

7 (p. 305). The diagram is often called the *de Bruijn* graph in honor of N. G. de Bruijn, the contemporary Dutch mathematician who discovered many of its properties.

8 (p. 308). Other writers call this generating function the code's *transmission gain* or *transfer function*.

9 (p. 310). Since the code is linear, the various bounds on the error probability we shall compute are in fact independent of the assumed transmitted path. (See the last two sentences in the proof of Theorem 7.5, p. 155.)

10 (p. 313). The material in this section is presented in the complete reverse of the historical order of discovery, which is as follows: Fano algorithm (1963), Stack algorithm (1969), analytic justification of the Fano metric (1972). The approach taken here is due to Massey [21].

11 (p. 317). Actually, the "stack" is better viewed as a "priority queue," a data structure defined in Chapter 4 of Aho, Hopcroft, and Ullman [1].

11

Variable-length source coding[1]

11.1 Introduction

Consider a discrete memoryless source with source statistics $\mathbf{p} = (p_0, p_1, \ldots, p_{r-1})$, that is, a sequence $U_1, U_2, \ldots$ of independent, identically distributed random variables with common distribution $P\{U = i\} = p_i$, $i = 0, 1, \ldots, r - 1$. According to the results of Chapter 3, it is in principle possible to represent this source faithfully using an average of $H_2(\mathbf{p}) = -\sum_{i=0}^{r-1} p_i \log_2 p_i$ bits per source symbol.[2] In this chapter we study an attractive constructive procedure for doing this, called *variable-length source coding*.

To get the general idea (and the reader is warned that the following example is somewhat meretricious), consider the particular source $\mathbf{p} = \left(\frac{1}{2}, \frac{1}{4}, \frac{1}{8}, \frac{1}{8}\right)$, whose entropy is $H_2(\frac{1}{2}, \frac{1}{4}, \frac{1}{8}, \frac{1}{8}) = 1.75$ bits. The source alphabet is $A_U = \{0, 1, 2, 3\}$; now let us encode the source sequence $U_1, U_2, \ldots$ according to Table 11.1. For example, the source sequence 03220100... would be encoded into 011111011001000 .... Clearly the average number of bits required per source symbol using this code is $\frac{1}{2} \cdot 1 + \frac{1}{4} \cdot 2 + \frac{1}{8} \cdot 3 + \frac{1}{8} \cdot 3 = 1.75$, the source entropy. This fact in itself is not surprising, since we could, for example, get the average down to only one bit per symbol by encoding everything into 0! The remarkable property enjoyed by the code of Table 11.1 is that it is *uniquely decodable*, that is, the source sequence can be reconstructed perfectly from the encoded stream. For example, the bit stream 1001011100010110... could have arisen only from the source sequence 101300012 .... (The fact that this particular code is uniquely decodable can be proved in several ways—for example by observing that no codeword is a prefix of any other codeword.) Hence the code of Table 11.1 does in fact represent the source $\left(\frac{1}{2}, \frac{1}{4}, \frac{1}{8}, \frac{1}{8}\right)$ faithfully using only an average of 1.75 bits per symbol.

330

Table 11.1 A variable-length code for the source $\left(\frac{1}{2}, \frac{1}{4}, \frac{1}{8}, \frac{1}{8}\right)$.

Source Symbol	Probability	Codeword
0	$\frac{1}{2}$	0
1	$\frac{1}{4}$	10
2	$\frac{1}{8}$	110
3	$\frac{1}{8}$	111

In the rest of this chapter we generalize the above example to arbitrary DMS's which are to be represented faithfully over arbitrary, not just binary, alphabets. In Section 11.2 we study the purely combinatorial problem of unique decodability. In Section 11.3 we match codes to sources. In Section 11.4 we present Huffman's famous algorithm for constructing optimal variable-length codes for a given source.

11.2 Uniquely decodable variable-length codes

Let S be a finite set containing s elements; unless otherwise specified, $S = \{0, 1, \ldots, s-1\}$. A *string of length* k over S is an ordered sequence $s_1 s_2 \ldots s_k$ of k elements from S. The *empty string*, denoted by $\varnothing$, is the string containing no symbols. If $\sigma = s_1 s_2 \ldots s_k$ and $\tau = t_1 t_2 \ldots t_l$ are strings, their *concatenation* $\sigma * \tau$ is the string $s_1 s_2 \ldots s_k t_1 t_2 \ldots t_l$. If $\sigma = \sigma_1 * \sigma_2 * \sigma_3$, σ_1 is called a *prefix*, σ_2 a *substring*, and σ_3 a *suffix* of σ. Thus the empty string $\varnothing$ is a prefix, suffix, and substring of every string. The *length* of a string σ is denoted by $|\sigma|$. It follows that $|\sigma * \tau| = |\sigma| + |\tau|$ for any pair of strings σ and τ.

A (variable-length) *code*[3] over S is a finite set of strings over S. The strings comprising a code C are called the *codewords* of C. Let C_1 and C_2 be two codes. Their *product*, denoted $C_1 * C_2$, consists of all strings of the form $\sigma_1 * \sigma_2$, $\sigma_1 \in C_1$, $\sigma_2 \in C_2$. The product of a code with itself k times will be denoted by C^k, that is, $C^k = C * C * \ldots * C$ (k factors).

The code C is called *uniquely decodable* (UD) if each string in each C^k arises in only one way as a concatenation of codewords. This means that if say $\tau_1 * \tau_2 * \cdots * \tau_k = \sigma_1 * \sigma_2 * \cdots * \sigma_k$ and each of the τ's and σ's is a codeword, then $\tau_1 = \sigma_1, \tau_2 = \sigma_2, \ldots, \tau_k = \sigma_k$. Thus every string in C^k can be uniquely decoded into a concatenation of codewords. (For an equivalent definition of UD, see Prob. 11.1)

Example 11.1 Let $s = 2$, and consider the codes $C_1 = \{0, 10, 110, 111\}$ and $C_2 = \{0, 10, 100, 101\}$. C_1 is UD (in fact it is even a prefix code; see the definition immediately below), but C_2 is not UD. For example, in C_2 $10 * 101 * 0 = 101 * 0 * 10$, so the string 101010 cannot be uniquely decoded with respect to C_2. $\qquad\square$

There is a special class of UD codes, called *prefix codes*,[4] which will be sufficient for our purposes. A code C is called a prefix code if no codeword is a prefix of any other. Prefix codes are necessarily UD (see Prob. 11.3), but the converse is not true. For example, $\{0, 01\}$ is UD in spite of the fact that 0 is a prefix of 01. The code C_1 of Example 11.1 is a prefix code, but C_2, not being UD, cannot be.

The problem to be posed and solved in this section is this: given a list $(n_0, n_1, \ldots, n_{r-1})$ of r nonnegative integers, does there exist a UD code $\{\sigma_0, \sigma_1, \ldots, \sigma_{r-1}\}$ over S with $|\sigma_i| = n_i$, $i = 0, 1, \ldots, r - 1$? The answer to this question turns out to be Yes, iff

$$\sum_{i=0}^{r-1} s^{-n_i} \leqslant 1. \tag{11.1}$$

This important inequality is called the *Kraft–McMillan* (KM) inequality. In Theorem 11.1 we shall show that if the KM inequality is not satisfied, no such code can exist. In Theorem 11.2 we shall show conversely that if KM is satisfied, not only a UD code but even a prefix code exists with lengths $n_0, n_1, \ldots, n_{r-1}$. It is because of these two results that nonprefix UD codes are rarely considered; prefix codes are quite easy to implement (see Prob. 11.3), and if a UD code exists with a certain set of lengths, a prefix code will also exist.

Theorem 11.1 (*McMillan*). *If* $C = \{\sigma_0, \sigma_1, \ldots, \sigma_{r-1}\}$ *is a UD code over* $S = \{0, 1, \ldots, s - 1\}$ *and if* $n_i = |\sigma_i|$, *then* $\sum_{i=0}^{r-1} s^{-n_i} \leqslant 1$.

Proof Let k be a positive integer. Then since C is UD,

$$\left(\sum_{i=0}^{r-1} s^{-n_i}\right)^k = \left(\sum_{\sigma \in C} s^{-|\sigma|}\right)^k = \sum_{\sigma \in C^k} s^{-|\sigma|} = \sum_{l=k \cdot n_{\min}}^{k \cdot n_{\max}} K_{l,k} s^{-1},$$

where $n_{\min}$ is the smallest n_i, $n_{\max}$ is the largest, and $K_{l,k}$ is the number of strings $\sigma_{i_1} * \sigma_{i_2} * \cdots * \sigma_{i_k}$ from C^k of length l. Since the set S has s elements, we know that $K_{l,k} \leqslant s^l$. Hence the final sum above is $\leqslant k \cdot n_{\max}$, that is,

$$\left(\sum_{i=0}^{r-1} s^{-n_i}\right)^k \leqslant k \cdot n_{\max},$$

for all k. Taking kth roots of both sides of this inequality and passing to the limit as $k \to \infty$, we obtain (11.1). □

Example 11.2 Take $s = 3$. Then no UD code with lengths $(1, 2, 2, 2, 2, 2, 3, 3, 3, 3)$ could exist since $\sum 3^{-n_i} = 28/27$. This fact would not be easy to prove directly. □

Theorem 11.2 (*Kraft*). *If $\sum_{i=0}^{r-1} s^{-n_i} \leqslant 1$, there exists a prefix (a fortiori UD) code with lengths n_i.*

Proof Let us reorder the n_i's so that $n_0 \leqslant n_1 \leqslant \cdots \leqslant n_{r-1}$. Define integers w_j, $j = 0, 1, \ldots, r-1$, by $w_0 = 0$, $w_j = \sum_{i=0}^{j-1} s^{n_j - n_i}$ for $j \geqslant 1$. Since $\sum_{i=0}^{r-1} s^{-n_i} \leqslant 1$, it follows that $w_j \leqslant s^{n_j} - 1$ for all j. Now define the string σ_j to be the s-ary representation of the integer w_j, with enough initial 0's added to make σ_j have length n_j. We claim that $\{\sigma_0, \sigma_1, \ldots, \sigma_{r-1}\}$ is a prefix code. For if σ_j were a prefix of σ_k for some $j < k$ it would follow that

Table 11.2 An application of Theorem 11.2.

j	n_j	w_j	σ_j
0	1	0	0
1	1	1	1
2	2	6	20
3	2	7	21
4	3	24	220
5	3	25	221
6	4	78	2220
7	4	79	2221
8	4	80	2222

$w_j = \lfloor w_k / s^{n_k - n_j} \rfloor.$ But $w_k / s^{n_k - n_j} = \sum_{i=0}^{k-1} s^{n_j - n_i} = w_j + \sum_{i=j}^{k-1} s^{n_j - n_i} \geqslant w_j + 1$, a contradiction. $\qquad\square$

Example 11.3 Take $s = 3$ with lengths $(1, 1, 2, 2, 3, 3, 4, 4, 4)$. Then $\sum 3^{-n_i} = 1$. The construction of Theorem 11.2 can be arranged as in Table 11.2. Notice that, if we try to apply this construction to the list in Example 11.2, we get $w_9 = 27 = 1000$, but $n_9 = 3$—so the construction fails simply because there is no string of length 3 that represents 27! (For a more transparent form of the construction of Theorem 11.2, see Probs. 11.6 and 11.7.) $\qquad\square$

11.3 Matching codes to sources

Having now learned something about the purely combinatorial properties of UD codes, we are prepared to match codes to sources for efficient communication.

Recall (p. 329) that a given DMS with source alphabet $A = \{0, 1, \ldots, r-1\}$ is completely specified by the probability vector $\mathbf{p} = (p_0, p_1, \ldots, p_{r-1})$. An *s-ary code* for the source $\mathbf{p}$ is defined to be a mapping of A onto a code $\{\sigma_0, \sigma_1, \ldots, \sigma_{r-1}\}$ over $S = \{0, 1, \ldots, s-1\}$. The idea is that if we wish to communicate the output of the source $\mathbf{p}$ over a noiseless channel with channel input alphabet S, we map the source symbol i into the codeword σ_i and transmit σ_i. If for example we use this strategy to transmit k source symbols, say $i_1, i_2, \ldots, i_k$, the receiver will obtain the string $\sigma = \sigma_{i_1} * \sigma_{i_2} * \cdots * \sigma_{i_k}$. If we want the receiver to be able to recover $i_1, \ldots, i_k$ from σ, we must assume that the code is UD.

The *average length* of the code $\{\sigma_0, \sigma_1, \ldots, \sigma_{r-1}\}$ with respect to the source $\mathbf{p} = (p_0, \ldots, p_{r-1})$ is defined to be $n = \sum_{i=0}^{r-1} p_i |\sigma_i|$. Its practical significance is that the number of s-ary symbols (i.e., symbols from S) required to encode the first k source symbols is likely to be about $k \cdot n$ for large k. Thus n is a measure of the average loading on the channel imposed by the use of the code. Theorem 11.3 (which actually follows from results in Part one) gives an interesting lower bound on n. We assume that n_i denotes the length of the ith codeword in C, that is, $n_i = |\sigma_i|$.

Theorem 11.3 *If the code C is UD, its average length must exceed the s-ary entropy of the source, that is,*

$$n \geqslant H_s(\mathbf{p}) = -\sum_{i=0}^{r-1} p_i \log_s p_i.$$

Proof

$$H_s(\mathbf{p}) - n = \sum_{i=0}^{r-1} p_i \log_s \frac{1}{p_i} - \sum_{i=0}^{r-1} p_i n_i$$

$$= \sum_{i=0}^{r-1} p_i \log_s \frac{s^{-n_i}}{p_i}$$

$$\leqslant \log_s \sum_{i=0}^{r-1} s^{-n_i},$$

the last step by Jensen's inequality (see Appendix B). But by Theorem 11.1 $\sum_{i=0}^{r-1} s^{-n_k} \leqslant 1$, and so $H_s(\mathbf{p}) - n \leqslant 0$. $\qquad\square$

Let us now denote the smallest possible average length for a UD s-ary code for $\mathbf{p}$ by $n_s(\mathbf{p})$. Theorem 11.3 asserts that $n_s(\mathbf{p}) \geqslant H_s(\mathbf{p})$. In fact, an examination of the proof shows that $n_s(\mathbf{p}) = H_s(\mathbf{p})$ iff each p_i is a negative integral power of s (Prob. 11.8). This result explains why the source $\left(\frac{1}{2}, \frac{1}{4}, \frac{1}{8}, \frac{1}{8}\right)$ discussed in Section 11.1 has such a nice binary code. In general, of course, we cannot expect to be so lucky, and it is natural to wonder about the actual value of $n_s(\mathbf{p})$. The exact calculation must be left to Section 11.4, but the following estimate is quite useful.

Theorem 11.4

$$H_s(\mathbf{p}) \leqslant n_s(\mathbf{p}) < H_s(\mathbf{p}) + 1.$$

[*Note. The lower bound of Theorem 11.3 is repeated here for aesthetic reasons.*]

Proof Define $n_i = \lceil \log_s p_i^{-1} \rceil$, $i = 0, 1, \ldots, r-1$. Then $s^{-n_i} \leqslant p_i$, and so $\sum_{i=0}^{r-1} s^{-n_i} \leqslant 1$. Hence by Theorem 11.2 there exists a UD code with lengths $n_0, n_1, \ldots, n_{r-1}$. The average length of such a code is $\sum_{i=0}^{r-1} p_i n_i$, but since $n_i < \log_s p_i^{-1} + 1$, this sum is $< \sum_{i=0}^{r-1} p_i(\log_s p_i^{-1} + 1) = H_s(\mathbf{p}) + 1$. $\qquad\square$

Example 11.4 Let $\mathbf{p} = (.1, .4, .5)$, $s = 2$. Then $H_s(\mathbf{p}) = 1.361$. The n_i's suggested in the proof of Theorem 11.4 are $(4, 2, 1)$, resulting in an average length of 1.7. However, the lengths $(2, 2, 1)$ also satisfy $\sum 2^{-n_k} \leqslant 1$, and so in fact $n_s(\mathbf{p}) \leqslant 1.5$. The point is that the upper bound of Theorem 11.4 is usually not sharp (but see Prob. 11.9); its merit lies in its great generality. As mentioned previously, a way to construct optimal UD codes for a given source will be given in Section 11.4.   □

At this point we have a problem. On one hand the results of Chapter 3 tell us that it should be possible to encode the source $\mathbf{p}$ faithfully using, on the average, only $H_s(\mathbf{p})$ s-ary symbols per source symbol. On the other hand, we have just seen that it is not possible to do this with a UD s-ary code for $\mathbf{p}$, since $n_s(\mathbf{p})$ is usually strictly larger than $H_s(\mathbf{p})$.

The solution to this problem is to consider codes for the *extended* sources $\mathbf{p}^m$, $m = 1, 2, \ldots$. The source $\mathbf{p}^m$ is defined to be the source with source alphabet A^m (recall that A is the source alphabet for $\mathbf{p}$), in which the source symbol $(u_1, u_2, \ldots, u_m)$ is assigned the probability $P\{U_1 = u_1, \ldots, U_m = u_m\} = p_{u_1} p_{u_2} \cdots p_{u_m}$. In effect, when we consider the source $\mathbf{p}^m$, we are blocking the source sequence $U_1, U_2, \ldots$ into consecutive blocks of m symbols from A, and regarding each such block as a single symbol from the alphabet A^m. Now a UD code for $\mathbf{p}^m$ with average length n_m will require an average of n_m s-ary symbols to represent one symbol from A^m. One symbol from A^m represents m symbols from the original source, and so this code faithfully represents the original source $\mathbf{p}$, using only n_m/m s-ary symbols per source symbol. The following theorem shows that by taking m large enough and using this technique, we can make the average number of symbols needed to represent $\mathbf{p}$ faithfully as close to $H_s(\mathbf{p})$ as desired.

Theorem 11.5

$$\lim_{m \to \infty} \frac{1}{m} n_s(\mathbf{p}^m) = H_s(\mathbf{p}).$$

Proof It is easy to see that $H_s(\mathbf{p}^m) = m H_s(\mathbf{p})$ (Prob. 11.12). Hence by Theorem 11.4 $m H_s(\mathbf{p}) \leq n_s(\mathbf{p}^m) < m H_s(\mathbf{p}) + 1$. Theorem 11.5 follows by dividing by m and taking limits.   □

Theorem 11.5 is satisfying theoretically, since it tells us that the source $\mathbf{p}$ can indeed be represented faithfully using (perhaps slightly more than) $H_s(\mathbf{p})$ s-ary symbols per source symbol. It leaves something to be desired from a constructive viewpoint, however, since it relies on the weak construction of Theorem 11.4. In the next section we shall remedy this situation by presenting a technique which will enable us to construct the best possible codes for the sources $\mathbf{p}^m$.

11.4 The construction of optimal UD codes (Huffman's algorithm)

According to Theorem 11.4, $n_s(\mathbf{p})$ lies somewhere between $H_s(\mathbf{p})$ and $H_s(\mathbf{p}) + 1$, and this estimate is adequate for some purposes (e.g., the proof of Theorem 11.5). But it is natural to wonder about the exact value of $n_s(\mathbf{p})$ for a fixed s and $\mathbf{p}$. In this section we present an alogrithm due to David Huffman that shows not only how to compute $n_s(\mathbf{p})$, but also how to construct a UD (indeed a prefix) code with average length $n_s(\mathbf{p})$.

Before giving a formal description of Huffman's algorithm, we shall work an example. Throughout this section we refer to a UD s-ary code for $\mathbf{p}$ whose average length is $n_s(\mathbf{p})$ as an *optimal* code for $\mathbf{p}$.

Example 11.5 Let $s = 4$, $\mathbf{p} = (.24, .21, .17, .13, .10, .07, .04, .03, .01)$. The first step in Huffman's algorithm is to replace the probability vector $\mathbf{p}$ with a simpler one $\mathbf{p}'$, which is obtained from $\mathbf{p}$ by combining the three smallest probabilities in $\mathbf{p}$. Thus $\mathbf{p}' = (.24, .21, .17, .13, .10, .08, .07)$ after the components are rearranged into decreasing order. Since $\mathbf{p}'$ is still complicated, we reduce $\mathbf{p}'$ still further by combining the four smallest probabilities in $\mathbf{p}'$ and obtain $\mathbf{p}'' = (.38, .24, .21, .17)$. Figure 11.1 gives a schematic diagram of these reductions. The reason why we combined three probabilities when going from $\mathbf{p}$ to $\mathbf{p}'$ and four when going from $\mathbf{p}'$ to $\mathbf{p}''$ will appear as the theory develops; for the moment just accept it.

It is of course clear how to construct an optimal code for $\mathbf{p}''$: the code $C'' = \{0, 1, 2, 3\}$ achieves $n_4(\mathbf{p}'') = 1$. Starting from this trivial code, we now can work backward and construct optimal codes C' and C for $\mathbf{p}'$ and $\mathbf{p}$ by "expanding" the code C'' in a simple way.

First we construct an optimal code for $\mathbf{p}'$. Notice that three of the probabilities (viz., .24, .21, .17) were not changed in the reduction from $\mathbf{p}'$ to $\mathbf{p}''$. The rule in this case is that in the expansion of C'' into C' the

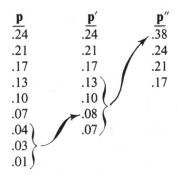

Figure 11.1 The successive reductions of $\mathbf{p}$.

corresponding codewords do not change, either. However, probability .38 in
p″ expands into four probabilities (.13, .10, .08, .07) in **p′**. Here the rule is
that the codeword (0) for .38 in $C″$ expands into four codewords
(00, 01, 02, 03) in $C′$ (see Fig. 11.2). The resulting code, according to
Theorem 11.7 below, is optimal for **p′**.

The construction of C from $C′$ proceeds similarly: every codeword but 02
corresponds directly to a single probability in **p**, so the corresponding codewords do not change. However, 02 expands into 020, 021, 022 (again see
Fig. 11.2). Thus allegedly $C = \{1, 2, 3, 00, 01, 03, 020, 021, 022\}$ is an
optimal code for **p**, and $n_4(\mathbf{p}) = 1 \cdot (.24 + .21 + .17) + 2(.13 + .10 + .07)$
$+ 3(.04 + .03 + .01) = 1.46.$ □

The preceding example is typical of the general Huffman algorithm: **p** is
successively reduced to **p′**, **p″**, etc., until a final reduction $\mathbf{p}^{(j)}$ with exactly s
probabilities is reached. The obvious optimal code $\{0, 1, \ldots, s - 1\}$ for $\mathbf{p}^{(j)}$
is then "expanded" in the above way until an optimal code for **p** is obtained.
The only mysterious feature of the algorithm is the computation of the number
of probabilities that must be combined in the reduction of **p** to **p′**. If we
denote this number by $s′$, it turns out that $s′$ is determined uniquely by the
following two conditions:

$$s' \in \{2, 3, \ldots, s\}, \tag{11.2}$$

$$s' \equiv r \pmod{s - 1}. \tag{11.3}$$

For example (cf. Example 11.5), of $s = 4$, $r = 9$, we get $s′ = 3$. If $s = 2$, then
$s′ = 2$ for all $r \geq 2$, and so this complication is absent for binary codes.[5] If
$s = 3$, then $s′ = 2$ if r is even and $s′ = 3$ if r is odd. Note that the number of
probabilities in **p′** is $r′ = r - s′ + 1$, which by Eq. (11.3) is congruent to 1

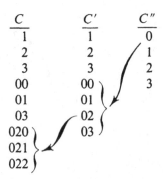

Figure 11.2 The synthesis of an optimal code for **p**. (Cf. Fig. 11.1.)

(mod $s - 1$). Hence after the first reduction s' will always be equal to s, and so it is necessary to compute s' using Eqs. (11.2) and (11.3) only once, no matter how many reductions of **p** are required.

Now we have described Huffman's algorithm; our remaining task—which is surprisingly tricky—is to show that it works, i.e. that it produces an s-ary prefix code for **p** whose average length is as small as possible. The following theorem is crucial; it guarantees that there is always an optimal prefix code for **p** in which the last s' words are of the same length, and agree except for their last component.

To fix our ideas, let us now assume that the probabilities $p_0, p_1, \ldots, p_{r-1}$ are arranged in decreasing order: $p_0 \geqslant p_1 \geqslant \cdots \geqslant p_{r-1}$. We consider prefix codes for **p** over the alphabet $\{0, 1, \ldots, s - 1\}$ whose words are denoted by $\sigma_0, \sigma_1, \ldots, \sigma_{r-1}$, with lengths $n_i = |\sigma_i|$, $i = 0, 1, \ldots, r - 1$.

Theorem 11.6 *If $r \geqslant 2$, there exists an optimal s-ary prefix code for **p** with the following two properties*:

(a) $n_0 \leqslant n_1 \leqslant \cdots \leqslant n_{r-1}$.
(b) *The last s' (see Eqs. (11.2) and (11.3)) codewords are identical except for their last component, that is, there exists a string σ of length $n_{r-1} - 1$ such that*:

$$\sigma_{r-s'} = \sigma * 0,$$

$$\sigma_{r-s'+1} = \sigma * 1,$$

$$\vdots$$

$$\sigma_{r-1} = \sigma * (s' - 1).$$

Proof For a given code $\{\sigma_0, \sigma_1, \ldots, \sigma_{r-1}\}$ for **p**, it is clear that if $i < j$ and $|\sigma_i| > |\sigma_j|$, interchanging σ_i and σ_j cannot increase the average length $\sum p_i |\sigma_i|$. Hence there exist optimal s-ary codes for **p** such that (a) holds. In the rest of the proof, "optimal" code will mean a code of minimal average length that also satisfies (a).

Now **p** may have several essentially different optimal codes (see Example 11.6). Among all such codes, let us select one for which $\sum n_i$ is as small as possible. By Theorem 11.1 we know that

$$s^{n_{r-1}} - \sum_{i=0}^{r-1} s^{n_{r-1}-n_i} = \Delta \geqslant 0. \tag{11.4}$$

If $\Delta \geqslant s - 1$, the lengths $(n_0, n_1, \ldots, n_{r-2}, n_{r-1} - 1)$ would satisfy the KM inequality, and so by Theorem 11.2 there would exist a prefix code with these lengths. This new code must also be optimal (and of course $p_{r-1} = 0$), but this would contradict the assumption that $\sum n_i$ is minimal. We conclude that $0 \leqslant \Delta \leqslant s - 2$, or, what is the same thing,

$$s - \Delta \in \{2, 3, \ldots, s\}. \tag{11.5}$$

Next, let r' denote the number of words of length n_{r-1} in the code. If $r' = 1$, we could shorten the unique longest codeword by deleting its last component without violating the prefix condition, again a contradiction. Hence

$$r' \geqslant 2. \tag{11.6}$$

If we examine Eq. (11.4) modulo s, we get $\Delta \equiv -r' \pmod{s}$, that is, $r' \equiv s - \Delta \pmod{s}$. But from Eqs. (11.5) and (11.6) it follows that

$$r' = ks + (s - \Delta) \qquad \text{for some } k \geqslant 0. \tag{11.7}$$

If we examine Eq. (11.4) modulo $s - 1$, we get $1 - r \equiv \Delta \pmod{s - 1}$, that is $s - \Delta \equiv r \pmod{s - 1}$. It follows from this congruence and (11.5) that $s - \Delta = s'$, where s' was defined by Eqs. (11.2) and (11.3). Thus from (11.7) we conclude that $r' \geqslant s'$, that is, there are at least s' longest codewords.

Next observe that if $\sigma' = \sigma * a$ is one of these r' codewords of length n_{r-1}, where σ is a string of length $n_{r-1} - 1$ and $a \in \{0, 1, \ldots, s - 1\}$, we can add each of the strings $\sigma * 0, \sigma * 1, \ldots, \sigma * (s - 1)$ not already present to the code without violating the prefix condition. Thus by replacing codewords of length n_{r-1} not of the form $\sigma * a$ with ones that are, we arrive at an optimal code which has $\sigma * 0, \sigma * 1, \ldots, \sigma * (s' - 1)$ among its codewords of length n_{r-1}. Finally, be rearranging the longest codewords if necessary, we may assume that conclusion (b) holds.   □

Before stating our main result, we need a definition. If $p_0 \geqslant p_1 \geqslant \cdots \geqslant p_{r-1}$ and $r \geqslant s$, the *s-ary Huffman reduction* of $\mathbf{p}$ is defined by

$$\mathbf{p}' = (p_0, p_1, \ldots, p_{r-s'-1}, p_{r-s'} + \cdots + p_{r-1}).$$

In words, $\mathbf{p}'$ is obtained from $\mathbf{p}$ by combining the smallest s' probabilities in $\mathbf{p}$, where s' is defined by Eqs. (11.2) and (11.3). The following theorem shows how an optimal code for $\mathbf{p}$ may be constructed from one for $\mathbf{p}'$. Since $\mathbf{p}'$ has fewer symbols than $\mathbf{p}$, it yields a recursive procedure for constructing optimal codes for $\mathbf{p}$.

Theorem 11.7 *If* $C' = \{\tau_0, \tau_1, \ldots, \tau_{r-s'-1}, \tau\}$ *is an optimal code for* $\mathbf{p}'$, *then*

$$C = \{\tau_0, \ldots, \tau_{r-s'-1}, \tau * 0, \tau * 1, \ldots, \tau * (s' - 1)\} \tag{11.8}$$

is an optimal code for $\mathbf{p}$. *Hence also*

$$n_s(\mathbf{p}) = n_s(\mathbf{p}') + p_{r-s'} + \cdots + p_{r-1}. \tag{11.9}$$

Proof We begin by proving (11.9). Let $(n_0, n_1, \ldots, n_{r-s'-1}, n)$ denote the lengths of the codewords in C'. By definition $p_0 n_0 + \cdots + p_{r-s'-1} n_{r-s'-1} + pn = n_s(\mathbf{p}')$, where $p = p_{r-s'} + \cdots + p_{r-1}$. If C is used as a code for $\mathbf{p}$, its average length is $n_s(\mathbf{p}') + p$; hence

$$n_s(\mathbf{p}) \leqslant n_s(\mathbf{p}') + p. \tag{11.10}$$

To prove the opposite inequality let $\{\sigma_0, \sigma_1, \ldots, \sigma_{r-1}\}$ be an optimal code for $\mathbf{p}$ that enjoys properties (a) and (b) of Theorem 11.6. The code $\{\sigma_0, \ldots, \sigma_{r-s'-1}, \sigma\}$, used for $\mathbf{p}'$, has average length $n_s(\mathbf{p}) - p$, that is,

$$n_s(\mathbf{p}') \leqslant n_s(\mathbf{p}) - p. \tag{11.11}$$

Combining Eqs. (11.10) and (11.11), we obtain (11.9). Hence the code C defined by (11.8) must be optimal, since its length is $n_s(\mathbf{p}') + p = n_s(\mathbf{p})$. $\square$

As promised, Theorem 11.7 yields a recursive alogrithm for constructing optimal s-ary prefix codes of any source $\mathbf{p} = (p_0, p_1, \ldots, p_{r-1})$. We can proceed inductively as follows. If $r \leqslant s$, the code $\{0, 1, \ldots, r-1\}$ is optimal for $\mathbf{p}$, and $n_s(\mathbf{p}) = 1$. If $r > s$, we assume inductively that we can construct an optimal code for any source with fewer that r symbols. Then we reduce $\mathbf{p}$ to $\mathbf{p}'$, which is a source with $r - s'$ symbols, inductively construct an optimal code for $\mathbf{p}'$, and expand it to an optimal code for $\mathbf{p}$ in the manner suggested by Theorem 11.7. As mentioned above, one comforting feature of this algorithm is that the annoying quantity s' need be computed only for the first reduction, $\mathbf{p} \to \mathbf{p}'$; in all future reductions it is equal to s. It is easy to show that the total number of source reductions required is $\lceil (r - s)/(s - 1) \rceil$; after this many reductions the source will have exactly s symbols and $\{0, 1, \ldots, s-1\}$ will be an optimal code (see Prob. 11.15).

Example 11.6 Let $s = 2$, $r = 4$, and $\mathbf{p} = (\frac{1}{3}, \frac{1}{3}, \frac{1}{6}, \frac{1}{6})$. The first reduction of the source is $(\frac{1}{3}, \frac{1}{3}, \frac{1}{3})$; the second is $(\frac{2}{3}, \frac{1}{3})$, for which $\{0, 1\}$ is an optimal code. The following sketch illustrates Huffman's construction:

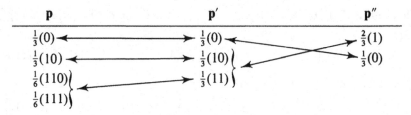

Alternatively, we could synthesize the code in the manner indicated below:

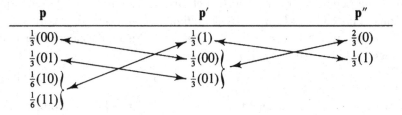

Thus $n_2\left(\frac{1}{3}, \frac{1}{3}, \frac{1}{6}, \frac{1}{6}\right) = 2$, although the ambiguity in the ordering of $\mathbf{p}'$ affords two essentially distinct optimal prefix codes for $\mathbf{p}$ (see Prob. 11.21). [Note. The binary source entropy is $H_2(\mathbf{p}) = 1.92$ bits.]

Example 11.7 Let $s = 2$, $r = 2$, $\mathbf{p} = (.9, .1)$. here $H_2(\mathbf{p}) = .469$, while Huffman's algorithm yields the following values:

$$n_2(\mathbf{p}) = 1.00,$$

$$\tfrac{1}{2}n_2(\mathbf{p}^2) = .645,$$

$$\tfrac{1}{3}n_3(\mathbf{p}^3) = .533,$$

$$\tfrac{1}{4}n_4(\mathbf{p}^4) = .493.$$

(These calculations should be verified by the reader; see Prob. 11.19.) The point here is that the convergence of $m^{-1}n_s(\mathbf{p}^m)$ to $H_s(\mathbf{p})$ (see Theorem 11.5) is usually quite rapid (but see Prob. 11.20).   □

Problems

11.1 The following definition of a uniquely decodable code is sometimes given: The code C is UD if $\tau_1 * \cdots * \tau_k = \sigma_1 * \cdots * \sigma_l$ (the τ's and σ's all being codewords) implies $k = l$ and $\tau_i = \sigma_i$ for $i = 1, 2, \ldots, k$. Show that this definition is equivalent to the one given in the text (see p. 330).

11.2 Let us call a UD code C *full* if it is impossible to adjoin a new codeword to C without violating the UD condition. A string σ is called *indecipherable* with respect to C if σ is not a substring of any finite concatenation of codewords of

C. If $C = \{\sigma_0, \sigma_1, \ldots, \sigma_{r-1}\}$, show that the following three conditions are equivalent:
(a) C is full.
(b) There exist no indecipherable strings with respect to C.
(c) $\sum_{i=0}^{r-1} s^{-|\sigma_i|} = 1$.

11.3 Prove that a prefix code C is necessarily UD, and design an efficient algorithm for decoding C, that is, for factoring a finite concatenation of codewords from C into its component codewords.

11.4 Show that the product of prefix codes is again a prefix code. Is the product of an arbitrary pair of UD codes necessarily UD? (See definition, p. 330.)

11.5 In this problem you are to try to construct UD codes over $S = \{0, 1, 2, 3\}$ with prescribed codeword lengths. In the following table k_i denotes the number of words of length i in the putative code:

		k_i		
i	Case 1	Case 2	Case 3	Case 4
1	3	2	1	0
2	3	7	7	7
3	3	3	3	3
4	3	3	7	11
5	4	5	4	3
6	0	0	0	4

In each of the four cases, construct the code or explain why such a code cannot exist.

11.6 A string σ over $S = \{0, 1, \ldots, s-1\}$ is said to be *lexicographically less* than another string τ (written $\sigma < \cdot \ \tau$) if σ and τ can be written as $\sigma = \rho_1 * i * \rho_2$, $\tau = \rho_1 * j * \rho_3$, with $i, j \in S$ and $i < j$. Show that lexicographic ordering is a total ordering, that is, for any pair (σ, τ) with $\sigma \neq \tau$, either $\sigma < \cdot \ \tau$ or $\tau < \cdot \ \sigma$.

11.7 (Continuation). Consider the following alternate fomulation of the construction in Theorem 11.2: Choose $\sigma_0 = 00 \ldots 0$ of length n_0, and inductively choose σ_{k+1} to be lexicographically least string of length n_{k+1} that has no prefixes in the set $\{\sigma_0, \sigma_1, \ldots, \sigma_k\}$. Show that this construction always yields the same code as the one given in Theorem 11.2.

11.8 Show that $n_s(\mathbf{p}) = H_s(\mathbf{p})$ iff there exist positive integers $(n_0, n_1, \ldots, n_{r-1})$ such that $p_i = s^{-n_i}$, $i = 0, 1, \ldots, r - 1$.

11.9 Fix the integer s, and show that the set $\{x : x = n_s(\mathbf{p}) - H_s(\mathbf{p}) \text{ for some } \mathbf{p}\}$ is dense in $[0, 1]$. In other words, for each $t \in [0, 1]$ and $\varepsilon > 0$, show that there exists a $\mathbf{p}$ such that $|n_s(\mathbf{p}) - H_s(\mathbf{p}) - t| < \varepsilon$.

11.10 The variable-length coding schemes of this chapter do not, as they stand, fit the model of Fig. 5.1, in which a fixed number k of source symbols is mapped onto a fixed number n of channel input symbols. This problem is designed to show how a variable-length code can be modified to fit that model. Throughout the source $\mathbf{p} = (p_0, p_1, \ldots, p_{r-1})$ will be fixed, and the channel will be a noiseless DMC with input and output alphabets both equal to $S = \{0, 1, \ldots, s-1\}$ and transition probabilities given by

$$p(y|x) = \begin{cases} 1 & \text{if } y = x, \\ 0 & \text{if } y \neq x. \end{cases}$$

Suppose we have found an optimal s-ary prefix code C with average length $n_s(\mathbf{p})$, and let n and k be fixed integers. Now consider the following strategy for mapping k source symbols $U_1, U_2, \ldots, U_k$ into n symbols from S. First, map each source symbol into the corresponding codeword from C, and form the concatenation σ of these codewords. The length of σ as a string over S is a random variable, but we modify σ so that it has length n as follows. If the length of σ exceeds n, we delete the last $|\sigma| - n$ symbols; if it is less that n, we adjoin $n - |\sigma|$ 0's. In this way every k-tuple of source symbols is mapped into an n-tuple of channel input symbols. Denote this n-tuple by σ'.

When the decoder receives σ', he decomposes it uniquely (see Prob. 11.3) into a string of the form $\sigma_1 * \sigma_2, * \cdots * \sigma_l * \tau$, where the σ_i are all codewords and τ is either the prefix of some codeword or a string of 0's. If $l \geq k$, well and good: the decoder has received all of the k source symbols perfectly. If, however, $l < k$, some of the source symbols did not arrive, that is, they were *erased* by the system. Denote by P_E the probability that at least one of the k symbols in the source block is erased. Let $\varepsilon > 0$ be arbitrary.

(a) Show that for any $\nu > n_s(\mathbf{p})$ there exists a scheme of the above sort with $n/k \leq \nu$ and $P_E < \varepsilon$.

(b) Using the results of part (a) and Theorem 11.5, show that for any $r < \lfloor H_s(\mathbf{p}) \rfloor^{-1}$ and $\varepsilon > 0$, there exists an (n, k) coding scheme of the type depicted in Fig. 5.1, with $k/n \geq r$ and $P\{V_i \neq U_i\} < \varepsilon$.

11.11 Let $\mathbf{U} = (U_1, U_2, \ldots, U_k)$ be a random vector, and let $n_s(\mathbf{U}) = n_s(\mathbf{p})$, where $\mathbf{p}$ is the probability distribution of $\mathbf{U}$. Now suppose $(U_1, U_2, \ldots)$ is a finite Markov chain (see Prob. 1.20). Show that $\lim_{k \to \infty} (1/k) n_s(U_1, \ldots, U_k)$ is equal to the s-ary entropy of the chain.

11.12 Prove that $H_s(\mathbf{p}^m) = m H_s(\mathbf{p})$ (see Theorem 11.5).

11.13 Define $n_s^{(\text{prefix})}(\mathbf{p})$ as the least possible average length for an s-ary prefix code for $\mathbf{p}$. Show that $n_s^{(\text{prefix})}(\mathbf{p}) = n_s(\mathbf{p})$.

11.14 Does Theorem 11.6 remain true if the restriction $r \geq 2$ is removed?

11.15 Show that, if $r \geq s$, after $\lceil (r-s)/(s-1) \rceil$ s-ary reductions of $\mathbf{p} = (p_0, p_1, \ldots, p_{r-1})$, the resulting source has exactly s symbols.

11.16 One popular way of avoiding the problem that in the first Huffman reduction of a source $\mathbf{p}$ the number of symbols combined may be less than s is to adjoin $s - s'$ extra "dummy" symbols to the source alphabet, each occurring with probability 0. If this is done, then every reduction, even the first, involves a collapsing of exactly s probabilities. (For example, in Example 11.5 $s - s' = 1$, and the source would become $\mathbf{p} = (.24, .21, .17, .13, .10, .07, .04, .03, .01, .00)$.) If one tries to prove that Huffman's construction works by using this artifice, however, a problem develops: an optimal code for $(\mathbf{p}, 0, \ldots, 0)$, when restricted to $\mathbf{p}$ itself, may no longer be optimal! Give an example of this pathological behavior.

11.17 Show that, for fixed s, the amount of computation required by Huffman's algorithm to find an optimal s-ary code for $\mathbf{p} = (p_0, \ldots, p_{r-1})$ is $O(r^2)$. How does the implied constant depend on s?

11.18 (Continuation). Suppose you are interested only in the vaule of $n_s(\mathbf{p})$ and not in finding an optimal code. Using Eq. (11.9), design an alogrithm for computing $n_s(\mathbf{p})$. Can you make its running time $o(r^2)$?

11.19 Verify the details of Example 11.7.

11.20 In the light of Theorem 11.5, answer the following:
  (a) Is it possible for $n_s(\mathbf{p}) < \frac{1}{2}n_s(\mathbf{p}^2)$?
  (b) Is it possible for $\frac{1}{2}n_s(\mathbf{p}^2) < \frac{1}{3}n_s(\mathbf{p}^3)$?

11.21 In Example 11.6 we noted that $\mathbf{p} = (\frac{1}{3}, \frac{1}{3}, \frac{1}{6}, \frac{1}{6})$ has two optimal binary prefix codes, namely, $\{00, 01, 10, 11\}$ and $\{0, 10, 110, 111\}$. Show that the set of all probability vectors (p_1, p_2, p_3, p_4) with $p_1 \geqslant p_2 \geqslant p_3 \geqslant p_4$ for which these same two codes are optimal is the convex hull of three fixed probability vectors; and find these three vectors.

11.22 Consider a game of "twenty questions" in which you are required to determine, after asking a certain number of questions that can be answered "yes" or "no," the outcome of one roll of a pair of dice. (The outcome is therefore one of the integers 2, 3, ..., 12.) What is the minimum number of questions you need to ask, on the average? [*Hint*: if you asked "Is it 2?" "Is it 3?" etc., you would average a little under six questions. It is possible to do better, however.]

11.23 The object of this problem is for you to learn how to design an optimal *binary* prefix code for the symmetric r-ary source $\mathbf{p}_r = (1/r, 1/r, \ldots, 1/r)$.
  (a) Construct such a code for $\mathbf{p}_2$, $\mathbf{p}_3$, $\mathbf{p}_4$, $\mathbf{p}_5$.
  (b) In general, what are the possible lengths of the words in an optimal binary prefix code for $\mathbf{p}_r$?
  (c) Compute $n_2(\mathbf{p}_r)$, in general.

Notes

1 (p. 330). It is unorthodox for the subject matter of this chapter to be presented so late in a text on information theory, since it is not particularly deep and really does not depend on any of the preceding material. However, we have chosen to place it here because it represents a constructive approach to the source coding problem, just as the material in Chapters 7–10 represents a constructive approach to the channel coding problem.

2 (p. 330). Although it is not necessary for an understanding of the material of this chapter, we pause to put our problem into the language of Chapter 3. We are dealing with a DMS relative to the *Hamming distortion*, that is, the source and destination alphabets are both equal to $\{0, 1, \ldots, s - 1\}$ and the distortion measure is

$$d(u, v) = \begin{cases} 0 & \text{if } u = v, \\ 1 & \text{if } u \neq v. \end{cases}$$

We are interested only in zero distortion, that is, we want to reproduce the source perfectly (technically we allow an error probability of ε for small ε) at the destination.

Since in this case the rate-distortion function $R(\delta)$ evaluated at the desired distortion ($\delta = 0$) is the source entropy $H_2(\mathbf{p})$ (cf. Fig. 3.3), it follows (nonconstructively!) from the source coding Theorem 3.4 that it is possible to represent the source faithfully at the destination using only $H_2(\mathbf{p})$ bits per source symbol.

3 (p. 331). Note that this kind of code is quite different from the block codes of Chapters 7–9, and the convolutional codes of Chapter 10. However, a variable-length code can be transformed, in a certain sense, into a block code (see Prob. 11.10).

4 (p. 332). There is a close connection between s-ary prefix codes and *s-ary trees*, which is emphasised in most texts, and we should mention it at least briefly. To illustrate the idea, consider the prefix code $\{0, 10, 110, 111\}$. The corresponding binary tree is depicted as follows:

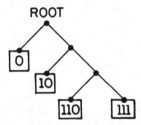

In general, each codeword represents a "leaf" on the tree, and the codeword's bit pattern describes the unique path from the root to the leaf—a leftward branch is denoted by a "0." and a rightward one by a "1." For details, consult Knuth [7], Vol. 1, Section 2.3.

5 (p. 338). Indeed, the entire proof that Huffman's algorithm works is comparatively trivial in the binary case. Most texts (e.g., Abramson [12], Gallager [17]) prove only the $s = 2$ case and leave the (nontrivial!) generalization as a problem. (See also Prob. 11.16.)

12
Survey of advanced topics for Part two

12.1 Introduction

This chapter serves the same function for Part two as Chapter 6 served for Part one, that is, it summarizes some of the most important results in coding theory which have not been treated in Chapters 7–11. In Sections 12.2, 12.3, and 12.4 we treat channel coding (block codes, convolutional codes, and a comparison of the two). Finally in Section 12.5 we discuss source coding.

12.2 Block codes

The theory of block codes is older and richer than the theory of convolutional codes, and so this section is much longer than Section 12.3. (This imbalance does not apply to practical applications, however; see Section 12.4.) In order to give this section some semblance of organization, we shall classify the results to be cited according to Berlekamp's [15] list of the three major problems of coding theory:

1. How good are the best codes?
2. How can we design good codes?
3. How can we decode such codes?

• *How good are the best codes?* One of the earliest problems which arose in coding theory was that of finding *perfect codes*. If we view a code of length n over the finite field F_q as a subset $\{x_1, x_2, \ldots, x_M\}$ of the vector space $V_n(F_q)$, the code is said to be perfect (or close packed) if for some integer e, the Hamming spheres of radius e around the M codewords completely fill $V_n(F_q)$ without overlap. By 1950 several kinds of perfect codes were known:

n	q	e	Remarks
$2e+1$	2	(arbitrary)	Repetition codes; see Prob. 7.18
$(q^m - 1)/(q - 1)$	(arbitrary prime power)	1	Hamming codes; see Section 7.4, Prob. 7.19
23	2	3	Binary Golay code; see Section 9.8
11	3	2	Ternary Golay code; see Probs. 9.30–9.33

Perfect codes (perhaps with the exception of the repetition codes) are beautiful combinatorial objects with remarkable error-correcting powers, and researchers naturally yearned to extend the above table. However, after a long series of extremely difficult investigations ending in the early 1970's the truth came out: *there are no perfect codes except those with the parameters listed above.* The full story can be read in MacWilliams and Sloane [19], Chapter 6. There do exist perfect codes with the same parameters as the Hamming codes which are not equivalent to them; the other three kinds, however, are unique. The question of the existence of perfect codes over alphabets with q symbols, where *q is not a prime power*, is at this writing not completely settled. It seems unlikely, however, that any exist.

Disappointed by the nonexistence of other perfect codes, researchers widened their search to codes that are almost as good: *nearly perfect* and *uniformly packed* codes. Over finite fields, such codes are now all classified for $e \geq 4$; for $e \leq 3$ only partial results are known; see van Tilborg [46].

The existence of the (23, 12) perfect Golay code tells us that the maximum number of vectors from $V_{23}(F_2)$ which can be chosen so that each pair of vectors is separated by a Hamming distance ≥ 7 is exactly $4096 = 2^{12}$. It is natural to ask what this maximum number is when 23 is replaced by n and 7 is replaced by d, and so we define

$A(n, d) =$ The largest integer M such that there exist M codewords
$\qquad \{x_1, \ldots, x_M\}$ from $V_n(F_2)$ such that $d_H(x_i, x_j) \geq d$ if $i \neq j$.
$\qquad$ Alternatively, $\qquad\qquad\qquad\qquad\qquad\qquad\qquad\qquad\qquad$ (12.1)

$A(n, d) =$ The largest number of codewords possible in a code of length n
$\qquad$ and minimum distance d.

The study of the numbers $A(n, d)$ is regarded by many as the central problem of coding theory. Although $A(n, d)$ is rarely known exactly unless n and d are relatively small, or $2d \geq n$, a great deal of first-rate research has gone into the problem of finding good upper and lower bounds for $A(n, d)$. In the next few pages we will summarize some of this research, but for the whole story the reader is referred to MacWilliams and Sloane [19], Chapter 17.

For relatively small values of n and d, essentially the only known way of obtaining lower bounds of the form $A(n, d) \geq M$ is to explicitly exhibit M vectors from $V_n(F_2)$ satisfying the conditions cited in (12.1). We discuss such code constructions below (see p. 349 ff.).

On the other hand, in order to obtain an upper bound of the form $A(n, d) < M$, it is necessary to show that every subset $\{\mathbf{x}_1, \ldots, \mathbf{x}_M\}$ of $V_n(F_2)$ contains at least one pair (i, j) such that $d_H(\mathbf{x}_i, \mathbf{x}_j) < d$. There are a great many possible ways of doing this, but currently the most powerful technique appears to be the *linear programming* approach, which we now describe.

If $C = \{\mathbf{x}_1, \ldots, \mathbf{x}_M\}$ is a binary code of length n, that is, a subset of $V_n(F_2)$, for each $i = 0, 1, \ldots, n$ and $\mathbf{x} \in C$, let $A_i(\mathbf{x})$ be the number of codewords in C at distance i from $\mathbf{x}$. The *distance distribution* of C is defined to be the $(n + 1)$-tuple $(A_0, A_1, \ldots, A_n)$ of nonnegative real numbers, where

$$A_i = \frac{1}{M} \sum_{\mathbf{x} \in C} A_i(\mathbf{x}).$$

Certain elementary properties of the A_i are immediate:

$$A_0 + A_1 + \cdots + A_n = M. \tag{12.2}$$

$$A_0 = 1. \tag{12.3}$$

And if the code has minimum distance d, it also follows that

$$A_1 = A_2 = \cdots = A_{d-1} = 0. \tag{12.4}$$

The key to the linear programming approach is the *Delsarte–MacWilliams (DM) inequalities*: If $P_j(i)$ denotes the coefficient of z^j in the polynomial $(1 - z)^i(1 + z)^{n-i}$, then the A_i's must satisfy

$$\sum_{i=0}^{n} A_i P_j(i) \geq 0, \qquad j = 0, 1, \ldots, n. \tag{12.5}$$

(For a proof of this result, see Prob. 7.28.) These facts suggest the following idea. Let $A_{LP}(n, d)$ denote the maximum possible value of the function $f(A_0, \ldots, A_n) = A_0 + \cdots + A_n$ of $n + 1$ nonnegative real numbers, subject to the linear constraints (12.3)–(12.5). Then clearly $A(n, d) \leq A_{LP}(n, d)$; this is called the *linear programming bound*.

At present, the best known upper bounds on $A(n, d)$ for almost every pair (n, d) come from the linear programming bound or a modification of it. For example, the bound on $A(13, 6)$ obtained by combining the DM inequalities (12.5) with the ad hoc inequality $A_{10} + 4A_{12} \leq 4$ (which is easy to obtain combinatorially) is $A(13, 6) \leq 32$. For several years before the discovery of the LP approach, $A(13, 6)$ was the simplest unknown value of $A(n, d)$, and researchers had struggled with it, using heroically complicated combinatorial arguments, but $32 \leq A(13, 6) \leq 35$ was the best they could do. Of course the old lower bound—which arises because there is a known code, the Nadler code, with $n = 13$, $d = 6$, $M = 32$—combined with the new upper bound, settles the matter: $A(13, 6) = 32$.

Perhaps the best known version of the $A(n, d)$ problem is the *asymptotic problem*: If $(d_n)_{n=1}^{\infty}$ is a sequence of integers with $d_n/n \to \delta$, $0 \leq \delta \leq 1$, how does $A(n, d_n)$ behave as $n \to \infty$? More formally, since the rate of a code of length n with M codewords is $1/n \log_2 M$, we define

$$\overline{R}(\delta) = \sup \varlimsup_{n \to \infty} \frac{1}{n} \log_2 A(n, d_n), \tag{12.6}$$

$$\underline{R}(\delta) = \inf \varliminf_{n \to \infty} \frac{1}{n} \log_2 A(n, d_n), \tag{12.7}$$

where the "sup" in (12.6) and the "inf" in (12.7) are both taken over all sequences $(d_n)_{n=1}^{\infty}$ satisfying $d_n/n \to \delta$. It is annoying that both an upper and a lower value of $R(\delta)$ must be defined; everybody believes (but nobody has proved) that $\overline{R}(\delta) = \underline{R}(\delta)$ for all δ. For simplicity in what follows, however, we shall refer only to $R(\delta)$, it being understood that an upper bound on $R(\delta)$ is an upper bound on $\overline{R}(\delta)$, and lower bound is a lower bound on $\underline{R}(\delta)$.

It is known, and relatively easy to prove, that $R(0) = 1$ and $R(\delta) = 0$ for $\frac{1}{2} \leq \delta \leq 1$, but $R(\delta)$ is unknown for $0 < \delta < \frac{1}{2}$. The best known lower bound for $R(\delta)$ in this range is

$$R(\delta) \geq 1 - H_2(\delta), \tag{12.8}$$

a result proved by Gilbert in 1952. (This bound is an asymptotic form of the bound obtained in Prob. 7.21.) No one has succeeded in raising Gilbert's lower bound, but there has been a whole series of steadily decreasing upper bounds on $R(\delta)$. The current record-holder, obtained by McEliece, Rodemich, Rumsey, and Welch using the linear programming approach, is

$$R(\delta) \leq \min_{0 \leq u \leq 1 - 2\delta} \{1 + g(u^2) - g(u^2 + 2\delta u + 2\delta)\}, \tag{12.9}$$

where $g(x) = H_2[(1 - \sqrt{1-x})/2]$. A plot of these two bounds is given in Fig. 12.1. It is widely conjectured that $R(\delta) = 1 - H_2(\delta)$, but the only real supporting evidence is the fact that over the years the upper bound has been decreasing while the lower bound has remained fixed!

• *How can we design good codes?* In the preceding few pages we have discussed some important nonconstructive results in coding theory. Much research has also been devoted to the problem of explicitly constructing powerful codes; and although nobody has yet explicitly constructed a sequence of codes achieving what the channel coding theorem promises (cf. remarks on p. 129), some remarkable results have nevertheless been achieved.

The most important class of codes yet discovered is undoubtedly the class of BCH codes described in Chapter 9. BCH codes are important not only because they are inherently powerful codes, but also because they possess a simple and efficient decoding algorithm. For most of the codes we shall describe, however, no good decoding algorithm is yet known. We shall describe some of the research that has been devoted to the design of decoding algorithms in the next subsection (see p. 353 ff.).

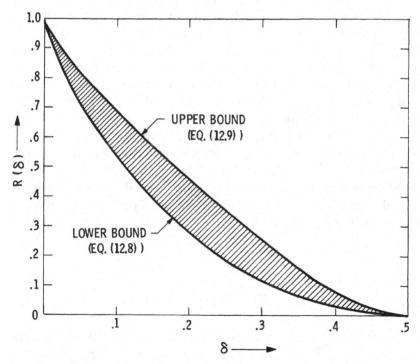

Figure 12.1 The best known upper and lower bounds on $R(\delta)$.

Probably the most impressive results in constructive coding theory deal with the class of *cyclic codes*. An (n, k) linear code C over F_q is said to be cyclic if, whenever $\mathbf{c} = (c_0, c_1, \ldots, c_{n-1})$ is in C, then so is the cyclic shift $\mathbf{c}' = (c_{n-1}, c_0, c_1, \ldots, c_{n-2})$. For any values of q and $n \geq 3$ there are always at least four cyclic codes of length n:

- The zero codeword $\mathbf{0} = (0, 0, 0, \ldots, 0)$ alone.
- The set of constant codewords $\boldsymbol{\alpha} = (\alpha, \alpha, \ldots, \alpha)$.
- The codewords satisfying $\sum_{i=0}^{n-1} c_i = 0$.
- The entire space $V_n(F_q)$.

These four cyclic codes have dimensions 0, 1, $n - 1$, and n, and for some pairs (q, n) there are no others (e.g., $q = 2$, $n = 11$; $q = 3$, $n = 7$). However, there are many extremely interesting cyclic codes. For example, the following codes from Chapter 9 are cyclic: BCH codes, Reed–Solomon codes, and the two Golay codes. In this subsection we shall describe several other cyclic codes, but we need a brief preliminary discussion.

Associated with each (n, k) cyclic code over F_q is a *generator polynomial* $g(x)$, which has coefficients in F_q, degree $n - k$, and which divides $x^n - 1$. The vector $\mathbf{c} = (c_0, \ldots, c_{n-1}) \in V_n(F_q)$ belongs to the code iff the polynomial $c_0 + c_1 x + \cdots + c_{n-1} x^{n-1}$ is a multiple of $g(x)$. The complementary polynomial $h(x) = (x^n - 1)/g(x)$ is called the *parity-check polynomial*. Since $g(x)$ divides $x^n - 1$, every zero of $g(x)$ must also be a zero of $x^n - 1$. If n and q are relatively prime (and cyclic codes for which this is not the case are usually uninteresting), the zeros of $x^n - 1$ are $\{1, \beta, \ldots, \beta^{n-1}\}$, where β is a primitive nth root of unity over F_q; thus the zeros of $g(x)$ are of the form $\{\beta^a : a \in A\}$, where $A \subseteq \{0, 1, \ldots, n - 1\}$. Since the coefficients of $g(x)$ are in F_q, it follows from Galois theory that A must be q-*closed*, that is, if $a \in A$, then also $qa \pmod{n} \in A$. If now $A \subseteq \{0, 1, \ldots, n - 1\}$ is arbitrary, we define its q-*closure* $\bar{A}$ to be the smallest q-closed subset containing A. Using this notation, we now factor the generator and parity-check matrices of an arbitrary (n, k) cyclic code over F_q:

$$g(x) = \prod_{a \in \bar{A}} (x - \beta^a),$$

$$h(x) = \prod_{b \in \bar{B}} (x - \beta^b).$$

Nearly every important result about cyclic codes can be stated in terms of the sets A, $\bar{A}$, B, and $\bar{B}$. For example, the BCH Theorem 9.3 can be stated as follows: If $n = 2^m - 1$, $q = 2$ and $A = \{1, 3, \ldots, 2t - 1\}$, then the code's minimum distance $d_{\min}$ satisfies $d_{\min} \geqslant 2t + 1$. We now list some of the most important results known about cyclic codes (all stated for $q = 2$, even though generalizations to larger q are usually known).

- If $n = 2^m - 1$ and $B = \{1, 3, \ldots, 2t - 1\}$, then $d_{\min} \geqslant 2^{m-1} - (t - 1)2^{m/2}$ (see van Lint [18], Theorem 6.3.6).
- If n is a prime congruent to $-1 \pmod 4$, and $\overline{A} = \{$quadratic residues $\pmod n\}$, then $k = (n + 1)/2$ and

$$d_{\min} \geqslant \begin{cases} \sqrt{n} & \text{if } n \equiv 1 \pmod 8, \\ \dfrac{1 + \sqrt{4n - 3}}{2} & \text{if } n \equiv -1 \pmod 8. \end{cases}$$

These are the *quadratic residue* codes; see MacWilliams and Sloane [19], Chapter 16.

- If there is no j-tuple $b_1, b_2, \ldots, b_j \in \overline{B}$ such that $b_1 + \cdots + b_j \equiv 0 \pmod n$, then every codeword has weight divisible by 2^j. (A disguised version of the special case $j = 2$ of this theorem appears in Lemma 5, p. 278. For proof and extensive generalization, consult Delsarte and McEliece [34].)
- If $B = \{1\}$, the codes are called *irreducible* cyclic codes (because the check polynomial $h(x)$ is irreducible) or *minimal* cyclic codes (because they contain no nontrivial cyclic subcodes). There is an elaborate theory for them; see MacWilliams and Sloane [19], Chapter 8, or McEliece [39].
- If $n = 2^m - 1$, $B = \{0 \leqslant b < 2^m - 1 : b$ can be written as a sum of $\leqslant d$ powers of 2$\}$, the resulting code is essentially equivalent to the Reed–Muller code RM(m, d) of Prob. 7.30. See MacWilliams and Sloane [19], Chapter 13, or Delsarte and McEliece [34], Section 5.

The foregoing sample should convince you that the theory of cyclic codes is very rich, but it is also somewhat disappointing, because the best codes are rarely cyclic. By that we mean roughly that for a randomly chosen pair (n, d) it is quite unlikely that there will exist a cyclic code with $A(n, d)$ codewords (see definition (12.1)). In fact, the best codes are often not even linear, and there is beginning to be quite a respectable collection of good nonlinear codes. Some examples (again, all codes have $q = 2$) follow:

- The best codes with $n \leqslant 2d$ are practically all nonlinear and are related to *Hadamard matrices*. See MacWilliams and Sloane [19], Chapter 2.

- The *Kerdock* nonlinear codes have parameters $n = 2^{2m}$, $M = 2^{4m}$ codewords, $d_{\min} = 2^{2m-1} - 2^{(2m-2)/2}$, $m \geqslant 2$. The simplest example ($m = 2$) is a code of length 16 with 256 codewords and $d_{\min} = 6$. The best linear code with $n = 16$, $d = 6$ has only 128 codewords. See MacWilliams and Sloane [19], Chapter 15.

- The *Preparata* codes have parameters $n = 2^{2m}$, $M = 2^{2^{2m}-4m}$, $d_{\min} = 6$, $m \geqslant 2$. They appear to be dual to the Kerdock codes; see MacWilliams and Sloane [19], Chapter 15.

Finally, we note that MacWilliams and Sloane [19], Appendix A, give a table of the best known codes for $d \leqslant 29$ and $n \leqslant 512$. Many of the codes listed there are clever ad hoc constructions that we cannot describe here.

- *How can we decode such codes?* The problem of finding the largest possible code with a given n and d is important and challenging. But to actually put such a code, or indeed any code, to use on a noisy channel it is necessary to have a practical decoding algorithm. If the number of codewords is small enough so that it is feasible to compare each of them to the received word, there is no problem. If the code is an (n, k) linear code and q^{n-k} is small enough so that the syndrome table approach of Sections 7.2 and 7.5 is possible, again no problem. However, in most practical situations it is unlikely that a code small enough to allow one of these primitive decoding algorithms will be powerful enough to do the job. For this reason considerable effort has been devoted to finding good decoding algorithms for large and powerful codes. In this subsection we shall describe some of these algorithms.

First and most important is the BCH–RS decoding algorithm that we described in detail in Chapter 9. Since this is such an important result, we pause here to pay tribute to its discoverers.

In 1959 and 1960 came the original papers by Hocquenghem ([15], pp. 72–74) and Bose and Ray-Chaudhuri ([15], pp. 75–78). It is important to note that these early papers concerned themselves only with the construction of the codes, not the decoding. In 1960 Peterson ([15], pp. 109–120) published a reasonably simple decoding algorithm which involved the solution of simultaneous linear equations over certain finite fields. One year later Gorenstein and Zierler ([15], pp. 87–89) generalized all the previous work to nonbinary codes and found that their generalization included the codes discovered by Reed and Solomon ([15], pp. 70–71) in 1960 before the Bose and Ray-Chaudhuri paper had been published! A real breakthrough in the decoding of BCH codes came in 1968, when Berlekamp ([15], pp. 145–148) (see also Chapter 7 in [14]) published his beautiful iterative decoding algorithm. A systems engineer who wishes to minimize absolutely the complexity of a

BCH decoder is still well advised to use Berlekamp's procedure. But certain important conceptual advances were yet to come.

In 1970 Goppa ([15], pp. 102–106) discovered a family of codes that bear his name and are a beautiful natural generalization of BCH codes. He showed that there are long Goppa codes that are asymptotically good in the sense that they lie above the Gilbert lower bound of Fig. 12.1. (This contrasts with the fact that BCH codes, in spite of their usefulness, are asymptotically "bad" in the sense that in any sequence of BCH codes with rates $\geq R > 0$ the ratio $d_{\min}/n$ must approach 0; see Berlekamp [31].) Goppa also gave a decoding procedure for his codes that was analogous to the old Peterson–Gorenstein–Zierler algorithm; he did not, however, generalize Berlekamp's iterative algorithm. In 1975 Sugiyama, Kasahara, Hirasawa, and Namekawa [45] discovered the beautiful fact, which we discussed at length in Chapter 9, that one can use Euclid's algorithm to decode BCH and Reed–Solomon codes. Actually, when applied to BCH codes Euclid's algorithm is inferior by a small constant factor to Berlekamp's. It is much easier to understand, however, and sheds considerable light on Berlekamp's algorithm. In fact, with hindsight it is now possible to view Berlekamp's algorithm as an improved version of Euclid's! Finally, in 1975, a circle that nobody knew was being formed was completed when Delsarte [33] and others showed that Goppa's codes can be obtained by modifying the ancient Reed–Solomon codes!

Although the algebraic decoding of BCH–RS codes is the most important result in the area, several other quite useful decoding methods are also available. We shall now describe some of them.

The earliest nontrivial decoding algorithm discovered was *threshold decoding*, which is a kind of generalization of the "majority vote" decoding mentioned in the Introduction, discovered by Reed in 1954 for decoding Reed–Muller codes (cf. Prob. 7.30. Muller invented the codes, and Reed the decoding algorithm, reported in separate 1954 papers.) We illustrate Reed's idea by considering the (7, 3) binary linear code with parity-check matrix

$$H = \begin{bmatrix} 1 & 1 & 0 & 1 & 0 & 0 & 0 \\ 1 & 0 & 1 & 0 & 1 & 0 & 0 \\ 0 & 1 & 1 & 0 & 0 & 1 & 0 \\ 1 & 1 & 1 & 0 & 0 & 0 & 1 \end{bmatrix}.$$

(This code is the dual of the (7, 4) Hamming code of the Introduction.) It is easy to verify that each codeword $\mathbf{C} = (C_0, C_1, \ldots, C_6)$ in this code satisfies the following three parity-check equations:

$$C_0 = C_1 + C_3,$$

$$C_0 = C_2 + C_4,$$

$$C_0 = C_5 + C_6.$$

Now assume that the codeword $\mathbf{C}$ is transmitted over a BSC with raw bit error probability p and is received as $\mathbf{R} = (R_0, \ldots, R_6) = \mathbf{C} + \mathbf{E}$, where $\mathbf{E} = (E_0, \ldots, E_6)$ is the error pattern. Then the decoder knows, from the above parity-check equations, that

$$
\begin{aligned}
C_0 &= R_1 + R_3 && \text{if } E_1 + E_3 = 0, \\
&= R_2 + R_4 && \text{if } E_2 + E_4 = 0, \\
&= R_5 + R_6 && \text{if } E_5 + E_6 = 0, \\
&= R_0 && \text{if } E_0 \quad\;\; = 0.
\end{aligned}
$$

Hence if the channel has made at most one error, it can affect at most one of the above estimates of C_0, and so C_0 can be obtained by taking a *majority vote* of these four estimates. If two channel errors occur and there is a 2–2 tie in the vote, the tie should be broken in favor of the estimate R_0, since $\{E_0 = 0\}$ is more probable than $\{E_i + E_j = 0\}$ if $0 < p < \frac{1}{2}$.

Reed obtained a generalization of the above technique applicable to the entire class of Reed–Muller codes. Later authors have discovered many classes of codes—whose constructions often depend on the existence of certain finite geometrics—that permit threshold decoding. The reader interested in pursuing this topic should begin with Chapter 13 of MacWilliams and Sloane [19], or with Massey [20].

The final decoding algorithm we wish to mention is a very restricted one, since it applies only to first-order Reed–Muller codes, which are essentially the duals of the Hamming codes. The first few of these have length and dimensions (8, 4), (16, 5), (32, 6), ..., and in general for each m there is a $(2^m, m+1)$ code. There is a decoding algorithm for these codes which is based on the *fast Fourier transform* (FFT) over the elementary Abelian group of order 2^m; see MacWilliams and Sloane [19], Chapter 14. The unique feature of this algorithm is that it is easily adapted to a Gaussian channel, without the need for binary output quantization. By this we mean the following. Any binary vector $\mathbf{x} = (x_1, \ldots, x_n) \in V_n(F_2)$ can be transmitted over a Gaussian channel with average power constraint $E(X^2) \leq \beta$ by mapping $\mathbf{x}$'s zero coordinates into $+\sqrt{\beta}$ and $\mathbf{x}$'s one coordinates into $-\sqrt{\beta}$. If say, $\mathbf{x}$ is from a BCH code and one wishes to use the decoding algorithm of Chapter 9 on the received vector $\mathbf{y}$, one will have to begin by transforming the coordinates of $\mathbf{y}$—which are real numbers—into 0's and 1's, by mapping positive coordinates into 0 and negative ones into 1. This is what we mean by binary output quantization. Unfortunately, this first step decreases the channel's capacity (by about 2dB; see Prob. 4.15), and for this reason codes with an inherently binary decoding algorithm are practically useless on noisy Gaussian channels.

However, as we have mentioned, FFT decoding of first-order Reed–Muller codes is not inherently binary, and so these codes are attractive for use on Gaussian channels. The deep-space communication channel is well modeled as a Gaussian channel; so it is not surprising that from 1969 to 1976 all of NASA's *Mariner* class deep-space probes were equipped with a (32, 6) Reed–Muller code. (Other deep-space probes have been equipped with convolutional codes; see Section 12.4.)

12.3 Convolutional codes

The theory of convolutional codes is much less well developed than the theory of block codes, partially because the subject is much younger, but mainly because it is intrinsically more difficult. For example, in 1970 Forney ([15], pp. 213–232) wrote a seminal paper that proved results for convolutional codes analogous to the simple results in Section 7.1 for block codes! This is not to say that convolutional coding theory has not produced important research. It does explain, however, why this section is much shorter than the preceding one. (Also, convolutional codes are often superior to block codes in practical situations; see Section 12.4.)

The *free distance* of a convolutional code (see p. 311) serves the same role as does the minimum distance of a block code: among codes of a fixed rate, the one with the biggest free distance is likely to be best. Much effort has been spent on the problem of finding convolutional codes with the largest possible free distances. Although no family of convolutional codes comparable to, say, the BCH block codes has yet been found, there exist extensive catalogues of specific good codes. For example, Larsen [37] gives a table of good (2, 1), (3, 1), and (4, 1) convolutional codes with memory $M \leqslant 13$; Johannesson [36] lists good (2, 1) codes up to $M = 35$. It is characteristic of the subject that these tables should be restricted to $(n, k) = (2, 1), (3, 1),$ (4, 1), especially (2, 1). The extra degree of freedom present in convolutional codes, namely, the memory M, makes the study of (2, 1) convolutional codes an extremely rich subject unto itself, whereas (2, 1) block codes are of course trivial. And although the possiblity of studying convolutional codes with fixed M and varying (n, k) suggests itself, nobody has yet pursued it very far.

The channel coding Theorem 2.4 was proved using block codes. It is natural to wonder whether it can be proved using convolutional codes. The answer so far is Not Quite. It can be proved using *tree codes* (codes obtained by assigning labels to a tree structure like the one in Fig. 10.20) or *trellis codes* (codes obtained by assigning labels to a trellis structure like the one in Fig. 10.7) or certain restricted subclasses of them. Such codes can be obtained from shift registers (see Figs. 10.3 and 10.4) in which the connections are allowed to vary with time; for this reason they are sometimes called *time-varying* convolutional codes, and ordinary convolutional codes are called *fixed* convolutional codes. The reader interested in studying coding theorems for time-varying convolutional codes should consult Massey [21] or Viterbi and Omura [26], Chapter 5.

Sequential decoding was the earliest known decoding algorithm for convolutional codes, and the analysis of sequential decoding has led to some of the deepest results in convolutional coding theory. We pointed out in Section 10.4 that a sequential decoding algorithm will practically always choose the same path through the code's tree as a maximum likelihood decoder. Unfortunately, the number of computations required to make its decision is a random variable which depends on the severity of the noise. Indeed, if C denotes the number of paths that the sequential decoder must investigate before it decodes the first L information bits, it is known that

$$P\{C > x\} \cong Lx^{-\alpha},$$

where α is a positive constant that depends on the channel and the code rate, but not on x. It turns out that if the code's rate is too close to channel capacity, the value of α lies between 1 and 2. This in turn implies that, although the random variable C has finite mean, its variance is infinite, and this has unpleasant practical implications. For remember that while the decoder is trying to find its way through the tree, more symbols are being received. These symbols waiting to be decoded must be stored in a buffer of finite length. If the decoder requires too many computations, this buffer will overflow and information will be lost. If the variance of C is infinite, this buffer will be overflowing all the time and sequential decoding is useless.

The rate at which this unpleasant phenomenon occurs is called the *computational cutoff rate*; it is just the rate R_0 we have already encountered in Probs. 2.21–2.26 and 4.3–4.9; R_0 is also sometimes denoted by the symbol R_{comp}. Even at rates below R_0, however, this buffer overflow problem often determines the code's performance. For example, by using a (2, 1) convolutional code with very large memory and free distance on a wideband Gaussian channel, one can easily make the bit error probability extremely small, but it is very difficult to decrease the *erasure* probability (i.e., the probability of there not being time to decode a given bit) below say 10^{-3}. The full story of sequential decoding computation distribution is told by Viterbi and Omura [26], Chapter 6.

In addition to Viterbi decoding and sequential decoding, there is a third important type of decoding for convolutional codes, called *threshold decoding*. It is analogous to the threshold decoding for block codes which we described in Section 12.2. It was developed in 1963 by Massey [20]; it is discussed by Gallager [17], Section 6.8; and a very clever application of threshold decoding to a concatenated convolutional coding system has recently been given by Wu [48].

Finally, we mention that in 1976 Schalkwijk and Vinck [44] invented a decoding algorithm that is an analogue for convolutional codes of syndrome decoding (see Section 7.2) for block codes. It also has much in common with Viterbi's decoding algorithm.

12.4 A comparison of block and convolutional codes

In this section we make a brief attempt to compare the relative merits of block and convolutional codes in practical applications.

It is fairly clear that the deepest and most impressive theoretical result in coding theory (block or convolutional) is the algebraic decoding of BCH–RS codes, which we described in Chapter 9. It is important to bear in mind, however, that these codes were designed for a very special class of channels, the q-ary symmetric channels (in particular the binary symmetric channel in the case of BCH codes), and that the decoding algorithms are not easily (if at all) adaptable to other channels.

On the other hand, the two most important decoding algorithms for convolutional codes, Viterbi and sequential decoding, although perhaps not as mathematically profound, are extremely robust and can rather easily be adapted to a very wide class of channels. And since a relatively small fraction of the communication channels that arise in practice are well modeled by a qSC, it is not surprising that a relatively large fraction of practical applications of coding theory involve convolutional and not block codes. We now illustrate this point with two specific examples.

Consider first a binary symmetric channel with raw bit error probability p, $0 < p < \frac{1}{2}$. In Figure 12.2 we give a graph of the performance of two practical coding systems for this channel, a (127, 64) 10-error-correcting BCH code

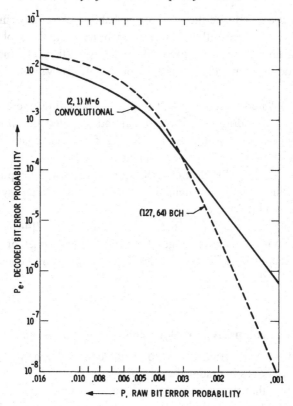

Figure 12.2 Comparison of a block and convolutional code on a BSC.

with the decoding algorithm described in Fig 9.1, and a $(2, 1)M = 6$ convolutional code with generator matrix $G = (x^6 + x^4 + x^3 + x + 1,$ $x^6 + x^5 + x^4 + x^3 + 1)$. For values of p greater than about $p = .03$ the convolutional code is slightly better, but for all smaller p the BCH code is better; for example, at $p = .001$ the BCH code is superior by more than two orders of magnitude. The essential reason the BCH code is superior is that its minimum distance (21) is much larger than the convolutional code's free distance (10). It would be quite difficult to implement a rate $1/2$ convolutional code with performance markedly superior to that shown in Fig. 12.2, since the only way to get better performance is to use a code with more memory; but for the Viterbi algorithm increasing M beyond 7 or 8 is extremely difficult, and if sequential decoding is used on a code with large M, the erasure probability (i.e., the probability that the algorithm will fail to make a decision because of the computational problems described in Section 12.3) will dominate the code's performance. On the other hand, it would not be difficult to go to a BCH code of rate approximately $1/2$ and length 255 or 511 and get performance markedly superior to that in Fig. 12.2 at a modest cost in increased complexity. We conclude then that for a binary symmetric channel, unless the channel is extremely noisy, a block code, in particular a BCH code, will probably be preferable to a convolutional code.

As a second example, consider the wideband Gaussian channel described in Chapter 4 (p. 95). In Fig. 12.3 we give performance curves for the same two codes as before; the quantity E_b is the ratio P/R, where P is the transmitter power (in watts) and R is the source rate (in bits per second). Thus E_b has dimensions joules per bit; it is usually called the *energy per bit*, and E_b/N_0 is called the *bit signal-to-noise ratio*. The notation "dB" means that what is

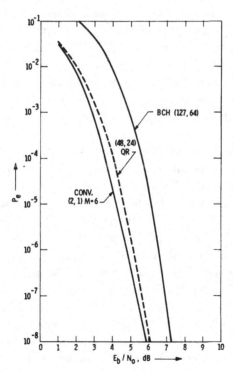

Figure 12.3 Comparison of block and convolutional codes on wideband Gaussian channel.

plotted is not E_b/N_0 directly, but $10\log_{10} E_b/N_0$. This normalization is the standard one for this channel; see, for example, Heller and Jacobs [35].

In Fig. 12.3 the BCH code not only has lost its superiority; it is now markedly inferior to the convolutional code. This is so because, in order to apply the algebraic decoding algorithm of Chapter 9 to the BCH code, it is necessary to use binary output quantization (cf. p. 335), whereas the Viterbi algorithm can be adapted to accept unquantized real numbers as inputs with relative ease (cf. Section 10.2). The penalty imposed by binary output quantization is simply too large to be compensated for by the fact that the BCH code is inherently more powerful than the convolutional code.

The important thing to know about these two examples is that the wideband Gaussian channel arises in a wide variety of practical communication situations, whereas a BSC hardly ever occurs in its pure form. So it should not be surprising that convolutional codes have been used in a majority of practical applications of error-correcting codes.

(An interesting example of practical applications: the communications link between a space vehicle and the earth is usually quite well modeled as a wideband Gaussian channel, and several sophisticated coding schemes have been used in deep-space applications since the late 1960's. In NASA's *Mariner*-class spacecraft from 1969 to 1976 a (32, 6) first-order Reed–Muller code with FFT decoding was used (see p. 356), but in 1977 this system was scrapped in favor of the $(2, 1)M = 6$ convolutional code described in this section. Other space missions (e.g., NASA's *Pioneer* missions and West Germany's *Helios*) have used sequential decoding on long $(M \geqslant 24)$ $(2, 1)$ convolutional codes.)

Let us also mention that in 1976 several of us at NASA's Jet Propulsion Laboratory devised a decoding algorithm for a powerful block code, the (48, 24) quadratic residue code, that does not require binary output quantization. The performance of this code has been plotted as the dashed curve in Fig. 12.3. Although its performance is not superior to that of the convolutional code, it suggests the following speculation: since the (127, 64) BCH code is a better code than the (48, 24) code ($d_{\min} = 21$ vs. $d_{\min} = 12$), if a good nonbinary decoding algorithm could be found for the BCH code it might well turn out to be significantly superior to the convolutional code.

Finally, we note that the performance of the convolutional code in Fig. 12.3 is the state of the art only for values of P_e of about 10^{-3} or more. For extremely small values of P_e, the best known practical system is a concatenation scheme (see Sections 6.2 and 9.6), in which the inner code is a convolutional code and the outer one is a Reed–Solomon (block) code (reference: Odenwalder [41]).

12.5 Source codes

Shannon's source coding Theorem 3.4 suffers from the same major flaw as his channel coding theorem: it is nonconstructive (see introductory remarks to Chapter 7). It is therefore natural to suppose that researchers may have developed a constructive theory for source coding analogous to the constructive theory for channel coding presented in Chapters 7–10. At present, however, no such theory exists. There are scattered results (e.g., the zero-distortion techniques in Chapter 10, and those to be described later in this section), but no unified theory. The reasons for this lack appear to be as follows: (a) source coding is apparently inherently much harder than channel coding; (b) the acceptable distortions in practice are so small that relatively little can be gained; (c) many or most sources that arise in practice are extremely difficult to model mathematically, as are the corresponding measures of distortion. The classic example of the last phenomenon is the case of visual data, for example, photographs of distant planets.

We now describe some of the more important results that have been obtained in constructive source coding.

We begin with the memoryless Gaussian source with mean-squared error fidelity criterion (cf. Section 4.2). It is important to have explicit source coding techniques for this source both because it arises frequently in practice and because perfect transmission is impossible, since $R(0) = \infty$ (cf. Fig. 4.2). Fortunately, there exists a very simple technique, called *quantization*, which is applicable here.

A quantization scheme is one that maps the source symbol u into $f(u)$, where f is a monotonic step function assuming only finitely many values. Thus there are real numbers $L_1 < \cdots < L_n$ (called quantization *levels*) and $T_1 < \cdots < T_{n-1}$ (called quantization *thresholds*) such that $f(u) = L_k$ iff $T_{k-1} < u \leq T_k$, where $T_0 = -\infty$ and $T_n = +\infty$. This mapping transforms the continuous memoryless source U (a mean 0, variance σ^2 Gaussian random variable) into a discrete memoryless source $f(U)$ that assumes the value L_k with probability $p_k = P\{T_{k-1} < U \leq T_k\}$. The average distortion that results from representing U by $f(U)$ is of course $\delta = E[U - f(U)]^2$. Furthermore, as we learned in Chapter 11, the average number of bits required to represent a symbol from the source $f(U)$ is $H[f(U)]$. Thus (referring to Fig. 4.2) the point $R = H[f(U)]$, $\delta = E[U - f(U)]^2$ is achievable by this quantization scheme. Surprisingly, Max and others have shown by numerical techniques that, if the levels and thresholds are properly chosen it is possible in this way to achieve performance that is quite close (within one bit) to the $R(\delta)$ curve. (Max's paper is on pp. 267–276 in [25]. Berger [13], Section 5.1, provides a more detailed summary of quantization techniques, including a discussion of their applicability to Gaussian sources with memory. See also papers 19–25 in [16].)

In order to describe some of the more sophisticated source coding techniques that are known, we pause to discuss the strong similarity between source coding and channel decoding. For clarity, we shall restrict ourselves to the binary symmetric source relative to the Hamming distortion (see pp. 82–83).

Recall that a source code of length k and rate R_s is a subset $C_s = \{\mathbf{v}_1, \mathbf{v}_2, \ldots, \mathbf{v}M_s\} \subseteq V_k(F_2) = V_k$ with $k^{-1} \log_2 M_s = R_s$. Its distortion is given by

$$\delta(C_s) = \frac{1}{2^k \cdot k} \sum_{\mathbf{u} \in V_k} d_{\mathrm{H}}[\mathbf{u}, f_s(\mathbf{u})],$$

where for each $\mathbf{u} \in V_k$ the encoding function $f_s(\mathbf{u})$ is defined to be the source codeword from C_s closest to $\mathbf{u}$ with respect to the Hamming distance d_{H} (cf. Eq. (3.18)).

Now recall that a channel code of length n with rate R_c is a subset $C_c = \{\mathbf{x}_1, \mathbf{x}_2, \ldots, \mathbf{x}_{M_c}\} \subseteq V_n$ with $R_c = n^{-1} \log_2 M_c$. Only the codewords $\mathbf{x}_i$ are transmitted, but because of channel noise $\mathbf{x}_i$ can be received as any $\mathbf{y} \in V_n$. The decoding function f_c must then map V_n into the code C_c; ideally f_c maps the received $\mathbf{y}$ onto the codeword closest to $\mathbf{y}$ with respect to the Hamming distance d_{H}.

In both cases the function f maps a vector from the whole space onto the closest codeword. This suggests that, given a good channel code and decoding algorithm, the same code can be used as a source code, where now the channel decoding algorithm is used as a source encoding algorithm. (Incidentally, the channel encoding algorithm should be used as the source decoding algorithm. Do you see why?) This idea is in fact a good one, but it is loaded with pitfalls, as we shall now explain.

We have already seen examples of running a channel code backward to get a source code in the Introduction: there we ran the repetition codes backward, and more important, we ran the (7, 4) Hamming code backward to get $R = 0.5714$ and $\delta = 0.1250$. The value of $R(\delta) = 1 - H_2(\delta)$ for $\delta = 0.1250$ is 0.4564, so the Hamming code performs tolerably well as a source code. Similarly, if the perfect (23, 12) Golay code (with syndrome table lookup decoding) is used as a source code, $R = 0.5217$ and $\delta = 0.1240$ (vs. $R(0.1240) = 0.4592$.) There is no problem so far.

It is natural, given the results of Chapter 9, to want to use BCH codes as source codes, but here serious problems do arise because the decoding algorithms of Chapter 9 are *incomplete* algorithms. By this we mean that these algorithms perform properly only if the received vector **y** lies fairly close to a codeword; for other vectors the performance is dismal. For example, consider the (31, 16) three-error-correcting BCH code. Its decoding algorithm (see Fig. 9.1) will succeed if the received vector lies within Hamming distance 3 of some codeword. But the total number of vectors in V_{31} this close to a codeword is only

$$2^{16} \cdot \left(1 + 31 + \binom{31}{2} + \binom{31}{3}\right) = 3.3 \times 10^8,$$

or about 15% of the $2^{31} = 2.1 \times 10^9$ vectors in V_{31}. For channel coding this presents no problem, since the code will normally be used only on a channel for which more than three errors is an extrememly improbable event. But for source coding, in which each of the vectors in V_{31} is equally likely to be emitted, it is a disaster; for 85% of the time the decoder/encoder will bomb out! (The usual failure mode will be that the error-locator polynomial $\sigma(x)$ will have degree 3 but will have fewer than 3 zeroes in the field F_{32}.)

The above considerations have led researchers to try to design practical decoding algorithms for BCH codes that will produce the nearest codeword, no matter what the decoder input is. For single-error-correcting BCH codes there is no problem: the decoding algorithm is already complete. Berlekamp [14], Chapter 16, has given a complete algorithm for the two-error-correcting case. And, finally, Vanderhorst and Berger [47] have studied the three-error-correcting case.

Although we have illustrated the concept of employing a channel code as a source code using block codes, one could just as well use convolutional codes, and indeed several researchers have done so. For example, Berger [13], Chapter 6, discusses certain modifications of sequential decoding that give good results. A more recent paper of Anderson [28] gives an adaptation of the stack algorithm to source coding. The anthology [16] edited by Davisson and Gray contains several more papers in the same area, including an interesting application to speech compression.

Appendix A
Probability theory

We emphasize at once that our object here is not to teach the reader probability theory. For that purpose we recommend either the short book by Lamperti [8] or the two-volume study by Feller [4]. Our object is really just to settle on standard notation, certain conventions, and so on, and to state the weak law of large numbers, which is the basic tool used in the proof of the coding theorems of Chapters 2 and 3.

The basic concept is that of a *probability space* $(\Omega, \mathcal{B}, P)$, where Ω is a nonempty set called the *sample space*, $\mathcal{B}$ is a Borel field of subsets of Ω, and P is a nonnegative function defined for all $A \in \mathcal{B}$ with the property that $P\{\Omega\} = 1$ and

$$P\left\{\bigcup_{n=1}^{\infty} A_n\right\} = \sum_{n=1}^{\infty} P\{A_n\},$$

provided each $A_n \in \mathcal{B}$, and the A_n's are disjoint. P is called the *probability measure*. For example, if $\Omega = \{\omega_1, \omega_2, \ldots\}$ is finite or countable, $\mathcal{B}$ is the collection of all subsets of Ω, $(p_1, p_2, \ldots)$ is a sequence of nonnegative numbers whose sum is 1, then the definition $P\{A\} = \sum\{p_n: \omega_n \in A\}$ makes $(\Omega, \mathcal{B}, P)$ a probability space; it is called a *discrete* probability space.

A *random variable* X is a function mapping Ω into some set R, called the *range of* X. (We shall denote random variables by upper case letters from near the end of the alphabet.) Normally we assume R is a subset of the real numbers but sometimes R will be another kind of set. For example, if R is a subset of n-dimensional Euclidean space, X will be called a *random vector* and be printed in boldface: $\mathbf{X}$. The components of $\mathbf{X}$ will be denoted by $X_1, X_2, \ldots, X_n$; thus a random vector $\mathbf{X} = (X_1, \ldots, X_n)$ can be viewed as a list of (one-dimensional) random variables.

Two random variables X and Y defined on the same probability space are said to be equal *almost everywhere* if the set $\{\omega: X(\omega) \neq Y(\omega)\}$ has P-measure 0. In symbols,

$$X = Y \text{ a.e. iff } P\{X \neq Y\} = 0.$$

If X is a random variable with range R equal to the real line, its *expectation*, or average, is defined by

$$E(X) = \int_{\Omega} X(\omega)\,dP;$$

in words, $E(X)$ is the Lebesgue integral of the function X with respect to the measure P.

There is another, equivalent definition of expectation that is often more convenient for computation. The measure P defined in Ω induces a probability measure P_X on R via the definition $P_X(S) = P\{\omega: X(\omega) \in S\}$ for any Borel set of reals. This measure is called the *distribution of X*, and the function S

$$F_X(x) = P\{\omega: X(\omega) \leq x\}$$

is called the *distribution function of X*. The expectation of X is equally given by

$$E(X) = \int_R x\,dP_X = \int_{-\infty}^{\infty} x\,dF_X(x), \tag{A.1}$$

in the sense of Lebesgue–Stieltjes. We pause to comment on two special cases of (A.1).

First, if Ω is discrete, and if for each $x \in R$ we define $p(x) = P_X(\{x\}) = P\{\omega: X(\omega) = x\}$, definition (A.1) becomes

$$E(X) = \sum_x p(x) \cdot x,$$

where the range of summation is purposely left unspecified; it can be over any discrete subset S of reals such that $P\{X \in S\} = 1$. More generally, if f is any real-valued function defined on S, $f(X)$ is a new random variable, and its expectation is given by

$$E[f(X)] = \sum_x p(x)f(x). \tag{A.2}$$

Sums of this kind appear throughout Part one of this book, and often $f(x)$ will assume the improper values $\pm\infty$, or will be undefined for certain values of x. However, it will always turn out that $f(x)$ will misbehave only on a set of measure 0, that is, only for values of x with $p(x) = 0$, and the corresponding contribution to the sum will always be assumed to be 0.

Second, if the random variable X has a density, that is, if the distribution function can be expressed as $F_X(x) = \int_{-\infty}^{x} p(u)\, du$ for some nonnegative function p, then the expectation of X can be computed from the formula

$$E(X) = \int_{-\infty}^{\infty} u \cdot p(u)\, du. \qquad (A.3)$$

Similarly, $E[f(X)] = \int f(u)p(u)\, du$.

We come now to the important concept of independence.

Let $(\Omega, \mathcal{B}, P)$ be a probability space, and let $A_1, A_2, \ldots, A_n$ be sets from $\mathcal{B}$. The "events" $A_1, A_2, \ldots, A_n$ are said to be *independent* if for every subset of them, $A_{i_1}, \ldots, A_{i_m}$, we have

$$P\{A_{i_1} \cap A_{i_2} \cap \cdots \cap A_{i_m}\} = P\{A_{i_1}\} \ldots P\{A_{i_m}\}.$$

More important, a collection $X_1, \ldots, X_n$ of random variables all defined on the same sample space is said to be independent if the events $A_i = \{\omega: X_i(\omega) \in S_i\}$ are independent for any choice of $S_1, S_2, \ldots, S_n \in \mathcal{B}$. This is not an easy condition to verify directly, so we give an alternate formulation in the case of only two random variables.

If X and Y are any pair of real random variables defined on the same sample space, independent or not, the mapping $\omega \rightarrow (X(\omega), Y(\omega))$ induces a measure P_{XY} on the field of two-dimensional Borel sets. This is called the *joint distribution* of X and Y, and the function

$$F_{XY}(x, y) = P\{\omega: X(\omega) \leqslant x, Y(\omega) \leqslant y\}$$

is called their *joint distribution function*. To say that X and Y are independent is to say that the measure P_{XY} is a product measure, that is, that for any pair S and T of one-dimensional Borel sets $P_{XY}(S \times T) = P_X(S)P_Y(T)$, where P_X and P_Y are the one-dimensional measures induced by X and Y. Equivalently, X and Y are independent iff

$$F_{XY}(x, y) = F_x(x)F_Y(y),$$

where F_X and F_Y are the distribution functions of X and Y.

In the discrete case, if we define for each x, y in (range of X, range of Y) $p(x, y) = P\{\omega: X(\omega) = x, Y(\omega) = y\}$, X and Y are independent iff $\dot{p}(x, y) = p(x)p(y)$, where $p(x) = P\{X(\omega) = x\}$, $p(y) = P\{Y(\omega) = y\}$. (Note the deliberate use of ambiguous notation: the same symbol $p(\cdot)$ is used for two completely different discrete distribution functions; they can be distinguished only by the fact that elements in the range of X are denoted by x's, elements in the range of Y by y's.)

In the continuous case, if X and Y both have densities $p(x)$ and $q(y)$, X and Y are independent iff

$$F_{XY}(x, y) = \int_{-\infty}^{x} \int_{-\infty}^{y} p(s)q(t) \, ds \, dt,$$

that is, X and Y have a *joint density* $p(x, y) = p(x)q(y)$.

We can now state the weak law of large numbers, which is vital in the proofs of the coding theorems.

Weak Law of Large Numbers. For each n, let $X_1, X_2, \ldots, X_n$ be independent random variables, each with finite expectation μ—indeed, each with the same distribution function. Then for each $\varepsilon > 0$,

$$\lim_{n \to \infty} P\left\{ \left| \frac{X_1 + \cdots + X_n}{n} - \mu \right| \geq \varepsilon \right\} = 0.$$

Appendix B
Convex functions and Jensen's inequality

References: Fleming [6], pp. 13–28 and 53–66; Feller [4], vol. II, pp. 153–155.

A subset $K \subseteq E^n$ (Euclidean n-space) is called *convex* if the line segment joining any two points of K is contained in K:

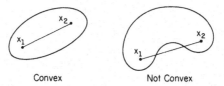

Convex   Not Convex

The line segment joining x_1 and x_2 is formally defined as $\{x : x = tx_1 + (1 - t)x_2,\ t \in [0, 1]\}$.

Although the definition of a convex set is given in terms of pairs of points, it can also be given in terms of *convex combinations* of any finite number of points, as follows. A point x is said to be a convex combination of x_1, $x_2, \ldots, x_m$ if there exist nonnegative scalars $\alpha_1, \alpha_2, \ldots, \alpha_m$ with $\sum \alpha_i = 1$, and $\sum \alpha_i x_i = x$. The set of all convex combinations of $x_1, x_2, \ldots, x_m$ is called the *convex hull* (Fig. B.1) of $x_1, x_2, \ldots, x_m$. It is easy to show that a set K is convex iff every convex combination of points in K is also in K.

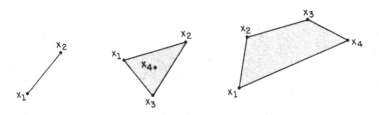

Figure B.1 Convex hulls in two dimensions.

Now let f be a real-valued function, and let K be a convex subset of the domain of f. Then f is said to be *convex* $\cup$ ("convex cup") if, for every $x_1, x_2 \in K$ and $t \in [0, 1]$,

$$f(tx_1 + (1 - t)x_2) \leqslant tf(x_1) + (1 - t)f(x_2). \tag{B.1}$$

Furthermore, if strict inequality holds in (B.1) whenever $x_1 \neq x_2$ and $0 < t < 1$, f is said to be *strictly convex* $\cup$. Examples in one dimension are as follows:

convex $\cup$ strictly convex $\cup$ not convex $\cup$

Geometrically, the definition says that f is convex $\cup$ iff all its chords lie above or on the graph of f.

Similarly, a function is called *convex* $\cap$ ("convex cap"),* or *strictly convex* $\cap$, if $-f$ is convex $\cup$, or strictly convex $\cup$, that is, if the inequality in (B.1) is reversed:

$$f(tx_1 + (1 - t)x_2) \geqslant tf(x_1) + (1 - t)f(x_2). \tag{B.1$'$}$$

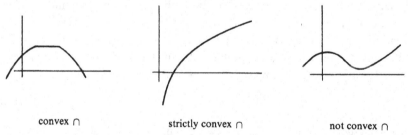

convex $\cap$ strictly convex $\cap$ not convex $\cap$

It is a remarkable fact that convex functions are continuous, in this sense: If K is an open convex set and if f is convex ($\cap$ or $\cup$) on K, then f is continuous on K. However, if K is not open, f may be discontinuous at boundary points of K: let $K = [0, 1]$, $f(x) = x$ for $0 < x \leqslant 1$, $f(0) = 1$. (See Fig. B.2).

If f is sufficiently smooth, we can test f for convexity by calculus. For example, in one dimension let f be a function that has a derivative $f'(x)$ for

---

*Sometimes *concave* is used instead of convex $\cap$, and *convex* mean convex $\cup$.

Figure B.2 f is convex $\cup$ (not convex $\cap$) on $[0, 1]$ but discontinuous at $x = 0$.

every point $x \in K$. Then f is convex $\cup$ iff f' is nondecreasing on K, and is strictly convex $\cup$ iff f' is increasing on K. Hence, if f'' exists, f is convex $\cup$ iff $f''(x) \geq 0$ in K; and if $f''(x) > 0$ except for a finite number of points, f is strictly convex $\cup$. The situation is similar for convex $\cap$ functions. (In higher dimensions there is a test for convexity that depends on the matrix $[D_{ij}f]$ of mixed second partial derivatives; see Fleming [6], §2–4, for details.)

Also, if f is differentiable on K, f is convex $\cup$ iff $f(x) \geq f(y) + \nabla f(y) \cdot (x - y)$ for all x, $y \in K$, and strictly convex $\cup$ iff strict inequality holds whenever $x \neq y$.

We come finally to *Jensen's inequality*, which deals with convex functions of only one variable. Let K be an interval in E^1, and let $F(x)$ be a probability distribution concentrated on K. Let X be the associated random variable, that is, $P\{X \leq x\} = F(x)$. If the expectation $E(X)$ exists, and if $f(x)$ is a convex $\cup$ function, Jensen's inequality says that

$$E(f(X)) \geq f(E(X)). \tag{B.2}$$

Furthermore, if f is strictly convex, inequality (B.2) is strict unless X is concentrated at a single point x_0, that is, $P\{X = x_0\} = 1$. Geometrically, Jensen's inequality (Fig. B.3) says that if a mass distribution is placed on the graph of f, the resulting center of mass will lie above (or on) the graph of f. Naturally, if f is instead convex $\cap$, the inequality reverses:

$$E(f(X)) \leq f(E(X)). \tag{B.3}$$

We conclude this appendix with two examples, in which the $f(x)$ is the

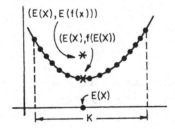

Figure B.3 Jensen's inequality for a discrete distribution.

convex ∩ function $\log x$ (base unspecified). In the first example X is discrete, and in the second X has a density.

Example B.1 Let $\Omega = \{\omega_1, \omega_2, \ldots\}$ be a discrete set of real numbers, and let $p(\omega_i)$ be a nonnegative function such that $\sum_i p(\omega_i) = 1$. Then Ω becomes in the obvious way (cf. Appendix A) a discrete sample space. Let $q(\omega_i)$ be any other nonnegative function defined on Ω, and define the random variable X by

$$X(\omega) = \frac{q(\omega)}{p(\omega)}.$$

(The values assumed by X at points where $p(\omega) = 0$ are unimportant; we may as well assume $X(\omega) = 0$ there.) Then by Jensen's inequality (B.3) $E(\log X) \leq \log E(X)$. But

$$E(\log X) = \sum_i p(\omega_i) \log \frac{q(\omega_i)}{p(\omega_i)},$$

and

$$E(X) = \sum_{i:\, p(\omega_i) \neq 0} q(\omega_i).$$

(see the discussion following Eq. (A.2).) Thus, if α denotes the above sum for $E(X)$, we get

$$\sum_i p(\omega_i) \log \frac{1}{p(\omega_i)} \leq \sum p(\omega_i) \log \frac{1}{q(\omega_i)} + \log \alpha.$$

Furthermore, since $\log x$ is strictly convex, equality holds iff $X = \beta$, a.e. for some constant β; but since $\sum p(\omega_i) = 1$ this constant must be α, and so equality holds iff $q(\omega_i) = \alpha p(\omega_i)$ for all i such that $p(\omega_i) \neq 0$. Clearly the underlying set Ω plays no part in this, and so we can state the following theorem.

Theorem B.1 *Let I be a discrete set of integers, and let p_i, $i \in I$, be a set of positive (NB) real numbers such that $\sum_{i \in I} p_i = 1$. If q_i is any other set of nonnegative (NB) real numbers with $\sum_{i \in I} q_i = \alpha$, then*

$$\sum_{i \in I} p_i \log \frac{1}{p_i} \leq \sum_{i \in I} p_i \log \frac{1}{q_i} + \log \alpha, \tag{B.4}$$

with equality iff $q_i = \alpha p_i$ for all i. □

Example B.2 Let Ω equal the real line, and let $p(x)$ be a density function, that is, a nonnegative function such that $\int_{-\infty}^{\infty} p(x)\,dx = 1$; $p(x)$ induces in the obvious way a probability measure on Ω (cf. Appendix A). Let $q(x)$ be any other nonnegative function defined on Ω, and define the random variable $X(x) = q(x)/p(x)$. Again applying Jensen's inequality, this time using Eq. (A.3) to compute the expectations, we get

$$\int_{-\infty}^{\infty} p(x) \log \frac{q(x)}{p(x)}\, dx \le \log \int_{I} q(x)\, dx,$$

where $I = \{x : p(x) > 0\}$. Arguing as in Example B.1, we can state another theorem.

Theorem B.2 *Let I be a measurable subset of the real line, and let $p(x)$ be a positive (NB) function defined in I, with $\int_{I} p(x)\, dx = 1$. If $q(x)$ is a nonnegative (NB) function defined on I, with $\int_{I} q(x)\, dx = \alpha$, then*

$$\int_{I} p(x) \log \frac{1}{p(x)}\, dx \le \int_{I} p(x) \log \frac{1}{q(x)}\, dx + \log \alpha, \qquad \text{(B.5)}$$

with equality iff $q(x) = \alpha p(x)$ a.e. $\square$

Note: The sum appearing on the left of Eq. (B.4) is called the *entropy* of the probability distribution $(p_1, p_2, \ldots)$; the integral on the left side of (B.5) is the *differential entropy* of the random variable whose density is $p(x)$. The study of entropy and differential entropy is central to information theory, and arguments similar to the above two are ubiquitous in Part one of this book.

Appendix C
Finite fields

In this appendix we shall state for reference purposes the main facts about finite fields, which are needed in Chapters 8 and 9. We recommend as references Berlekamp [14], Chapters 2, 3, 4, 6, and Niederreiter [11].

C.1 Construction

The basic building blocks are the prime fields F_p, where p is a prime number. F_p is the field whose elements are $\{0, 1, \ldots, p-1\}$, and arithmetic is performed modulo p.

The m-dimensional vector space $V_m(F_p)$ over F_p is the set of m-tuples $\mathbf{a} = (a_0, a_1, \ldots, a_{m-1})$, with addition defined componentwise. $V_m(F_p)$ can be made into a field by defining a multiplication in the following way. Let $f(x) = f_0 + f_1 x + \ldots + f_m x^m$ be an mth degree polynomial *irreducible* over F_p. Such polynomials exist for all m and p; indeed there are exactly

$$I_m(p) = \frac{1}{m} \sum_{d|m} \mu(d) p^{m/d}$$

monic irreducible polynomials of degree m over F_p, where μ is Möbius' mu function and the summation is extended over all (positive) divisors of m. Extensive tables of irreducible polynomials have been computed, e.g. [22], [23]. Now in $V_m(F_p)$, define the product of $\mathbf{a} = (a_0, a_1, \ldots, a_{m-1})$ and $\mathbf{b} = (b_0, b_1, \ldots, b_{m-1})$ to be $\mathbf{c} = (c_0, c_1, \ldots, c_{m-1})$, where $\mathbf{c}$ is uniquely determined by the equation

$$(a_0 + a_1 x + \cdots + a_{m-1} x^{m-1})(b_0 + b_1 x + \cdots + b_{m-1} x^{m-1})$$
$$\equiv (c_0 + c_1 x + \cdots + c_{m-1} x^{m-1}) \pmod{f(x)}.$$

The multiplication defined in this way, combined with the componentwise addition, makes $V_m(F_p)$ into a field with p^m elements. According to the above construction, there are as many ways to define this multiplication as there are degree m irreducible polynomials; but it turns out that all these fields are isomorphic, and so the field is usually called *the* field with p^m elements, and denoted by $GF(p^m)$ (GF = Galois fields in honor of their discoverer) or by F_{p^m}.

The above construction shows how to construct a field with p^m elements for all p, m. Conversely it is known that any field with a finite number of elements must be of this kind, so, for example, there is no field with six elements. The prime p is called the *characteristic* of the field.

More generally, if F_q is a finite field with q elements (and now q need not be a prime; it can be a prime power), the vector space $V_m(F_q)$ can be made into a larger finite field by the use of an mth degree polynomial that is irreducible over F_q. By taking $q = p^d$ and $m = n/d$, this shows that F_{p^d} can be viewed as a subfield of F_{p^n} if d divides n. Conversely, if d does not divide n, F_{p^d} is not a subfield of F_{p^n}.

C.2 Multiplicative structure

Let F_q be a finite field with $q = p^m$ elements. The nonzero elements of F_q form a commutative group F_q^* of order $q - 1$, which is in fact a cyclic group. It follows then that F_q^* contains elements of order d iff d is a divisor of $q - 1$, and in fact there are exactly $\phi(d)$ elements of order d, if d divides $q - 1$ (ϕ = Euler's phi function). An element of mulitplicative order $q - 1$, that is, a generator of the group F_q^* is called a *primitive root*.

It thus follows that every element in F_q^* satisfies $\alpha^{q-1} = 1$, and so every element in F_q satisfies $\alpha^q = \alpha$. If F_q is viewed as a subfield of F_{q^m} for some m, then this equation characterizes F_q, that is,

$$\alpha^q = \alpha \text{ iff } \alpha \in F_q.$$

C.3 Conjugation and minimal polynomials

If F_{q^m} is regarded as an mth degree extension of F_q, the mapping $\alpha \to \alpha^q$ is called *conjugation*. Conjugation is linear, that is,

$$(\alpha + \beta)^q = \alpha^q + \beta^q.$$

The *conjugates* of α are the distinct elements in the sequence α, $\alpha^q \alpha^{q^2}$, ... Thus, if k is the least integer such that $\alpha^{q^k} = \alpha$. the conjugates of α are

$\{\alpha, \alpha^q, \ldots, \alpha^{q^{k-1}}\}$. Here k is called the *degree* of α; it follows that k is a divisor of m, and that $\alpha \in F_{q^k}$ but no smaller field.

The *minimal polynomial* of α is defined to be the monic polynomial $f(x)$ of least degree with coefficients in F_q such that $f(\alpha) = 0$. Over F_q, $f(x)$ is irreducible, but in the larger field F_{q^m}, $f(x)$ factors linearly:

$$f(x) = (x - \alpha)(x - \alpha^q) \ldots (x - \alpha^{q^{k-1}}).$$

Thus the degree of $f(x)$ is the same as the degree of α.

If α is a *primitive root* in F_{q^m} the minimal polynomial of α is called a *primitive polynomial* over F_q. It is often convenient to reverse this process and use a primitive polynomial to construct the field. For example, by consulting tables (e.g., [22] or [23]) we find that $f(x) = x^4 + x + 1$ is a primitive (and hence also irreducible) polynomial over F_2. Thus $f(x)$ is the minimal polynomial of a primitive root in F_{16}, which we denote by α. Then F_{16} consists of the elements $\{0, 1, \alpha, \alpha^2, \ldots, \alpha^{14}\}$, and each power α^i can be represented as a polynomial of degree $\leqslant 3$ in α by repeated use of the equation $\alpha^4 = \alpha + 1$. In this way we can compute the following table:

i	α^i
0	1
1	α
2	α^2
3	α^3
4	$\alpha + 1$
5	$\alpha^2 + \alpha$
6	$\alpha^3 + \alpha^2$
7	$\alpha^3 + \alpha + 1$
8	$\alpha^2 + 1$
9	$\alpha^3 + \alpha$
10	$\alpha^2 + \alpha + 1$
11	$\alpha^3 + \alpha^2 + \alpha$
12	$\alpha^3 + \alpha^2 + \alpha + 1$
13	$\alpha^3 + \alpha^2 + 1$
14	$\alpha^3 + 1$

This kind of table is convenient for making calculations. For example, if F_{16} is represented by the vectors from $V_4(F_2)$, to multiply 1101 by 0111 we do the following:

$$1101 = \alpha^3 + \alpha^2 + 1 = \alpha^{13},$$

$$0111 = \alpha^2 + \alpha + 1 = \alpha^{10},$$

$$\alpha^{13} \cdot \alpha^{10} = \alpha^{23} = \alpha^8 = \alpha^2 + 1 = 0101.$$

Hence $(1101) \cdot (0111) = (0101)$ if F_{16} is realized in this way.

C.4 Factorization of $x^n - 1$ over F_q

If F_q has characteristic p and $n = p^i n_0$ for some $i \geqslant 1$, then $x^n - 1 = (x^{n_0} - 1)^{p^i}$ in F_q. Hence we assume n and p are relatively prime. If this is so, there will be at least integer m such that $q^m \equiv 1 \pmod{n}$, that is, such that n divides $q^m - 1$. Hence the field F_{q^m}, but no smaller field, will contain a primitive nth root of unity, which we denote by β. In the field F_{q^m},

$$x^n - 1 = \prod_{i=0}^{n-1} (x - \beta^i),$$

but in the smaller field F_q, $x^n - 1$ is the product of the (distinct) minimal polynomials of the powers of β. We illustrate with an example (cf. p. 276).

Let $q = 2$, $n = 23$. Since $2^{11} \equiv 1 \pmod{23}$, $F_{2^{11}}$ contains a primitive 23rd root of unity. The minimal polynomial of $\beta^0 = 1$ is clearly $x - 1$. The minimal polynomial of β is

$$g(x) = \prod_{i=0}^{10} (x - \beta^{2^i}),$$

since $\beta^{2^{11}} = \beta$, and so (using $\beta^{23} = 1$),

$$g(x) = (x - \beta)(x - \beta^2)(x - \beta^4)(x - \beta^8)(x - \beta^{16})(x - \beta^9)(x - \beta^{18})$$

$$\cdot (x - \beta^{13})(x - \beta^3)(x - \beta^6)(x - \beta^{12}).$$

Similarly, the minimal polynomial of β^5 is

$$\tilde{g}(x) = (x - \beta^5)(x - \beta^{10})(x - \beta^{20})(x - \beta^{17})(x - \beta^{11})(x - \beta^{22})(x - \beta^{21})$$

$$\cdot (x - \beta^{19})(x - \beta^{15})(x - \beta^7)(x - \beta^{14}).$$

Since every power of β is now accounted for, the complete factorization of $x^{23} - 1$ into irreducible factors over F_2 is

$$(x^{23} - 1) = (x - 1)g(x)\tilde{g}(x).$$

By factoring $x^{23} - 1$ over F_2 directly, it can be shown that we may take

$$g(x) = x^{11} + x^9 + x^7 + x^6 + x^5 + x + 1,$$

$$\tilde{g}(x) = x^{11} + x^{10} + x^6 + x^5 + x^4 + x^2 + 1.$$

(See [22] or [23].)

Appendix D
Path enumeration in directed graphs

(For more detailed treatments of this subject, consult Aho, Hopcroft, and Ullman [1], Chapter 5, or Mason and Zimmerman [10], Chapter 4.)

Let $G = (V, E)$ be a directed graph, where $V = \{v_1, v_2, \ldots\}$ is a finite set of *vertices* and E is a subset of ordered pairs from V, called *edges*. G can be represented pictorially by a set of points in the plane corresponding to the elements of V; if $(v_i, v_j) \in E$, a directed line is drawn from v_i to v_j. For example, in Fig. D.1 we have drawn the graph G, where $V = \{a_0, a_1, b, c, d\}$ and $E = \{(a_0, c),\ (b, a_1),\ (b, c),\ (c, b),\ (c, d),\ (d, b),\ (d, d)\}$. [*Note.* The graph in Fig. D.1 is identical to the graph in Fig. 10.14.]

A *path* in such a graph is a sequence of edges of the form (v_0, v_1), $(v_1, v_2), \ldots, (v_{n-1}, v_n)$. Such a path is said to be a path of length n from v_0 to v_n, and is usually represented by the string $v_0 v_1 \ldots v_n$. For example, in the graph of Fig. D.1 $P = a_0 cbcddba_1$ is a path of length 7 from a_0 to a_1.

Suppose now that each edge in G is assigned a label. For definiteness we will assume that the labels are real numbers between 0 and 1. However, the

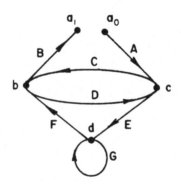

Figure D.1 The graph G.

380

discussion that follows remains valid when the labels come from much more general algebraic domains. (See [1], Section 5.6.) In Fig. D.1 the seven edges have been labeled $A, B, \ldots, G$. Let us define the label of a path to the product of the labels of the component edges. The path P cited above has label $ABCDEFG$, for example. The problem we wish to consider is to compute the sum of the labels of all paths joining a specified pair of vertices in G. This quantity we shall call the *transmission* between the two vertices.

Before tackling the graph of Fig. D.1, let us consider a somewhat simpler graph, depicted in fig. D.2, where $V = \{v_i, v_j, v_s\}$, and $E = \{(v_i, v_j),$ $(v_i, v_s), (v_s, v_s), (v_s, v_j)\}$. In this graph there are infintely many paths between v_i and v_j, namely $(v_i v_j)$, $(v_i v_s v_j)$, $(v_i v_s v_s v_j)$, $(v_i v_s v_s v_s v_j)$, etc. The sum of the labels of all paths between v_i and v_j is clearly

$$L_{ij} + L_{is}L_{sj} + L_{is}L_{ss}L_{sj} + L_{is}L_{ss}^2 L_{sj} + \cdots$$

$$= L_{ij} + L_{is}L_{sj}(1 + L_{ss} + L_{ss}^2 + L_{ss}^3 + \cdots)$$

$$= L_{ij} + \frac{L_{is}L_{sj}}{1 - L_{ss}}.$$

Hence the transmission between v_i and v_j in Fig. D.2 is given by

$$T(V_i, V_j) = L_{ij} + \frac{L_{is}L_{sj}}{1 - L_{ss}}.$$

In other words, the transmission between v_i and v_j in Fig. D.2 is the same as the transmission between v_i and v_j in Fig. D.3, the graph G' of Fig. D.3 having been obtained from the graph G of Fig. D.2 by deleting the vertex v_s and replacing the label L_{ij} according to the rule

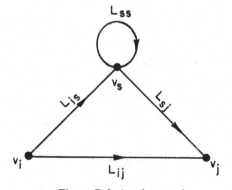

Figure D.2 Another graph.

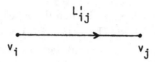

Figure D.3 The graph of G'.

$$L'_{ij} = L_{ij} + \frac{L_{is}L_{sj}}{1 - L_{ss}}. \tag{D.1}$$

What this means is that it is possible to compute the transmission between a given pair of vertices in a complicated graph by removing intermediate vertices from the graph, one by one, and redefining the edge labels by means of Eq. (D.1). One will eventually reach a graph with only two vertices, at which point the transmission between these vertices can be read directly from the graph (see Prob. 10.9).

Let us use this technique to compute the transmission between a_0 and a_1 in the graph of Fig. D.1. We first remove vertex d; the resulting graph appears in Fig. D.4.

Next we remove vertex c. The result is shown in Fig. D.5. The graph of Fig. D.5 has the form of Fig. D.2, and so we can compute the transmission between a_0 and a_1 directly from Eq. (D.1). The result is

$$T(a_0, a_1) = \frac{AB(C - CG + EF)}{1 - G - CD + CDG - DEF}. \tag{D.2}$$

In Chapter 10 we needed two special cases of Eq. (D.2), in which the labels are given by the following table (see Figs. 10.14 and 10.15).

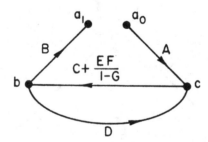

Figure D.4 The graph of Fig. D.1 after vertex d is removed.

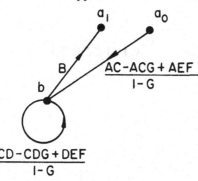

Figure D.5 The graph of Fig. D.4 after vertex c is removed.

Label	Case 1	Case 2
A	x^2	x^2yz
B	x^2	x^2y
C	x	xy
D	1	yz
E	x	xyz
F	x	xy
G	x	xyz

Substituting these values into Eq. (D.2), we obtain

$$A(x) = \frac{x^5}{1 - 2x},\tag{D.3}$$

$$A(x, y, z) = \frac{x^5 y^3 z}{1 - xyz(1 + y)}.\tag{D.4}$$

(cf. Eqs. (10.11) and (10.13)).

References

1. General reference textbooks

1. Aho, Alfred, Hopcroft, John, and Ullman, Jeffrey. *The Analysis and Design of Computer Algorithms*. Reading, Mass.: Addison-Wesley Publishing Co., 1974.
2. Apostol, Tom. *Mathematical Analysis*, 2nd ed. Reading, Mass.: Addison-Wesley Publishing Co., 1974.
3. Birkhoff, Garrett, and MacLane, Saunders. *A Survey of Modern Algebra*, rev. ed. New York: The Macmillan Co., 1953.
4. Feller, William. *An Introduction to Probability Theory and Its Applications*, 2 vols. Vol. 1, 3rd ed. rev. Vol. 2, 2nd ed. New York: John Wiley and Sons, 1968 and 1971.
5. Feynman, Richard, Leighton, Robert, and Sands, Matthew. *The Feynman Lectures on Physics*, 3 vols. Reading Mass.: Addison-Wesley Publishing Co., 1963.
6. Fleming, Wendell. *Functions of Several Variables*. Reading, Mass.: Addison-Wesley Publishing Co., 1965.
7. Knuth, Donald. *The Art of Computer Programming*, 3 vols. Reading, Mass.: Addison-Wesley Publishing Co., 1968, 1969, 1973.
8. Lamperti, John. *Probability*. New York: W. A. Benjamin, 1966.
9. Lang, Serge. *Algebra*. Reading, Mass.: Addison-Wesley Publishing Co., 1965.
10. Mason, Samuel, and Zimmermann, Henry. *Electronic Circuits, Signals, and Systems*. New York: John Wiley and Sons, 1960.
11. Lidl, R., and Niederreiter, Harold. *Finite Fields*. Encyclopedia of Mathematics and Its Applications, Vol. 20. Reading, Mass.: Addison-Wesley Publishing Co.

2. An annotated bibliography of the theory of information and coding

12. Abramson, Norman. *Information Theory and Coding*. New York: McGraw-Hill Book Co., 1963.

A short, elementary, and highly readable introduction to information theory, culminating with a proof of the channel coding theorem for discrete memoryless channels.

13. Berger, Toby. *Rate Distortion Theory*. Englewood Cliffs, N.J.: Prentice-Hall, 1971.

An advanced book dealing wholly with the source coding theorem, its generalizations, and its practical applications.

14. Berlekamp, Elwyn. *Algebraic Coding Theory*. New York: McGraw-Hill Book Co., 1968.

Begins slowly, then rapidly accelerates into the furthest reaches of the theory of block codes. A *tour de force* by an authentic post-Shannon genius.

15. Berlekamp, Elwyn, ed. *Key Papers in the Development of Coding Theory*. New York: IEEE Press, 1974.

    An anthology of 44 important papers. Also notable for the extensive editorial comments by Berlekamp.

16. Davisson, Lee, and Gray, Robert, eds. *Data Compression*. Stroudsberg, Pa.: Dowden, Hutchinson, and Ross, 1976.

    An anthology of 46 important papers about rate-distortion theory and source coding techniques.

17. Gallager, Robert. *Information Theory and Reliable Communication*. New York: John Wiley and Sons, 1968.

    For some years the standard advanced textbook on the subject. Includes extensive treatment of block coding error exponents and waveform channels.

18. Van Lint, Jacobus. *Coding Theory*. Lecture Notes in Mathematics, No. 201. Berlin: Springer-Verlag, 1971.

    A brief and clearly written introduction to the theory of block codes, from a combinatorial and algebraic viewpoint.

19. MacWilliams, Jessie, and Sloane, Neil. *The Theory of Error-Correcting Codes*. Amsterdam: North-Holland Publishing Co., 1977.

    An encyclopedia of the rich combinatorial mathematics of block codes. Includes a bibliography containing over 1400 entries.

20. Massey, James. *Threshold Decoding*. Cambridge, Mass.: M.I.T. Press, 1963.

    A thin monograph, still the best introduction to this interesting decoding technique.

21. Massey, James. *Error Bounds for Tree Codes, Trellis Codes, and Convolutional Codes with Encoding and Decoding Procedures.*, in *Coding and Complexity*, G. Longo, Ed., CISM Courses and Lectures No. 216. Wien-New York: Springer-Verlag, 1977.

    A brief and lucid introduction to the theory and application of convolution codes.

22. Peterson, Wesley. *Error-Correcting Codes*. Cambridge, Mass.: M.I.T. Press, 1961.

    The classic text on coding theory. A little dated now but still notable for its treatment of shift-register implementation and its tables of irreducible polynomials. (See next entry.)

23. Peterson, Wesley, and Weldon, Edward. *Error-Correcting Codes*, 2nd ed. Cambridge, Mass.: M.I.T. Press, 1972.

    A completely revised version of the preceding entry, containing much new information, notably on the theory of polynomial codes.

24. Pinsker, M. S. *Information and Information Stability of Random Variables and Processes*. Translated by A. Feinstein. San Francisco: Holden-Day, 1964.

An English translation of the classic Russian text, containing the only available complete treatment of mutual information for arbitrary random variables.

25. Slepian, David, ed. *Key Papers in the Development of Information Theory.* New York: I.E.E.E. Press, 1974.

An anthology of 49 important papers in information theory, 12 by Shannon, including his classics, "A mathematical theory of communication" and "Coding theorems for a discrete source with a fidelity criterion."

26. Viterbi, Andrew, and Omura, Jim. *Digital Communication and Coding.* New York: McGraw-Hill Book Co., 1978.

Encyclopedic coverage of information theory, especially recent advances in error exponents and rate-distortion theory. Emphasis is on convolutional codes throughout.

27. Wolfowitz, Jacob. *The Coding Theorems of Information Theory,* 3rd Ed. Berlin: Springer-Verlag, 1978.

Contains a wide variety of generalizations of the channel coding theorem and its converse.

3. Original papers cited in the text

[*Note.* Most of these papers appeared in the *IEEE Transactions on Information Theory*, which we hereafter abbreviate as *IT*.]

28. Anderson, John. "A stack algorithm for source encoding with a fidelity criterion," *IT* **20** (1974), 211–226.
29. Arimoto, Suguru. "An algorithm for computing the capacity of arbitrary discrete memoryless channels," *IT* **18** (1972), 14–20.
30. Arimoto, Suguru. "On the converse to the coding theorem for discrete memoryless channels," *IT* **19** (1973), 357–359.
31. Berlekamp, Elwyn. "Long primitive BCH codes have distance $d \sim 2nlnR^{-1}/\log n$," *IT* **18** (1972), 415–426.
32. Blahut, Richard. "Computation of channel capacity and rate-distortion functions," *IT* **18** (1972), 460–473.
33. Delsarte, Philippe. "On subfield subcodes of modified Reed–Solomon codes," *IT* **21** (1975), 575–576.
34. Delsarte, Philippe, and McEliece, Robert. "Zeros of functions in finite Abelian group algebras," *Amer. J. of Math.* **98** (1976), 197–224.
35. Heller, Jerold, and Jacobs, Irwin. "Viterbi decoding for satellite and space communication," *IEEE Trans. Comm. Technol.* **COM-19** (1971), 835–848.
36. Johannesson, Rolf. "Robustly optimal rate one-half binary convolutional codes," *IT* **21** (1975), 464–468.
37. Larsen, Knud. "Short convolutional codes with maximal free distance for rates 1/2, 1/3, and 1/4," *IT* **19** (1973), 371–372.
38. Massey, James. "On the fractional weight of distinct binary n-tuples," *IT* **20** (1974), 131.
39. McEliece, Robert. "Irreducible cyclic codes and Gauss sums." Pp. 185–202 in *Combinatorics*, M. Hall, Jr., and J. H. van Lint, eds. Dordrecht, Holland: D. Reidel Publishing Co., 1975.
40. McEliece, Robert, and Omura, Jim. "An improved upper bound on the block coding error exponent for binary-input discrete memoryless channels," *IT* **23** (1977), 611–613.

41. Odenwalder, John. "Optimal decoding of convolutional codes." Ph.D. dissertation, University of California, Los Angeles, 1970.

42. Patterson, Nicholas. "The algebraic decoding of Goppa codes." *IT* **21** (1975), 203–207.

43. Pilc, R. J. "Coding theorems for discrete source-channel pairs." Ph.D. dissertation, Massachusetts Institute of Technology, Cambridge, 1967.

44. Schalkwijk, J. P. M., and Vinck, A. J. "Syndrome decoding of binary rate-1/2 convolutional codes," *IEEE Trans. Commun.* **COM-24** (1976), 977–985.

45. Sugiyama, Yasuo, Kasahara, Masao, Hirasawa, Shigeichi, and Namekawa, Toshihiko. "A method for solving key equation for decoding Goppa codes," *Information and Control* **27** (1975), 87–99.

46. Van Tilborg, Henricus. "Uniformly packed codes." Ph.D. dissertation, Technische Hogeschool Eindhoven, The Netherlands, 1976.

47. Vanderhorst, Jose, and Berger, Toby. "Complete decoding of triple-error-correcting binary BCH codes," *IT* **22** (1976), 138–147.

48. Wu, William. "New convolutional codes—Part III," *IEEE Trans. Commun.* **COM-24** (1976), 946–955.

49. Wyner, Aaron. "Recent results in the Shannon theory," *IT* **20** (1974), 2–10.

50. Wyner, Aaron, and Ziv, Jacob. "The rate-distortion function for source coding with side information at the decoder," *IT* **22** (1976), 1–10.

51. Lovász L. "On the Shannon capacity of a graph," *IT* **25** (1979), 1–7.

Index of theorems

Index

Printed in the United States
by Baker & Taylor Publisher Services